An Introduction to
Mechanical
Engineering

Jonathan Wickert
Iowa State University

Kemper Lewis
University at Buffalo—SUNY

CENGAGE

Australia • Brazil • Mexico • Singapore • United Kingdom • United States

An Introduction to Mechanical Engineering,
Enhanced Fourth Edition
Jonathan Wickert and Kemper Lewis

Product Director, Global Engineering:
Timothy L. Anderson

Learning Designer:
MariCarmen Constable

Associate Content Manager:
Alexander Sham

Product Assistant: Andrew Reddish

Associate Marketing Manager:
Shannon Hawkins

Intellectual Property Analyst:
Deanna Ettinger

Production Service: RPK Editorial
Services, Inc.

Compositor: MPS Limited

Designer: Chris Doughman

Cover Designer: Felicia Bennett

Cover Image: Mario Cea Sanchez/Robert
Harding

Manufacturing Planner: Doug Wilke

For product information and technology assistance, contact us at
Cengage Customer & Sales Support, 1-800-354-9706 or
support.cengage.com.

For permission to use material from this text or product,
submit all requests online at
www.cengage.com/permissions.

Library of Congress Control Number: 2019910598

Student Edition:
ISBN: 978-0-357-38229-5

Loose-leaf Edition:
ISBN: 978-0-357-38247-9

Cengage
200 Pier 4 Boulevard
Boston, MA 02210
USA

Cengage is a leading provider of customized learning solutions
with employees residing in nearly 40 different countries and sales
in more than 125 countries around the world. Find your local
representative at **www.cengage.com**.

Cengage products are represented in Canada by Nelson
Education, Ltd.

To learn more about Cengage platforms and services, register or
access your online learning solution, or purchase materials for
your course, visit **www.cengage.com**.

Printed at CLDPC, USA, 08-19

Contents

▶ Purpose

This textbook will introduce you to the ever-emerging field of mechanical engineering and help you appreciate how engineers design the hardware that builds and improves societies all around the world. As the title implies, this textbook is neither an encyclopedia nor a comprehensive treatment of the discipline. Such a task is impossible for a single textbook, and, regardless, our perspective is that the traditional four-year engineering curriculum is just one of many steps taken during a lifelong education. By reading this textbook, you will discover the "forest" of mechanical engineering by examining a few of its "trees," and along the way you will be exposed to some interesting and practical elements of the profession called mechanical engineering.

▶ Approach and Content

This textbook is intended for students who are in the first or second years of a typical college or university program in mechanical engineering or a closely related field. Throughout the following chapters, we have attempted to balance the treatments of technical problem-solving skills, design, engineering analysis, and modern technology. The presentation begins with a narrative description of mechanical engineers, what they do, and the impact they can have (Chapter 1). Seven "elements" of mechanical engineering are emphasized subsequently in Chapter 2 (Mechanical Design), Chapter 3 (Professional Practice), Chapter 4 (Forces in Structures and Machines), Chapter 5 (Materials and Stresses), Chapter 6 (Fluids Engineering), Chapter 7 (Thermal and Energy Systems), and Chapter 8 (Motion and Power Transmission). Some of the applications that you will encounter along the way include commercial space travel, 3-D printing, Boeing's 787 Dreamliner, medical device design, nanomachines, internal combustion engines, robotics, sports technology, advanced materials, micro-fluidic devices, automatic transmissions, and renewable energy.

What should you be able to learn from this textbook? First and foremost, you will discover who mechanical engineers are; what they do; and what technical, social, and environmental challenges they solve with the technologies they create. Section 1.3 details a "top ten" list of the profession's achievements. By looking at this list, you will recognize how the profession has contributed to your day-to-day life and society around the world in general. Second, you will find that engineering is a practical endeavor with the objective of designing things that work, that are cost-effective to manufacture, that are safe to use, and that are responsible in terms of their environmental impact. Third, you will learn some of the calculations, estimates, and approximations that mechanical engineers can perform as they solve technical problems and communicate their results. To accomplish their jobs better and faster, mechanical engineers combine mathematics, science, computer-aided engineering tools, experience, and hands-on skills.

You will not be an expert in mechanical engineering after having read this textbook, but that is not our intention, and it should not be yours. If our objective has been met, however, you will set in place a solid foundation of problem-solving, design, and analysis skills, and those just might form the basis for your own future contributions to the mechanical engineering profession.

▶ Approach

This textbook is intended for a course that provides an introduction to mechanical engineering during either the freshman or sophomore years. Over the past decade, many colleges and universities have taken a fresh look at their engineering curricula with the objective of positioning engineering content earlier in their programs. Particularly for the freshman year, the formats vary widely and can include seminars on "who are mechanical engineers" and "what do they do," innovative design experiences, problem-solving skills, basic engineering analysis, and case studies. Courses at the sophomore level often emphasize design projects, exposure to computer-aided engineering, principles of engineering science, and a healthy dose of mechanical engineering hardware.

Core engineering-science courses (for example, strength of materials, thermodynamics, fluid mechanics, and dynamics) have evolved since the post–World War II era into their present, relatively mature states. On the other hand, little if any standardization exists among introductory mechanical engineering courses. With limited discipline-specific instructional materials available for such courses, we believe that an important opportunity remains for attracting students, exciting them with a view of what to expect later in their program of study and in their future careers, and providing them with a foundation of sound engineering analysis, technical problem-solving, and design skills.

▶ Objectives

While developing the fourth edition of this textbook, our objective has been to provide a resource that others can draw upon when teaching introductory mechanical engineering to first-year and second-year students. We expect that most such courses would encompass the bulk of material presented in Chapter 1 (The Mechanical Engineering Profession), Chapter 2 (Mechanical Design), and Chapter 3 (Technical Problem Solving and Communication Skills). Based on the level and contact hours of their particular courses, instructors can select additional topics from Chapter 4 (Forces in Structures and Machines), Chapter 5 (Materials and Stresses), Chapter 6 (Fluids Engineering), Chapter 7 (Thermal and Energy Systems), and Chapter 8 (Motion and Power Transmission). For instance, Section 5.5 on materials selection is largely self-contained, and it provides an introductory-level student with an overview of the different classes of engineering materials. Similarly, the descriptions in Sections 7.6 and 7.7 of internal-combustion engines and electrical power plants are expository in nature, and that material can be incorporated in case studies to demonstrate the operation of some important mechanical engineering hardware. Rolling-contact bearings, gears, and belt and chain drives are similarly discussed in Sections 4.6, 8.3, and 8.6.

This textbook reflects our experiences and philosophy for introducing students to the vocabulary, skills, applications, and excitement of the mechanical engineering profession. Our writing has been motivated in part by teaching introductory mechanical engineering courses at our respective universities. Collectively, these courses have included lectures, computer-aided design and manufacturing projects, product dissection laboratories (an example of which is discussed in Section 2.1), and team design projects (many of which have been adapted into the new open-ended design problems at the end of each chapter). A number of vignettes and case studies are also discussed to demonstrate for students the realism of what they are learning, including the "top ten" list of previous mechanical engineering achievements and a list of the top emerging fields in mechanical engineering both developed by the American Society of Mechanical Engineers (Section 1.3); the fourteen "grand challenges" from the National Academy of Engineering (NAE) (Section 2.1); design innovation, patents, and a summary of the recently updated patent protection system in the United States (Section 2.2); the design of the Boeing 787 Dreamliner (Section 2.3); the loss of the *Mars Climate Orbiter* spacecraft and the refueling error on Air Canada Flight 143 (Section 3.1); the Deepwater Horizon oil spill disaster (Section 3.6); the design of a heart implant (Section 4.5); the design of products and materials for extreme environments (Section 5.2); the design of advanced materials for innovation applications (Section 5.5); microfluidic devices (Section 6.2); blood flow in the human body (Section 6.5); sports technology (Sections 6.6 and 6.7); global energy consumption (Section 7.3); renewable energies (Section 7.5); internal combustion engines (Section 7.6); solar power generation and the development of innovative energy solutions through crowdsourcing (Section 7.7); nanomachines (Section 8.3); and advanced geartrains for next generation engines (Section 8.5).

The "Focus on ..." boxes in each chapter are used to highlight some of these interesting topics and other emerging concepts in mechanical engineering.

▶ Content

We certainly have not intended this textbook to be an exhaustive treatment of mechanical engineering, and we trust that it will not be read in that light. Quite the contrary: In teaching first-year and second-year students, we are ever conscious of the mantra that "less really is more." To the extent possible, we have resisted the urge to add just one more section on a particular subject, and we have tried to keep the material manageable and engaging from the reader's perspective. Indeed, many topics that are important for mechanical engineers to know are simply not included here; this is done intentionally (or, admittedly, by our own oversight). We are confident, however, that students will be exposed to those otherwise omitted subjects in due course throughout the remainder of their engineering curricula.

In Chapters 2 through 8, we have selected a subset of mechanical engineering "elements" that can be sufficiently covered for early students to develop useful design, technical problem-solving, and analysis skills. The coverage has been chosen to facilitate the textbook's use within the constraints of courses having various formats. While there is more material here than can be comfortably covered in a single semester, instructors should find a reasonable menu from which to choose. In particular, we have selected content that we have found to

1. Match the background, maturity, and interests of students early in their study of engineering

2. Expose students to the significance of mechanical design principles in the development of innovative solutions to technical challenges that face our global societies

3. Help students think critically and learn good problem-solving skills, particularly with respect to formulating sound assumptions, making order-of-magnitude approximations, performing double-checks, and bookkeeping proper units

4. Convey aspects of mechanical engineering science and empiricism that can be applied at the freshman and sophomore levels

5. Expose students to a wide range of hardware, innovative designs, engineering technology, and the hands-on nature of mechanical engineering

6. Generate excitement through applications encompassing space flight, 3-D printing, Boeing's 787 Dreamliner, medical device design, nanomachines, engines, robotics, sports technology, consumer products, advanced materials, micro-fluidic devices, automotive transmissions, renewable energy generation, and more

To the extent possible at the freshman and sophomore levels, the exposition, examples, and homework problems have been drawn from realistic applications. You will find no masses on inclined planes or block-and-tackle systems in this textbook. Because we find engineering to be a visual and graphical activity, we have placed particular emphasis on the quality and breadth of the nearly three hundred photographs and illustrations, many of which were provided by our colleagues in industry, federal agencies, and academe. Our view has been to leverage that realism and motivate students through interesting examples that offer a glimpse of what they will be able to study in later courses and, subsequently, practice in their own careers.

▶ New in the Fourth Edition

In preparing this fourth edition, we have made many of the types of changes that one would expect: Sections have been rewritten and reorganized, new material has been added, some material has been removed, new example problems have been created, and small mistakes have been corrected. Over 20 new homework problems have been developed and over 30 new figures have been included. We are excited about the new homework problems, as they are all open-ended problems whose solutions depend upon the set of assumptions made. While these problems do not have a single correct answer, there are better answers and worse answers. Therefore, students are challenged to consider their problem-solving approach, the validity of their assumptions, and the appropriateness of their answers. These new problems are included as the last homework problems in each chapter and have been developed to be used in group settings, including flipped classroom environments. These larger open-ended problems are denoted with an asterisk "*".

We have attempted to remain faithful to the philosophy of the first three editions by emphasizing the importance of the mechanical engineering profession to solving global problems, including new information in Chapter 1 on recent professional trends, technology development, traditional and emerging mechanical engineering career paths, and knowledge areas. Also, in Chapter 1, we update, in Figure 1.2, the energy range that mechanical engineers are creating devices or machines to produce and/or consume. We tighten the presentation of the top past accomplishments in mechanical engineering and add a discussion about the top emerging fields within mechanical engineering adapted from a recent report from ASME.

In Chapter 2, new material is included on global design patents, and the new patent law in the United States. The previous case studies from Chapter 2 and one from Chapter 7 are now located on the student companion website.

Throughout the book, we have continued the use of the improved pedagogical format comprising the problem's statement, approach, solution, and discussion. In particular, the discussion portion is intended to highlight why the numerical answer is interesting or why it makes intuitive sense. Symbolic equations are written alongside the numerical calculations. Throughout the textbook, the dimensions appearing in these calculations are explicitly manipulated and canceled in order to reinforce good technical problem-solving skills.

The "Focus on …" boxes contain topical material, either conceptual or applied, that broadens the textbook's coverage without detracting from its flow. New topics in the "Focus on …" boxes include the emerging career opportunities for mechanical engineers; product archaeology "digs"; global design teams; the types of engineering estimations used in predicting the oil flow rates during the Deepwater Horizon disaster; ineffective communication practices using illustrative technical charts; innovative design opportunities that arise from engineering failure analysis; design of devices for extreme environments; development of new engineering materials; the crowdsourcing of innovative solutions to global energy challenges; and the design of advanced automotive geartrains to address fuel economy standards.

As was the intent with the first three editions, we have attempted to make the fourth edition's content readily accessible to any student having a conventional secondary school background in mathematics and physics. We have not relied on any mathematics beyond algebra, geometry, and trigonometry (which is reviewed in Appendix B), and in particular, we have not used any cross-products, integrals, derivatives, or differential equations. Consistent with that view, we have intentionally not included a chapter that addresses the subjects of dynamics, dynamic systems, and mechanical vibration (ironically, my own areas of specialization). We remain focused on the earliest engineering students, many of whom will be studying calculus concurrently. Keeping those students in mind, we feel that the added mathematical complexity would detract from this textbook's overall mission.

▶ Supplements

Supplements for instructors are available on the Instructor's Resource Center at http://login.cengage.com.

- Case Studies: We have shifted some of the case studies in previous editions to the student website without any loss of content in the text. The case studies were longer design applications of the chapter concepts and as such, they function well as effective reference information for the students who want to explore the chapter concepts further.

- Instructor's Solutions Manual: As was the case with the first three editions, the enhanced fourth edition is also accompanied by a detailed Instructor's Solutions Manual. With the exception of some "open-ended" problems at the end of each chapter, the manual contains a solution to each of the numerical problems in this textbook. The description and style of these solutions (stating the problem, writing a brief approach, making appropriate assumptions, making sketches, carrying out calculations, keeping track of units and significant figures, checking one's work, discussing the validity of the solution, and so forth) are intended to guide students with respect to the formatting of their own work.

- Lecture Note PowerPoints: We have created PowerPoint presentations for each chapter that capture the primary teaching points, along with a number of the figures and tables from the chapters.

▶ Acknowledgments

It would have been impossible to develop the four editions of this textbook without the contributions of many people and organizations, and at the outset, we would like to express our appreciation to them. Generous support was provided by the Marsha and Philip Dowd Faculty Fellowship, which encourages educational initiatives in engineering. Adriana Moscatelli, Jared Schneider, Katie Minardo, and Stacy Mitchell, who are now alumni of Carnegie Mellon University, helped to get this project off the ground by drafting many of the illustrations. The expert assistance provided by Ms. Jean Stiles in proofreading the textbook and preparing the *Instructor's Solutions Manual* was indispensable. We very much appreciate the many contributions she made.

Our colleagues, graduate students, and teaching assistants at Iowa State University, Carnegie Mellon, and the University at Buffalo—SUNY provided many valuable comments and suggestions as we wrote the editions. We would specifically like to thank Adnan Akay, Jack Beuth, Paul Steif, Allen Robinson, Shelley Anna, Yoed Rabin, Burak Ozdoganlar, Parker Lin, Elizabeth Ervin, Venkataraman Kartik, Matthew Brake, John Collinger, Annie Tangpong, Matthew Iannacci, James Lombardo, Phil Odonkor, Erich Devendorf, Phil Cormier, Aziz Naim, David Van Horn, Brian Literman, and Vishwa Kalyanasundaram for their comments and help. We are likewise indebted to the students in our courses: Fundamentals of Mechanical Engineering

(Carnegie Mellon), Introduction to Mechanical Engineering Practice (University at Buffalo—SUNY), and Design Process and Methods (University at Buffalo—SUNY). Their collective interest, feedback, and enthusiasm have always provided much-needed forward momentum. Joe Elliot and John Wiss kindly offered the engine dynamometer and cylinder pressure data to frame the discussion of internal-combustion engines in Chapter 7. Solutions to many of the homework problems were drafted by Brad Lisien and Albert Costa, and we appreciate their hard work and conscientious effort.

In addition, the following reviewers of the first, second, third, and fourth editions were kind enough to let us benefit from their perspectives and teaching experience: Terry Berreen, Monash University; John R. Biddle, California State Polytechnic University at Pomona; Terry Brown, University of Technology (Sydney); Peter Burban, Cedarville University; David F. Chichka, George Washington University; Scott Danielson, Arizona State University; Amirhossein Ghasemi, University of Kentucky; William Hallett, University of Ottawa; David W. Herrin, University of Kentucky; Robert Hocken, University of North Carolina (at Charlotte); Damir Juric, Georgia Institute of Technology; Bruce Karnopp, University of Michigan; Kenneth A. Kline, Wayne State University; Pierre M. Larochelle, Florida Institute of Technology; Steven Y. Liang, Georgia Institute of Technology; Per Lundqvist, Royal Institute of Technology (Stockholm); William E. Murphy, University of Kentucky; Petru Petrina, Cornell University; Anthony Renshaw, Columbia University; Oziel Rios, University of Texas-Dallas; Hadas Ritz, Cornell University; Timothy W. Simpson, Pennsylvania State University; K. Scott Smith, University of North Carolina (at Charlotte); Michael M. Stanisic, University of Notre Dame; Gloria Starns, Iowa State University; David J. Thum, California Polytechnic State University (San Luis Obispo); and David A Willis, Southern Methodist University. We are grateful for their detailed comments and helpful suggestions.

On all counts, we have enjoyed interacting with the editorial staff at Cengage. As the textbook's publisher, Tim Anderson was committed to developing a high-quality product, and it has been a continuing pleasure to collaborate with him. We would also like to acknowledge MariCarmen Constable, Alexander Sham, and Andrew Reddish for their contributions to the enhanced fourth edition, along with Rose Kernan of RPK Editorial Services. They managed production with skill and professionalism and with a keen eye for detail. To each, we express our thanks for a job well done.

Colleagues at the following industrial, academic, and governmental organizations were remarkably helpful and patient in providing us with photographs, illustrations, and technical information: General Motors, Intel, Fluent, General Electric, Enron Wind, Boston Gear, Mechanical Dynamics, Caterpillar, NASA, NASA's Glenn Research Center, W. M. Berg, FANUC Robotics, the U.S. Bureau of Reclamation, Niagara Gear, Velocity11, Stratasys, National Robotics Engineering Consortium, Lockheed-Martin, Algor, MTS Systems, Westinghouse Electric, Timken, Sandia National Laboratories, Hitachi Global Storage Technologies, Segway LLC, the U.S. Department of Labor, and the U.S. Department of Energy. We've surely not listed everyone who has helped us with this endeavor, and we apologize for any inadvertent omissions that we may have made.

▶ New Digital Solution for Your Engineering Classroom

WebAssign is a powerful digital solution designed by educators to enrich the engineering teaching and learning experience. With a robust computational engine at its core, WebAssign provides extensive content, instant assessment, and superior support.

WebAssign's powerful question editor allows engineering instructors to create their own questions or modify existing questions. Each question can use any combination of text, mathematical equations and formulas, sound, pictures, video, and interactive HTML elements. Numbers, words, phrases, graphics, and sound or video files can be randomized so that each student receives a different version of the same question.

In addition to common question types such as multiple choice, fill-in-the-blank, essay, and numerical, you can also incorporate robust answer entry palettes (mathPad, chemPad, calcPad, physPad, pencilPad, Graphing Tool) to input and grade symbolic expressions, equations, matrices, and chemical structures using powerful computer algebra systems. You can even use Camtasia to embed "clicker" questions that are automatically scored and recorded in the GradeBook.

▶ WebAssign Offers Engineering Instructors the Following

- The ability to create and edit algorithmic and numerical exercises.
- The opportunity to generate randomized iterations of algorithmic and numerical exercises. When instructors assign numerical WebAssign homework exercises (engineering math exercises), the WebAssign program offers them the ability to generate and assign their students differing versions of the same engineering math exercise. The computational engine extends beyond and provides the luxury of solving for correct solutions/answers.
- The ability to create and customize numerical questions, allowing students to enter units, use a specific number of significant digits, use a specific number of decimal places, respond with a computed answer, or answer within a different tolerance value than the default.

Visit https://www.webassign.com/instructors/features/ to learn more. To create an account, instructors can go directly to the signup page at http://www.webassign.net/signup.html.

▶ WebAssign Features for Students

- **Review Concepts at Point of Use**

Within WebAssign, a "Read It" button at the bottom of each question links students to corresponding sections of the textbook, enabling access to the MindTap Reader at the precise moment of learning. A "Watch It" button allows a short video to play. These videos help students understand and review the problem they need to complete, enabling support at the precise moment of learning.

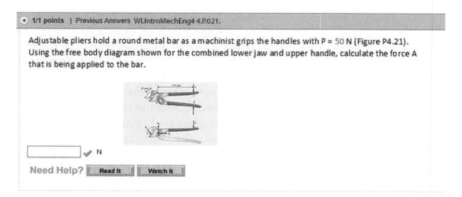

- **My Class Insights**

WebAssign's built-in study feature shows performance across course topics so that students can quickly identify which concepts they have mastered and which areas they may need to spend more time on.

- **Ask Your Teacher**

This powerful feature enables students to contact their instructor with questions about a specific assignment or problem they are working on.

▶ MindTap Reader

Available via WebAssign and our digital subscription service, Cengage Unlimited, **MindTap Reader** is Cengage's next-generation eBook for engineering students.

The MindTap Reader provides more than just text learning for the student. It offers a variety of tools to help our future engineers learn chapter concepts in a way that resonates with their workflow and learning styles.

- **Personalize their experience**

Within the MindTap Reader, students can highlight key concepts, add notes, and bookmark pages. These are collected in My Notes, ensuring they will have their own study guide when it comes time to study for exams.

6.2 Properties of Fluids

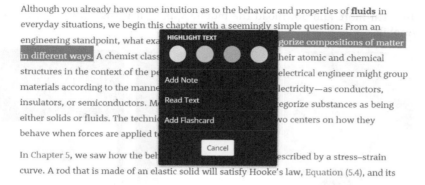

Although you already have some intuition as to the behavior and properties of **fluids** in everyday situations, we begin this chapter with a seemingly simple question: From an engineering standpoint, what exa...   gorize compositions of matter in different ways. A chemist class...   heir atomic and chemical structures in the context of the pe...   electrical engineer might group materials according to the manne...   lectricity—as conductors, insulators, or semiconductors. M...   egorize substances as being either solids or fluids. The techni...   o centers on how they behave when forces are applied t...

In Chapter 5, we saw how the beh...   escribed by a stress–strain curve. A rod that is made of an elastic solid will satisfy Hooke's law, Equation (5.4), and its

HIGHLIGHT TEXT

Add Note

Read Text

Add Flashcard

Cancel

- **Flexibility at their fingertips**

With access to the book's internal glossary, students can personalize their study experience by creating and collating their own custom flashcards. The Readspeaker feature reads text aloud to students, so they can learn on the go— wherever they are.

Chapter 6 - Fluids Engineering

3 of 27 | Flip Card

fluids

Next Card

Previous Card

Shuffle Deck

▶ The Cengage Mobile App

Available on iOS and Android smartphones, the Cengage Mobile App provides convenience. Students can access their entire textbook anyplace and anytime. They can take notes, highlight important passages, and have their text read aloud whether they are online or off.

To learn more and download the mobile app, visit https://www.cengage.com /mobile-app/.

CENGAGE
UNLIMITED

▶ All-You-Can-Learn Access with Cengage Unlimited

Cengage Unlimited is the first-of-its-kind digital subscription that gives students total and on-demand access to all the digital learning platforms, eBooks, online homework, and study tools Cengage has to offer—in one place, for one price. With Cengage Unlimited, students get access to their WebAssign courseware, as well as content in other Cengage platforms and course areas from day one. That's 70 disciplines and 675 courses worth of material, including engineering.

With Cengage Unlimited, students get **unlimited access** to a library of more than 22,000 products. To learn more, visit https://www.cengage.com/unlimited.

Photo courtesy of Dave Gieseke

Jonathan Wickert

Jonathan Wickert serves as Senior Vice President and Provost at Iowa State University and is responsible for the university's educational, research, and extension and outreach programs. He previously served as department chair and dean at Iowa State, and before then was a member of the faculty at Carnegie Mellon University. Dr. Wickert received his B.S., M.S., and Ph.D. degrees in mechanical engineering from the University of California, Berkeley, and he was a postdoctoral fellow at the University of Cambridge. The Society of Automotive Engineers, the American Society for Engineering Education, and the Information Storage Industry Consortium have recognized Dr. Wickert for his teaching and research, and he was elected a fellow of the American Society of Mechanical Engineers and the National Academy of Inventors. As a researcher and consultant in the field of mechanical vibration, he has worked with companies and federal agencies on a diverse range of technical applications including computer data storage; the manufacture of sheet metal, fiberglass, polymers, and industrial chemicals; automotive brakes; radial flow gas turbines; and consumer products.

© 2011 University at Buffalo/Douglas Levere

Kemper Lewis

Kemper Lewis serves as Chair of the Mechanical and Aerospace Engineering Department at the University at Buffalo – SUNY where he also holds a faculty appointment as Professor. He teaches and conducts research in the areas of mechanical design, system optimization, and decision modeling. As a researcher and consultant, he has worked with companies and federal agencies on a wide range of engineering design problems including turbine engine product and process design; optimization of industrial gas systems; air and ground vehicle design; innovation in consumer product design; and manufacturing process control for thin film resistors, heat exchangers, and medical electronics. Dr. Lewis received his B.S. in mechanical engineering and B.A. in Mathematics from Duke University, and his M.S. and Ph.D. degrees in mechanical engineering from the Georgia Institute of Technology. He has served as associate editor of the ASME *Journal of Mechanical Design*, on the ASME Design Automation Executive Committee, and on the National Academies Panel on Benchmarking the Research Competitiveness of the United States in Mechanical Engineering. Dr. Lewis has received awards in recognition of his teaching and research from the Society of Automotive Engineers, the American Society for Engineering Education, the American Institute of Aeronautics and Astronautics, and the National Science Foundation. He was also elected fellow of the American Society of Mechanical Engineers.

CHAPTER 1

The Mechanical Engineering Profession

- Describe some of the differences among engineers, mathematicians, and scientists.
- Discuss the type of work that mechanical engineers do, list some of the technical issues they address, and identify the impact that they have in solving global, social, environmental, and economic problems.
- Identify some of the industries and governmental agencies that employ mechanical engineers.

- List some of the products, processes, and hardware that mechanical engineers design.
- Recognize how the mechanical engineering profession's "top ten" list of achievements has advanced our society and improved day-to-day lives.
- Understand the objectives and format of a typical curriculum for mechanical engineering students.

▶ 1.1 Overview

I n this introductory chapter, we describe who mechanical engineers are, what they do, what their challenges and rewards are, what their global impact can be, and what their notable accomplishments have been. Engineering is the practical endeavor in which the tools of mathematics and science are applied to develop cost-effective solutions to the technological problems facing our society. Engineers design many of the consumer products that you use every day. They also create a large number of other products that you do not necessarily see or hear about because they are used in business and industrial settings. Nevertheless, they make important contributions to our society, our world, and our planet. Engineers develop the machinery that is needed to manufacture most products, the factories that make them, and the quality control systems that guarantee the product's safety and performance. Engineering is all about making useful things that work and impact lives.

The Elements of Mechanical Engineering

The discipline of mechanical engineering is concerned in part with certain "elements":

- Design (Chapter 2)
- Professional Practices (Chapter 3)
- Forces (Chapter 4)
- Materials (Chapter 5)
- Fluids (Chapter 6)
- Energy (Chapter 7)
- Motion (Chapter 8)

Mechanical engineers invent machines and structures that exploit those elements in order to serve a useful purpose and solve a problem. Original design and the practical issue of making something that works are the themes behind any engineering endeavor. An engineer creates a machine or product to help someone solve a technical problem. The engineer might start from a blank sheet of paper, conceive something new, develop and refine it so that it works reliably, and—all the while—satisfy the constraints of safety, cost, and manufacturability.

Robotic welding systems (Figure 1.1), internal combustion engines, sports equipment, computer hard disk drives, prosthetic limbs, automobiles, aircraft, jet engines, surgical tools, and wind turbines are some of the thousands of technologies that mechanical engineering encompasses. It would not be much of an exaggeration to claim that, for every product you can imagine, a mechanical engineer was involved at some point in its design, materials selection, temperature control, quality assurance, or production. Even if a mechanical engineer didn't conceive or design the product per se, it's still a safe bet that a mechanical engineer designed the machines that built, tested, or delivered the product.

Mechanical engineering has been defined as the profession in which power-producing and power-consuming machines are researched, designed, and manufactured. In fact, mechanical engineers devise machines that produce or consume power over the remarkably wide scale shown in Figure 1.2, ranging from

Figure 1.1

Robots are used in environments that require precise and repetitive tasks such as industrial assembly lines and in extreme environments like this deep sea repair on a corroded pipe.

Paul Fleet/Shutterstock.com

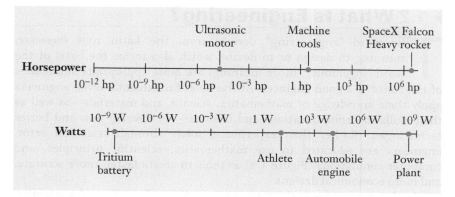

Figure 1.2

Mechanical engineers work with machines that produce or consume power over a remarkably wide range.

nanowatts (nW) to gigawatts (GW). Few professions require a person to deal with physical quantities across so many orders of magnitude (one quintillionfold or a factor of 1,000,000,000,000,000,000), but mechanical engineering does. At the lower end of the power range, batteries powered by the decay of the hydrogen isotope tritium can generate nanowatts of power for sensors, electronic equipment, and small precision ultrasonic motors, such as those used in a camera's autofocus lens. Moving upward in power level, an athlete using exercise equipment, such as a rowing machine or a stair climber, can produce up to several hundred watts (about 0.25–0.5 horsepower [hp]) over an extended period of time. The electric motor in an industrial drill press might develop 1000 W, and the engine on a sport utility vehicle is capable of producing about 100 times that amount of power. At the upper end of the scale, SpaceX's Falcon Heavy rocket (Figure 1.3) produces approximately 1,800,000 hp at liftoff. Finally, a commercial electrical power plant can generate one billion watts of power, which is an amount sufficient to supply a community of 800,000 households with electricity.

Figure 1.3

The Falcon Heavy rocket from SpaceX, which is able to lift into orbit the equivalent of a Boeing 737 jetliner loaded with passengers, crew, luggage, and fuel.

▶ 1.2 What Is Engineering?

The word "engineering" derives from the Latin root *ingeniere*, meaning to design or to devise, which also forms the basis of the word "ingenious." Those meanings are quite appropriate summaries of the traits of a good engineer. At the most fundamental level, engineers apply their knowledge of mathematics, science, and materials—as well as their skills in communications and business—to develop new and better technologies. Rather than experiment solely through trial and error, engineers are educated to use mathematics, scientific principles, and computer simulations (Figure 1.4) as tools to create faster, more accurate, and more economical designs.

In that sense, the work of an engineer differs from that of a scientist, who would normally emphasize the discovery of physical laws rather than apply those phenomena to develop new products. Engineering is essentially a bridge between scientific discovery and product applications. Engineering does not exist for the sake of furthering or applying mathematics, science, and computation by themselves. Rather, engineering is a driver of social and economic growth and an integral part of the business cycle. With that perspective, the U.S. Department of Labor summarizes the engineering profession as follows:

Engineers apply the theories and principles of science and mathematics to research and develop economical solutions to technical problems. Their work is the link between perceived social needs and commercial applications. Engineers design products, machinery to build those products, plants in which those products are made, and the systems that ensure the quality of the products and the efficiency of the workforce and manufacturing process. Engineers design, plan, and supervise the construction of buildings, highways, and transit systems. They develop and implement improved ways to extract, process, and use raw materials, such as petroleum and natural gas. They develop new materials that both improve the performance of products and take advantage of advances in technology. They harness the power of the sun, the Earth, atoms, and electricity for use in supplying

Figure 1.4

On a day-to-day basis, mechanical engineers use state-of-the-art cyber-enabled tools to design, visualize, simulate, and improve products.

Copyright © Kevin C. Hulsey.

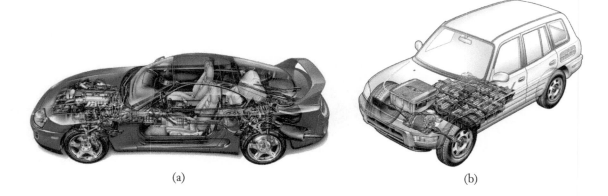

(a)   (b)

the Nation's power needs, and create millions of products using power. They analyze the impact of the products they develop or the systems they design on the environment and on people using them. Engineering knowledge is applied to improving many things, including the quality of healthcare, the safety of food products, and the operation of financial systems.

Many students begin to study engineering because they are attracted to the fields of mathematics and science. Others migrate toward engineering careers because they are motivated by an interest in technology and how everyday things work or, perhaps with more enthusiasm, how not-so-everyday things work. A growing number of others are impassioned by the significant impact that engineers can have on global issues such as clean water, renewable energy, sustainable infrastructures, and disaster relief.

Regardless of how students are drawn to it, engineering is distinct from the subjects of mathematics and science. At the end of the day, the objective of an engineer is to have built a device that performs a task that previously couldn't have been completed or couldn't have been completed so accurately, quickly, or safely. Mathematics and science provide some of the tools and methods that enable an engineer to test fewer mock-ups by refining designs on paper and with computer simulations, before any metal is cut or hardware is built. As suggested by Figure 1.5, "engineering" could be defined as the intersection of activities related to mathematics, science, computer simulation, and hardware.

Approximately 1.5 million people are employed as engineers in the United States. The vast majority work in industry, and fewer than 10% are employed by federal, state, and local governments. Engineers who are federal employees are

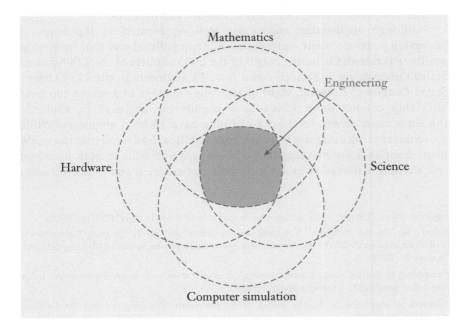

Figure 1.5

Engineers combine their skills in mathematics, science, computers, and hardware.

often associated with such organizations as the National Aeronautics and Space Administration (NASA) or the Departments of Defense (DOD), Transportation (DOT), and Energy (DOE). About 3–4% of all engineers are self-employed, working mostly in consulting and entrepreneurial capacities. Further, an engineering degree prepares students to work in a wide range of influential fields. In a recent list of the CEOs from the Standard & Poor's 500, 33% have undergraduate degrees in engineering, which is almost three times the number as those who earned business administration or economics degrees. Similar surveys showed that 28% of the CEOs in the Fortune 50 had an undergraduate engineering degree. Of the 13 major industry sectors, engineering was the most popular major for CEOs in nine of them:[1]

- Business services
- Chemicals
- Communications
- Electricity, gas, and sanitary
- Electronic components
- Industrial and commercial machinery
- Measuring instruments
- Oil and gas extraction
- Transportation equipment

This is understandable since engineers know that successful problem solving starts with effective information gathering and sound assumptions. They know how to process information to make decisions while taking into account unknown parameters. They also know when to isolate facts and emotions in their decisions while being incredibly innovative and intuitive.

Although engineering majors are well represented in top business leadership positions, their representation in top political and civic leadership positions is mixed. Currently, only 9 of the 540 members of the 114th United States Congress are engineers[2] down from 11 engineers in the 113th United States Congress.[3] However, eight of the nine members of a recent top civic leadership committee in China have engineering degrees.[4] In addition, the three most recent Presidents of China have all been engineers. While government service may not be your career ambition, leaders all over the world in all disciplines are realizing that a broad range of skills in both hard and soft sciences is necessary in a world where globalization and communication

---

[1] Spencer Stuart, "Leading CEOs: A Statistical Snapshot of S&P 500 Leaders" (Chicago, 2008).

[2] Lucey, B., "By the Numbers: How well do you know the 114th Congress?" http://www.dailynewsgems.com/2015/01/by-the-numbers-how-well-do-you-know-the-114th-congress.html, January 25, 2015.

[3] "Engineers in Politics," https://www.asme.org/career-education/early-career-engineers/me-today/me-today-march-2013-issue/engineers-in-politics

[4] Norman R. Augustine, *Is America Falling off the Flat Earth*? (Washington, DC: The National Academies Press, 2007).

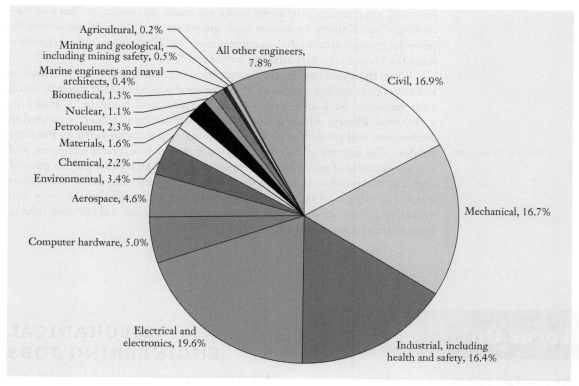

Agricultural, 0.2%
Mining and geological, including mining safety, 0.5%
Marine engineers and naval architects, 0.4%
Biomedical, 1.3%
Nuclear, 1.1%
Petroleum, 2.3%
Materials, 1.6%
Chemical, 2.2%
Environmental, 3.4%
Aerospace, 4.6%
Computer hardware, 5.0%
Electrical and electronics, 19.6%
Industrial, including health and safety, 16.4%
Mechanical, 16.7%
Civil, 16.9%
All other engineers, 7.8%

Figure 1.6

Percentages of engineers working in the traditional engineering fields and their specializations.

Based on data from the United States Department of Labor.

technologies are making geographic divisions increasingly irrelevant. As a result, the field of engineering is changing and this textbook encompasses many of these changes in how engineers need to view, model, analyze, solve, and disseminate the technical, social, environmental, economic, and civic challenges from a global perspective.

Most engineers, while earning a degree in one of the major branches, end up specializing. Though 17 engineering specialties are covered in the Federal Government's Standard Occupational Classification (SOC) system, numerous other specialties are recognized by professional societies. Further, the major branches of engineering have many subdivisions. For example, civil engineering includes the subdivisions of structural, transportation, urban, and construction engineering; electrical engineering includes the subdivisions of power, control, electronics, and telecommunications engineering. Figure 1.6 depicts the distribution of engineers in the major branches, as well as several other specializations.

Engineers develop their skills first through formal study in an accredited bachelor's degree program and later through advanced graduate studies and/or practical work experience under the supervision of accomplished and senior engineers. When starting a new project, engineers often rely on their reasoning, physical intuition, hands-on skills, and the judgment gained through previous technical experiences. Engineers routinely make approximate "back-of-the-envelope" calculations to answer such questions as, "Will a 10-hp engine be powerful enough to drive that air compressor?" or "How many g's of acceleration must the blade in the turbocharger withstand?"

Lifelong learning

When the answer to a particular question isn't known or more information is needed to complete a task, an engineer conducts additional research using such resources as books, professional journals, and trade publications in a technical library; sites such as Google Scholar or CiteSeer; engineering conferences and product expositions; patents; and data provided by industry vendors. The process of becoming a good engineer is a *lifelong* endeavor, and it is a composite of education and experience. One can make a good argument that it is not possible to build a lifelong career on only the material that was learned in college. As technologies, markets, and economies quickly grow and evolve, engineers are constantly learning new approaches and problem-solving techniques and informing others of their discoveries.

Focus On ⊕

MECHANICAL ENGINEERING JOBS

As you begin your formal mechanical engineering education, keep the outcome of your degree in mind. As your education process continues, either formally with more degrees or informally with on the job training, the immediate outcome is a job that matches your skills, passions, and education. While you may have some perceptions about where mechanical engineers work, you may be surprised to find mechanical engineering opportunities in almost every company. For instance, a quick search reveals the following positions for candidates with a bachelor level degree in mechanical engineering.

CAD models, part sourcing, and assembly of support equipment. You will take responsibility of some subset of our environmental tests which include sine and random vibration testing, acoustics testing, shock testing, and mechanism characterization. You will contribute detailed design documentation including test plans and reports, test procedures, assembly drawings and assembly instructions as needed to the satellite or supporting equipment designs. You will own production design-work such as CAD updates incorporating lessons learned, drawing fixes, and clarification documents.

Google/Skybox Imaging

Description Summary: You will design in CAD our next generation satellite (a large, complex assembly) making sure parts fit properly together and the CAD model matures efficiently with the program; early layout in a lightweight model up through detailed models representing "as built." You will be responsible for small-to-medium

General Requirements:

- Collaborate with other engineers to establish the best solution or design
- Design mechanical parts with an acute awareness of manufacturability and strive for simplicity
- Plan complex tests on a critical piece of hardware

- Work hands-on in the machine shop to prototype, assemble completed designs, and run tests
- Experience with mechanical engineering fundamentals (thermodynamics, fluid dynamics, mechanics of materials)

Apple, Inc.

Description Summary: Lead the design, development and validation of sensor technologies including owning the end-to-end mechanical development and integration of a sensor module into a product, with responsibilities including:

- Brainstorming design concepts and executing design concepts in all phases of a development cycle
- Generating innovative designs with cross-functional teams while driving design towards Apple's cosmetic requirements
- Defining mechanical component outlines and assembly schemes
- Generating dimensional and tolerance analysis
- Participating in the development of new manufacturing processes
- Design validation and characterization from prototype bring-up to product testing

General Requirements:

- 3D CAD experience required
- Scientific method, experimental process, root cause analysis
- Applied knowledge in flex circuits, printed circuit boards, material science and basic chemistry is beneficial
- Excellent written and verbal communication skills and people skills; ability to interact with management, team members and external vendors
- Teamwork: the candidate must be able to communicate well with cross-functional team members, be able to efficiently collaborate with team members to achieve project goals, and contribute positively to the engineering community

Amazon

Description Summary: Given the rapid growth of our business, we can achieve Earth's biggest selection and still manage to offer lower prices every day to customers by providing cutting-edge automation technology and excellent decision support tools/services. If you are seeking an environment where you can drive innovation, want to apply state-of-the-art technologies to solve extreme-scale real world problems, and want to provide visible benefit to end-users in an iterative fast paced environment, the Amazon Prime Air Team is your opportunity. You will work with an interdisciplinary team to execute product designs from concept to production, including design, prototyping, validation, testing and certification. You will also work with manufacturing, supply chain, quality and outside vendors to ensure a smooth transition to production.

General Requirements:

- Experience designing and analyzing robust, mechanical systems
- Enjoy problem solving and possess practical knowledge of prototype design as well as production run manufacturing methods
- Experience with CREO with knowledge of robust part design, managing large assemblies and creating detailed documentation
- Strong hands on experience with the ability to craft simple proof-of-concept models in-house
- Thorough understanding and use of principles, theories and concepts in mechanical, aerospace, or robotics engineering and design

In this textbook, we cover a number of these skills to help you prepare to be a successful professional in the dynamic field of mechanical engineering.

▶ 1.3 Who Are Mechanical Engineers?

The field of mechanical engineering encompasses the properties of forces, materials, energy, fluids, and motion, as well as the application of those elements to devise products that advance society and improve people's lives. The U.S. Department of Labor describes the profession as follows:

> Mechanical engineers research, develop, design, manufacture and test tools, engines, machines, and other mechanical devices. They work on power-producing machines such as electricity-producing generators, internal combustion engines, steam and gas turbines, and jet and rocket engines. They also develop power-using machines such as refrigeration and air-conditioning equipment, robots used in manufacturing, machine tools, materials handling systems, and industrial production equipment.

Mechanical engineers are known for their broad scope of expertise and for working on a wide range of machines. Just a few examples include the microelectromechanical acceleration sensors used in automobile air bags; heating, ventilation, and air-conditioning systems in office buildings; land, ocean, and space robotic exploration vehicles; heavy off-road construction equipment; hybrid gas-electric vehicles; gears, bearings, and other machine components (Figure 1.7); artificial hip implants; deep-sea research vessels; robotic manufacturing systems; replacement heart valves; noninvasive equipment for detecting explosives; and interplanetary exploration spacecraft (Figure 1.8).

Based on employment statistics, mechanical engineering is one of the largest engineering fields, and it is often described as offering the greatest flexibility of career choices. In 2013, approximately 258,630 people were employed as mechanical engineers in the United States, a population representing over 16% of all engineers. The discipline is closely related to the technical areas of industrial (254,430 people), aerospace (71,500), and nuclear (16,400)

Figure 1.7

Mechanical engineers design machinery and power–transmission equipment using various types of gears as building-block components.

Figure 1.8

The *Mars Exploration Rover* is a mobile geology laboratory used to study the history of water on Mars. Mechanical engineers contributed to the design, propulsion, thermal control, and other aspects of these robotic vehicles.

NASA/JPL/Cornell University

engineering, since each of those fields evolved historically as a spin-off from mechanical engineering. Together, the fields of mechanical, industrial, aerospace, and nuclear engineering account for almost 39% of all engineers. Emerging fields like biotechnology, materials science, and nanotechnology are expected to create new job opportunities for mechanical engineers. The U.S. Bureau of Labor Statistics predicts an increase of nearly 20,000 mechanical engineering jobs by the year 2022. A degree in mechanical engineering can also be applied to other engineering specialties, such as manufacturing engineering, automotive engineering, civil engineering, or aerospace engineering.

While mechanical engineering often is regarded as the broadest of the traditional engineering fields, there are many opportunities for specialization in the industry or technology that interests you. For example, an engineer in the aviation industry might focus her career on advanced technologies for cooling turbine blades in jet engines or fly-by-wire systems for controlling an aircraft's flight.

Above all else, mechanical engineers make hardware that works. An engineer's contribution to a company or another organization ultimately is evaluated based on whether the product functions as it should. Mechanical engineers design equipment, it is produced by companies, and it is then sold to the public or to industrial customers. In the process of that business cycle, some aspect of the customer's life is improved, and society as a whole benefits from the technical advances and additional opportunities that are offered by engineering research and development.

Mechanical Engineering's Top Ten Achievements

Mechanical engineering isn't all about numbers, calculations, computers, gears, and grease. At its heart, the profession is driven by the desire to advance society through technology. One of the most important professional

Table 1.1

Top Ten Achievements of the Mechanical Engineering Profession Compiled by the American Society of Mechanical Engineers

1. The automobile
2. The Apollo program
3. Power generation
4. Agricultural mechanization
5. The airplane
6. Integrated-circuit mass production
7. Air conditioning and refrigeration
8. Computer-aided engineering technology
9. Bioengineering
10. Codes and standards

organizations in the field is ASME, founded as the American Society of Mechanical Engineers, which currently "promotes the art, science, and practice of multidisciplinary engineering and allied sciences around the globe." ASME surveyed its members to identify the major accomplishments of mechanical engineers. This top ten list of achievements, summarized in Table 1.1, can help you better understand who mechanical engineers are and appreciate the contributions they have made to your world. In descending order of the accomplishment's perceived impact on society, the following milestones were recognized in the survey:

1. **The automobile.** The development and commercialization of the automobile were judged as the profession's most significant achievement in the twentieth century. Two factors responsible for the growth of automotive technology have been high-power, lightweight engines and efficient processes for mass manufacturing. In addition to engine improvements, competition in the automobile market has led to advances in the areas of safety, fuel economy, comfort, and emission control (Figure 1.9). Some of the newer technologies include hybrid gas-electric vehicles, antilock brakes, run-flat tires, air bags, widespread use of composite materials, computer control of fuel-injection systems, satellite-based navigation systems, variable valve timing, and fuel cells.

Figure 1.9

Mechanical engineers design, test, and manufacture advanced automotive systems, such as this (a) suspension system, (b) automatic transmission, and (c) six-cylinder gas-electric hybrid engine.

Copyright © Kevin C. Hulsey.

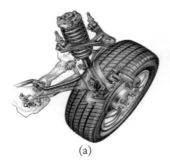
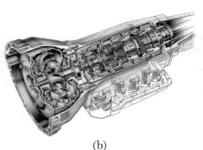
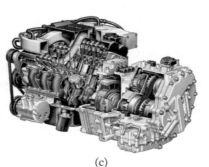

(a)   (b)   (c)

Figure 1.10

Astronaut John Young, commander of the *Apollo 16* mission, leaps from the lunar surface at the Descartes landing site as he salutes the United States flag. The roving vehicle is parked in front of the lunar module.

NASA/Charlie Duke

2. **The Apollo program.** In 1961, President John F. Kennedy challenged the United States to land a man on the Moon and return him safely to Earth. The first portion of that objective was realized fewer than ten years later with the July 20, 1969 landing of *Apollo 11* on the lunar surface. The three-man crew of Neil Armstrong, Michael Collins, and Buzz Aldrin returned safely several days later. Because of its technological advances and profound cultural impact, the Apollo program was chosen as the second most influential mechanical engineering achievement of the twentieth century (Figure 1.10).

   The Apollo program was based on three primary engineering developments: the huge three-stage *Saturn V* launch vehicle that produced some 7.5 million pounds of thrust at liftoff, the command and service module, and the lunar excursion module, which was the first vehicle ever designed to be flown only in space.

3. **Power generation.** One aspect of mechanical engineering involves designing machinery that can convert energy from one form to another. Abundant and inexpensive energy is recognized as an important factor behind economic growth and prosperity, and the generation of electrical power is recognized as having improved the standard of living for billions of people across the globe. In the twentieth century, entire societies changed as electricity was produced and routed to homes, businesses, and factories.

   Although mechanical engineers are credited with having developed efficient technologies to convert various forms of stored energy into more easily distributed electricity, the challenge to bring power to every man, woman, and child around the globe still looms for mechanical engineers. As the supply of natural resources diminishes and as fuels become more expensive in terms of both cost and the environment, mechanical engineers will become even more involved in developing advanced power-generation technologies, including solar, ocean, and wind power systems (Figure 1.11, see page 14).

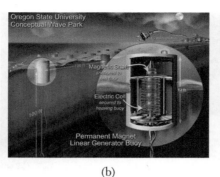

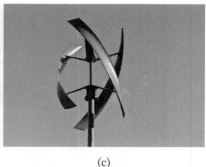

(a)   (b)   (c)

Figure 1.11

Mechanical engineers design machines for producing energy from a variety of renewable sources, such as (a) solar power towers, (b) wave energy power plants , and (c) innovative wind turbines.

(a) © PM photos/Alamy. (b) Science Source. (c) eldeiv/Shutterstock.com

4. **Agricultural mechanization.** Mechanical engineers have developed technologies to improve significantly the efficiency in the agricultural industry. Automation began in earnest with the introduction of powered tractors in 1916 and the development of the combine, which greatly simplified harvesting grain. Decades later, research is underway to develop the capability for machines to harvest a field autonomously, without any human intervention using advanced machinery, GPS technology, and intelligent guidance and control algorithms (Figure 1.12). Other advances include improved weather observation and prediction, high-capacity irrigation pumps, automated milking machines, and the digital management of crops and the control of pests.

5. **The airplane.** The development of the airplane and related technologies for safe powered flight were also recognized by the American Society of Mechanical Engineers as a key achievement of the profession.

Mechanical engineers have developed or contributed to nearly every aspect of aviation technology. Advancements in high-performance military aircraft include vectored turbofan engines that enable the pilot to redirect the engine's thrust for vertical takeoffs and landings. Mechanical

Figure 1.12

Robotic vehicles under development can learn the shape and terrain of a field of grain and harvest it with essentially no human supervision.

Reprinted with permission of the National Robotics Engineering Consortium.

Figure 1.13

This prototype of the X-48B, a blended wing-body aircraft, is being tested at the full-scale wind tunnel at NASA Langley Research Center in Virginia.

NASA/Jeff Caplan

engineers design the combustion systems, turbines, and control systems of such advanced jet engines. By taking advantage of testing facilities such as wind tunnels (Figure 1.13), they have also spearheaded the design of turbines, development of control systems, and discovery of lightweight aerospace-grade materials, including titanium alloys and graphite-fiber-reinforced epoxy composites.

6. **Integrated-circuit mass production.** The electronics industry has developed remarkable technologies for miniaturizing integrated circuits, computer memory chips, and microprocessors. The mechanical engineering profession made key contributions during the twentieth century to the manufacturing methods involved in producing integrated circuits. While the vintage 8008 processor that was first sold by the Intel Corporation in 1972 had 2500 transistors, the current SPARC M7 processor from Oracle has over 10 billion transistors (Figure 1.14, see on page 16).

   Mechanical engineers design the machinery, alignment systems, advanced materials, temperature control, and vibration isolation that enable integrated circuits to be made at the nanometer scale.

   The same manufacturing technology can be used to produce other machines at the micro or nano level. Using these techniques, machines with moving parts can be made so small that they are imperceptible to the human eye and can be viewed only under a microscope. As shown in Figure 1.15 (see page 16), individual gears can be fabricated and assembled into geartrains that are no bigger than a speck of pollen.

7. **Air conditioning and refrigeration.** Mechanical engineers invented the technologies of efficient air conditioning and refrigeration. Today, these systems not only keep people safe and comfortable, but also preserve food and medical supplies in refrigeration systems. Like other infrastructures, we typically do not recognize the value of air conditioning until it is gone. In a record European heat wave during the summer of

Figure 1.14

Mechanical engineers have been instrumental in developing the manufacturing technologies that are necessary to build billions of electronic components on devices such as the Oracle SPARC M7.

Courtesy of Intel.

2003, for instance, over 10,000 people—many elderly—died in France as a direct result of the searing temperatures.

8. **Computer-aided engineering technology.** The term "computer-aided engineering" (CAE) refers to a wide range of automation technologies in mechanical engineering, and it encompasses the use of computers for performing calculations, preparing technical drawings, simulating performance (Figure 1.16), and controlling machine tools in a factory. Over the past several decades, computing and information technologies have changed the manner in which mechanical engineering is practiced.

As an example, the Boeing 777 was the first commercial airliner to be developed through a paperless computer-aided design process. The 777's design began in the early 1990s, and a new computer infrastructure had to be created specifically for the design engineers. Conventional paper-and-pencil drafting services were nearly eliminated. Computer-aided design, analysis, and manufacturing activities were integrated across some 200 design teams that were spread over 17 time zones. Because the plane had over 3 million individual components, making everything fit together was a remarkable challenge. Through the extensive usage of CAE

Figure 1.15

Mechanical engineers design and build machines that are microscopic in size. These tiny gears are dwarfed by a spider mite, and the entire geartrain is smaller than the diameter of a human hair.

Courtesy of Sandia National Laboratories.

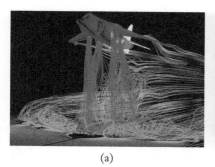
(a)

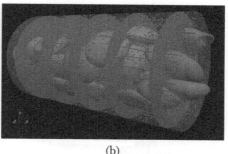
(b)

Figure 1.16

(a) Mechanical engineers use computer simulations to analyze and visualize the flow of air around aircraft including the Harrier Jet.

Science Source

(b) A dynamic simulation of the flow of blood through an artery of the brain is used to observe the interaction between plasma and blood, helping engineers design medical devices and helping doctors understand disease diagnosis and treatment.

Joseph A. Insley and Michael E. Papka, Argonne National Laboratory.

tools, designers were able to check part-to-part fits in a virtual, simulated environment before any hardware was produced. Current CAE tools are being developed for diverse computing platforms including leveraging mobile devices, cloud computing technologies, and virtual machines.

9. **Bioengineering.** The discipline of bioengineering links traditional engineering fields with the life sciences and medicine. Although bioengineering is considered an emerging field, it ranked in the American Society of Mechanical Engineer's top ten list not only for the advances that have already been made, but also for its future potential in addressing medical and health-related problems.

One objective of bioengineering is to create technologies to expand the pharmaceutical and healthcare industries, including drug discovery, genomics (Figure 1.17), ultrasonic imaging, artificial joint replacements, cardiac pacemakers, artificial heart valves, and robotically assisted surgery (Figure 1.18, see page 18). For instance, mechanical engineers apply the principles of heat transfer to assist surgeons with cryosurgery, a technique in which the ultralow temperature of liquid nitrogen is used to destroy malignant tumors. Tissue engineering and the development of artificial organs are other fields where mechanical engineers contribute, and they often work with physicians and scientists to restore damaged skin, bone, and cartilage in the human body.

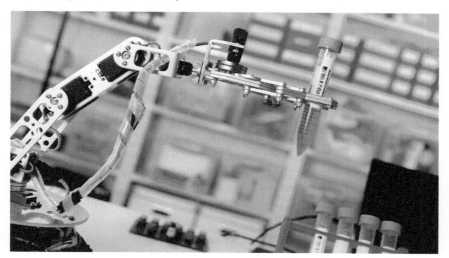

Figure 1.17

Mechanical engineers design and build automated test equipment that is used in the biotechnology industry.

science photo/Shutterstock.com

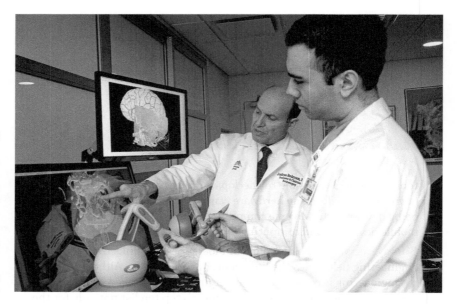

Figure 1.18

Neurosurgeons plan their case using patient-specific augmented reality and virtual reality simulation technology before performing the operation.

Joshua B. Bederson, MD and Mount Sinai Health System

10. **Codes and standards.** The products that engineers design must connect to, and be compatible with, the hardware that is developed by others. Because of codes and standards, you can have confidence that the gasoline purchased next month will work in your car just as well as the fuel purchased today, and that the socket wrench purchased at an automobile parts store in the United States will fit the bolts on a vehicle that was manufactured in Germany. Codes and standards are necessary to specify the physical characteristics of mechanical parts so that others can clearly understand their structure and operation.

Many standards are developed through consensus among governments and industry groups, and they have become increasingly important as companies compete internationally for business. Codes and standards involve a collaboration among trade associations, professional engineering societies such as the American Society of Mechanical Engineers, testing groups such as Underwriters Laboratories, and organizations such as the American Society of Testing and Materials.

The Future of Mechanical Engineering

While the ASME's list of the top ten accomplishments captures the past achievements of the field, ASME also released a study identifying a number of emerging fields within mechanical engineering.[5] In Figure 1.19, the emerging fields are sorted according to how often they were mentioned in the respondent surveys. Not surprisingly, the top emerging fields are related to solving heath care and energy issues. Collectively, these fields will become prominent opportunities for mechanical engineers to have significant impact on global health, social, and environmental issues.

---

[5]*The State of Mechanical Engineering: Today and Beyond*, ASME, New York, NY, 2012.

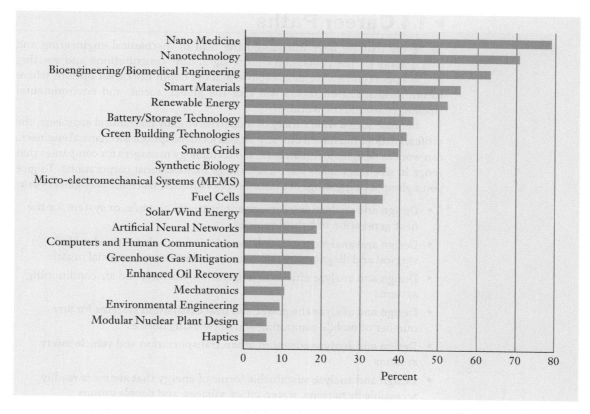

Figure 1.19

The top emerging fields of mechanical engineering identified by an ASME study.

Source: Based on data from *The State of Mechanical Engineering: Today and Beyond*, ASME, New York, NY, 2012.

Part of our intent is to help prepare you for a successful and significant career in any mechanical engineering field, emerging or traditional. As a result, in the subsequent chapters we cover the fundamental principles of mechanical engineering that will allow you to progress in your continued study and practice of the discipline. In fact, in the same ASME study, the most important enduring fields/tools in mechanical engineering were also identified. The principles covered in the following chapters correspond to many of these enduring fields including the following:

- Design Engineering (Chapter 2)
- Finite Element Analysis (Chapters 4 and 5)
- Fluid Mechanics (Chapter 6)
- Stress of Heat Flow Analysis (Chapter 7)
- Electronic Motor Development (Chapter 8)

In addition, ASME identified the "most needed skills" for mechanical engineers to succeed in the next 10 to 20 years. At the top of the list was *the ability to communicate*, which is the focus of Chapter 3. So whether your career path takes you into an emerging field or an enduring field, mechanical engineering will prepare you to be able solve problems using a blend of analytical models, computational tools, and personal skills.

▶ 1.4 Career Paths

Now that we have introduced the field of mechanical engineering and some of the profession's most significant contributions and exciting opportunities, we next explore in more depth the career options where future mechanical engineers will face the global, social, and environmental challenges around the world.

Because such a wide variety of industries employ mechanical engineers, the profession does not have a one-size-fits-all job description. Mechanical engineers can work as designers, researchers, and technology managers for companies that range in size from small start-ups to large multinational corporations. To give you a glimpse of the range of available opportunities, mechanical engineers can:

- Design and analyze any component, material, module, or system for the next generation of hybrid vehicles
- Design and analyze medical devices, including aids for the disabled, surgical and diagnostic equipment, prosthetics, and artificial organs
- Design and analyze efficient refrigeration, heating, and air-conditioning systems
- Design and analyze the power and heat dissipation systems for any number of mobile computing and networking devices
- Design and analyze advanced urban transportation and vehicle safety systems
- Design and analyze sustainable forms of energy that are more readily accessible by nations, states, cities, villages, and people groups
- Design and analyze the next generation of space exploration systems
- Design and analyze revolutionary manufacturing equipment and automated assembly lines for a wide range of consumer products
- Manage a diverse team of engineers in the development of a global product platform, identifying customer, market, and product opportunities
- Provide consultant services to any number of industries, including chemical, plastics, and rubber manufacturing; petroleum and coal production; computer and electronic products; food and beverage production; printing and publishing; utilities; and service providers
- Work in public service for such governmental agencies as the National Aeronautics and Space Administration, Department of Defense, National Institute of Standards and Technology, Environmental Protection Agency, and national research laboratories
- Teach mathematics, physics, science, or engineering at the high school, 2-year college, or 4-year university level
- Pursue significant careers in law, medicine, social work, business, sales, or finance

Historically, mechanical engineers could take either a technical track or a management track with their careers. However, the lines between these tracks are blurring as emerging product development processes are demanding knowledge not only about technical issues but also about economic, environmental, customer, and manufacturing issues. What used to be done in colocated teams using centrally located engineering expertise is now done by globally distributed teams taking advantage of engineering expertise in multiple geographic regions, lower-cost processes, global growth opportunities, and access to leading technologies.

Job openings historically labeled as "mechanical engineer" now include a number of diverse titles that reflect the changing nature of the profession. For example, the following job position titles all required a degree in mechanical engineering (taken from a leading job website):

Career opportunities

- Product engineer
- Systems engineer
- Manufacturing engineer
- Renewable energy consultant
- Applications engineer
- Product applications engineer
- Mechanical device engineer
- Process development engineer
- Principal engineer
- Sales engineer
- Design engineer
- Power engineer
- Packaging engineer
- Electro-mechanical engineer
- Facilities design engineer
- Mechanical product engineer
- Energy efficiency engineer
- Mechatronics engineer
- Project capture engineer
- Plant engineer

Aside from requiring technical knowledge and skills, landing a job, keeping a job, and progressing upward through one's career will depend on a number of skills that, at first glance, might appear to be nontechnical in nature. Mechanical engineers must be capable of taking initiative when handling work assignments, efficiently finding answers to problems, and accepting additional responsibility with success. A quick survey of engineering positions on any job website will demonstrate that employers place significant value on the ability of a mechanical engineer to communicate to a wide range of backgrounds and in all forms of verbal and written media. In fact, companies looking to hire engineers routinely note effective communication as the most important nontechnical attribute for aspiring engineers. The reason is quite simple—at each stage of a product's development, mechanical engineers work with a wide range of people: supervisors, colleagues, marketing, management, customers, investors, and suppliers. An engineer's ability to discuss and explain technical and business concepts clearly, and to interact well with coworkers, is critical.

Communication skills

After all, if you have an outstanding and innovative technical breakthrough, but you are unable to convey the idea to others in a convincing manner, your idea is not very likely to be accepted.

▶ 1.5 Typical Program of Study

As you begin to study mechanical engineering, your program will most likely include the following four components:

- General education courses in the humanities, social sciences, and fine arts
- Preparatory courses in mathematics, science, and computer programming
- Core courses in fundamental mechanical engineering subjects
- Elective courses on specialized topics that you find particularly interesting

Innovation and design

After completing the core curriculum, you often will have the flexibility to build an individualized program of study through elective courses in such fields as aerospace engineering, automotive engineering, computer-aided design, manufacturing, biomedical engineering, and robotics, among other fields.

Figure 1.20

Hierarchy of topics and courses studied in a typical mechanical engineering curriculum.

The major topics in a typical mechanical engineering curriculum are shown in Figure 1.20. While the topics are allocated into separate branches, the mechanical engineering curriculum is becoming an integrated system with interrelationships among many of the courses, topics, and knowledge areas. At the heart of being a mechanical engineer are *innovation and design*. An important place to start your study is to understand that the design of products,

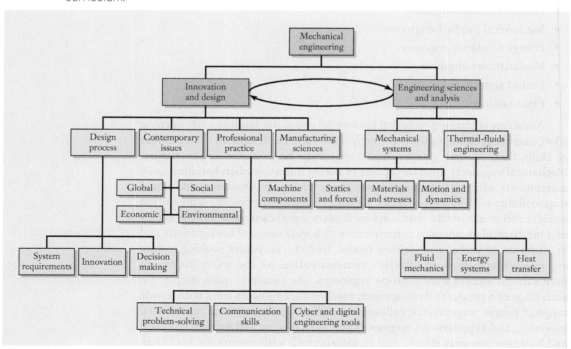

systems, and processes is how mechanical engineers impact the social, global, environmental, and economic challenges in the world. Engineers are relied on to be creative not only in solving technical problems in innovative ways, but to find and to pose these problems in novel ways.

Knowledge of innovation and design will require the study of how a design process is structured, including the following topics:

- The development of system requirements from a variety of system stakeholders
- The generation of innovative concept alternatives and the effective selection and realization of a final design
- Principles of sound decision making applied to the multitude of trade-offs involved in a product development process

In addition, knowledge of contemporary and emerging issues is critical to design products and systems that will sustain and transform lives, communities, economies, nations, and the environment. Of course, because of the direct impact mechanical engineers have on potentially billions of lives, they must be outstanding professionals of the highest character. To become such a professional, you will learn the following skills:

- Sound technical problem-solving skills
- Effective practices in technical communications (oral presentations, technical reports, e-mails)
- The latest digital and cyber-enabled tools to support engineering design processes

Instruction on innovation and design would not be complete without some fundamental understanding of the processes required to physically realize products. This includes course materials focused on the manufacturing sciences and on how products actually get built, produced, and assembled.

Providing the foundation for the curricular components of innovation and design are the core *engineering sciences and analysis*. A series of courses focuses on mechanical systems, including modeling and analyzing the components of mechanical devices (e.g., gears, springs, mechanisms). These core courses usually include the following issues:

Engineering sciences and analysis

- Understanding the forces that act on machines and structures during their operation, including components that move and those that do not
- Determining whether structural components are strong enough to support the forces that act on them and what materials are the most appropriate
- Determining how machines and mechanisms will move and the amount of force, energy, and power that is transferred between them

Another series of courses focuses on thermal fluid principles, including modeling and analyzing the behavior and properties of thermodynamic and fluidic systems. These core courses usually include the following issues:

- The physical properties of liquids and gases and the drag, lift, and buoyancy forces present between fluids and structures

- The conversion of energy from one form to another by efficient power generation machinery, devices, and technologies
- Temperature control and the management of heat through the processes of conduction, convection, and radiation

Gaining experience   Along with formal study, it is also important to *gain experience* through summer employment, internships, research projects, co-op programs, and study-abroad opportunities. Those experiences, as well as courses completed outside the formal engineering program, will greatly broaden your perspective of the role that engineering plays in our global societies. Increasingly, employers are looking for engineering graduates who have capabilities and experiences above and beyond the traditional set of technical and scientific skills. Knowledge of business practices, interpersonal relationships, organizational behavior, international cultures and languages, and communication skills are important factors for many engineering career choices. For instance, a corporation with overseas subsidiaries, a smaller company that has customers in foreign countries, or a company that purchases instrumentation from an overseas vendor will each value engineers who are conversant in foreign languages. As you plan your engineering degree, choose electives, and perhaps prepare for a minor degree, pay attention to those broader skills.

The Accreditation Board for Engineering and Technology (ABET, www.abet.org) is an organization formed by over two dozen technical and professional societies, including the American Society of Mechanical Engineers. ABET endorses and certifies almost 3000 engineering programs at more than 600 colleges and universities across the United States through their accreditation process. ABET has also begun accrediting international engineering programs. The board has identified a set of skills that new engineering graduates are expected to have, which are useful benchmarks for you to consider while monitoring progress during your studies:

a. *An ability to apply knowledge of mathematics, science, and engineering.* Since World War II, science has been a mainstay of engineering education, and mechanical engineering students have traditionally studied mathematics, physics, and chemistry.

b. *An ability to design and conduct experiments, as well as to analyze and interpret data.* Mechanical engineers set up and perform experiments, use state-of-the-art measurement equipment, and interpret the physical implications of the results.

c. *An ability to design a system, component, or process to meet desired needs within realistic constraints such as economic, environmental, social, political, ethical, health and safety, manufacturability, and sustainability.* This skill is the core of mechanical engineering. Engineers are trained to conceive solutions to open-ended technical problems and to prepare a detailed, functional, safe, environmentally friendly, and profitable design.

d. *An ability to function on multidisciplinary teams.* Mechanical engineering is not an individual activity, and you will need to demonstrate the skills necessary to interact effectively with others in the business community.

e. *An ability to identify, formulate, and solve engineering problems.* Engineering is firmly based on mathematical and scientific principles, but it also involves creativity and innovation to design something new. Engineers are often described as problem solvers who can confront an unfamiliar situation and develop a clear solution.

f. *An understanding of professional and ethical responsibility.* Through your courses and personal experiences, you will see that engineers have a responsibility to act professionally and ethically. Engineers need to recognize ethical and business conflicts and resolve them with integrity when they arise.

g. *An ability to communicate effectively.* Engineers are expected to be competent in both written and verbal communication, including the presentation of engineering calculations, computations, measurement results, and designs.

h. *The broad education necessary to understand the impact of engineering solutions in a global, economic, environmental, and societal context.* Engineers create products, systems, and services that potentially impact millions of people across the globe. A mechanical engineer who is aware of that context is able to make sound technical, ethical, and career decisions.

i. *A recognition of the need for and an ability to engage in lifelong learning.* "Educate" does not mean to fill up with facts; rather, it means to "bring out." Therefore, your intellectual growth should continue to bring out new knowledge and understanding long after you graduate.

j. *A knowledge of contemporary issues.* Engineers need to be aware of social, global, environmental, economic, and political developments that are of current importance, since they provide the context for the technical problems that a society faces and that engineers are expected to solve.

k. *An ability to use the techniques, skills, and modern engineering tools necessary for engineering practice.* This skill is based in part on using computer-aided engineering software tools and the ability to think critically about numerical results.

Outcomes (a) and (b) are attained by learning the core engineering science and mathematics fundamentals throughout the mechanical engineering curriculum, which are introduced for you in Chapters 4–8. Outcomes (c), (h), and (j) are addressed in Chapters 1 and 2 and will also be part of other mechanical engineering courses, including upper-level design courses. Some of the computer-aided design and manufacturing tools that are discussed in Chapter 2 are also relevant to outcome (k). In Chapter 3, we focus directly on preparing you to attain outcomes (e) and (g), which are critical to you becoming a successful engineering professional, ready to design, create, innovate, study, analyze, produce, and impact lives in a dynamic and global society. You will have opportunities to develop skills and understanding to address outcomes (d), (f), and (g) throughout your curriculum.

Summary

This chapter is intended to give you a perspective on the purpose, challenges, responsibilities, opportunities, rewards, and satisfaction of being a mechanical engineer. Simply stated, engineers conceive, design, and build things that work and impact lives. Engineers are regarded as being good problem solvers who can clearly communicate the results of their work to others through drawings, written reports, and verbal presentations. Mechanical engineering is a diverse discipline, and it is generally regarded as being the most flexible of the traditional engineering fields. In Section 1.3, the top ten contributions of the mechanical engineering profession were described as having improved the day-to-day lives of literally billions of people. To accomplish those achievements, mechanical engineers use computer-aided software tools for designing, simulating, and manufacturing. Technologies that you may have previously taken for granted—such as abundant and inexpensive electricity, refrigeration, and transportation— take on new meanings as you reflect on their importance to our society and on the remarkable hardware that makes them possible. These enduring technologies and other emerging technologies are now expanding the impact that mechanical engineers continue to have designing and implementing solutions around the world.

Self-Study and Review

1.1. What is engineering?

1.2. What are the differences among engineers, mathematicians, and scientists?

1.3. What is mechanical engineering?

1.4. Compare mechanical engineering to the other traditional engineering fields.

1.5. Describe a half dozen products that mechanical engineers design, improve, or produce, and list some technical issues that must be solved.

1.6. Describe several of the top ten achievements of the mechanical engineering profession listed in Section 1.3.

1.7. Discuss the career options and job titles that are available to mechanical engineers.

1.8. Describe some of the main subjects that comprise a typical mechanical engineering curriculum.

Problems

P1.1

For each of the following systems, give two examples of how a mechanical engineer would be involved in its design.

(a) Passenger automobile engine

(b) Escalator

(c) Computer hard disk drive

(d) Artificial hip implant

(e) Baseball pitching machine

P1.2

For each of the following systems, give two examples of how a mechanical engineer would be involved in its analysis.

(a) Jet engine for a commercial airliner

(b) Rover robot for planetary exploration

(c) Computer ink-jet printer

(d) Smart phone

(e) Can of soda from a vending machine

P1.3

For each of the following systems, give two examples of how a mechanical engineer would be involved in its manufacture.

(a) Graphite-epoxy skis, tennis racket, or golf club

(b) Elevator

(c) Blu-ray player

(d) Automatic teller banking machine

(e) Automotive child safety seat

P1.4

For each of the following systems, give two examples of how a mechanical engineer would be involved in its testing.

(a) Hybrid gas-electric passenger vehicle

(b) Motors for remote-control cars, planes, and boats

(c) Bindings for snowboards

(d) Global positioning system (GPS) satellite receiver

(e) Motorized wheelchair

P1.5

For each of the following systems give an example of how a mechanical engineer would have to address global issues in its design.

(a) Dialysis machine

(b) Microwave oven

(c) Aluminum mountain bike frame

(d) Automotive antilock brake system

(e) Lego toys

P1.6

For each of the following systems give an example of how a mechanical engineer would have to address social issues in its design.

(a) Dishwasher

(b) E-book reader

(c) Coffee maker

(d) Cordless electric drill

(e) Infant high chair

P1.7

For each of the following systems give an example of how a mechanical engineer would have to address environmental issues in its design.

(a) Wetsuit

(b) Refrigerator

(c) Space tourism vehicle

(d) Automobile tires

(e) Infant jogging stroller

P1.8

For each of the following systems give an example of how a mechanical engineer would have to address economic issues in its design.

(a) Clothes dryer

(b) Robotic lawnmower

(c) Retractable stadium roof

(d) Toothpaste tube

(e) Paper clip

P1.9

Read one of the *Mechanical Engineering* articles listed in the References section at the end of this chapter, describing a top ten achievement. Prepare a technical report of at least 250 words summarizing the interesting and important aspects of the achievement.

P1.10

What other product or device do you think should be on the list of the top mechanical engineering achievements? Prepare a technical report of at least 250 words detailing your rationale and listing the interesting and important aspects of the achievement.

P1.11

Looking forward by 100 years, what technological advance do you think will be regarded at that time as having been a significant achievement of the mechanical engineering profession during the twenty-first century? Prepare a technical report of at least 250 words that explains the rationale for your speculation.

P1.12

What do you think are the three most significant issues facing engineers today? Prepare a technical report of at least 250 words that explains your rationale.

P1.13

Interview someone whom you know or contact a company and learn some of the details behind a product that interests you. Prepare a technical report of at least 250 words of the product, the company, or the manner in which a mechanical engineer contributes to the product's design or production.

P1.14

Interview someone whom you know or contact a company and learn about a computer–aided engineering software tool that is related to mechanical engineering. Prepare a technical report of at least 250 words that describes what the software tool does and how it can help an engineer work more efficiently and accurately.

P1.15

Find a recent engineering failure in the news and prepare a technical report of at least 250 words explaining how a mechanical engineer could have prevented the failure. Explain clearly how your proposed prevention is a design, manufacturing, analysis, and/or testing issue.

P1.16

Think of a product you personally have used that you think failed to do what it was supposed to do. Prepare a technical report of at least 250 words describing why you think this product failed and then explain what a mechanical engineer could have done differently to prevent the product from not living up to your expectation.

P1.17*

Google to find the report from *ASME, The State of Mechanical Engineering: Today and Beyond.* Read the report and as a group choose one of the tools and techniques on page 15 of the report that you are unfamiliar with and prepare a short (oral or written) description of it with a clear explanation of what kind of mechanical engineering problems it could be used for.

P1.18*

From the same report in P1.17*, as a group take a look at the chart on page 6 regarding the future view of engineering. Select three of the chart entries (rows) and find a current article that specifically highlights the principle being expressed in the entry. Prepare a short presentation of your articles and the impact your group thinks the three selected principles should have on a mechanical engineering curriculum.

References

Armstrong, N. A., "The Engineered Century," *The Bridge,* National Academy of Engineering, Spring 2000, pp. 14–18.

Gaylo, B. J., "That One Small Step," *Mechanical Engineering,* ASME International, October 2000, pp. 62–69.

Ladd, C. M., "Power to the People," *Mechanical Engineering,* ASME International, September 2000, pp. 68–75.

Lee, J. L., "The Mechanics of Flight," *Mechanical Engineering,* ASME International, July 2000, pp. 55–59.

Leight, W., and Collins, B., "Setting the Standards," *Mechanical Engineering,* ASME International, February 2000, pp. 46–53.

Lentinello, R. A., "Motoring Madness," *Mechanical Engineering,* ASME International, November 2000, pp. 86–92.

Nagengast, B., "It's a Cool Story," *Mechanical Engineering,* ASME International, May 2000, pp. 56–63.

Petroski, H., "The Boeing 777," *American Scientist,* November–December 1995, pp. 519–522.

Rastegar, S., "Life Force," *Mechanical Engineering,* ASME International, March 2000, pp. 75–79.

Rostky, G., "The IC's Surprising Birth," *Mechanical Engineering,* ASME International, June 2000, pp. 68–73.

Schueller, J. K., "In the Service of Abundance," *Mechanical Engineering,* ASME International, August 2000, pp. 58–65.

Weisberg, D. E., "The Electronic Push," *Mechanical Engineering,* ASME International, April 2000, pp. 52–59.

CHAPTER 2

Mechanical Design

- Outline the major steps involved in a mechanical design process.
- Recognize the importance of mechanical design for solving the technical, global, and environmental challenges that society faces.
- Recognize the importance of innovation in designing effective engineered products, systems, and processes.
- Recognize the importance of multidisciplinary teams, collaboration, and technical communication in engineering.

- Be familiar with some of the processes and machine tools used in manufacturing.
- Understand how patents are used to protect a newly developed technology in the business side of engineering.
- Describe the role played by computer-aided engineering tools in linking mechanical design, analysis, and manufacturing.

▶ 2.1 Overview

The National Academy of Engineering (NAE) has identified 14 Grand Challenges facing the global engineering community and profession in the twenty-first century. These challenges are reshaping how engineers view themselves, how and what they learn, and how they think. The challenges are also broadening the perspective of engineers and how they view the communities they impact. The 14 challenges are as follows:

- Make solar energy economical
- Provide energy from fusion
- Develop carbon sequestration methods
- Manage the nitrogen cycle
- Provide access to clean water
- Restore and improve urban infrastructure

- Advance health informatics
- Engineer better medicines
- Reverse-engineer the brain
- Prevent nuclear terror
- Secure cyberspace
- Enhance virtual reality
- Advance personalized learning
- Engineer the tools of scientific discovery

Not only will mechanical engineers play important roles in each of these challenges, but they will also take on significant technical and global leadership roles in a number of the challenges. While reading this list, you may even resonate with one or more of the challenges, perhaps envisioning yourself creating innovative solutions that impact millions of lives. Although the challenges span many scientific and engineering disciplines, the principle that connects them all is *design*. In fact, a recent report from the American Society of Mechanical Engineers[1] states that design engineering will be the most enduring field in mechanical engineering over the next one to two decades. Many multidisciplinary teams will need to design innovative and effective solutions to meet the myriad of subchallenges that each challenge embodies. The focus of this chapter is on understanding the fundamental principles and learning the skills necessary to be part of, to contribute to, or to lead a successful design process.

Element 1: Mechanical design

While discussing the differences between engineers, scientists, and mathematicians in Chapter 1, you saw that the word "engineering" is related to both "ingenious" and "devise." In fact, the process of developing something new and creative lies at the heart of the engineering profession. The ultimate goal, after all, is to build hardware that solves one of the global society's technical problems. The objective of this chapter is to introduce you to some of the issues arising when a new product is designed, manufactured, and patented. The relationship of this chapter to the hierarchy of mechanical engineering disciplines is shown by the blue shaded boxes in Figure 2.1.

You don't need a formal education in engineering to have a good idea for a new or improved product. In fact, your interest in studying mechanical engineering may have been sparked by your own ideas for building hardware. The elements of mechanical engineering that we examine in the remaining chapters of this book—forces in structures and machines, materials and stresses, fluids engineering, thermal and energy systems, and motion and power transmission—are intended to set in place a foundation that will enable you to bring your ideas to reality effectively and systematically. In that respect, the approach taken in this textbook is an analog of the

---

[1] *The State of Mechanical Engineering: Today and Beyond*, ASME, New York, NY, 2012.

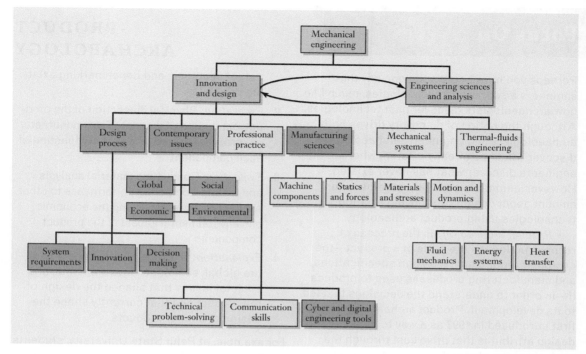

Figure 2.1

Relationship of the topics emphasized in this chapter (blue shaded boxes) relative to an overall program of study in mechanical engineering.

traditional mechanical engineering curriculum: estimation, mathematics, and science are applied to design problems to reduce trial and error. You can use the types of calculations described in Chapters 3–8 to answer many of the questions that might arise during the design process. Those calculations are not just academic exercises; rather, they will enable you to design better, smarter, and faster.

In this chapter, we present an overview of the product development process, beginning with the definition of a design problem, proceeding to the development of a new concept, continuing to production, and culminating in the patenting of the new technology. We begin with a discussion of the high-level steps in a design process that engineers follow when they transform a new idea into reality. Once the new product has been designed and built, an engineer or company generally wants to obtain a competitive advantage in the marketplace by protecting the new technology and preventing others from using it. The United States Constitution contains provisions that enable inventions to be patented, an important aspect of the business side of engineering. Once the details of the product have been determined, the hardware needs to be built economically. Mechanical engineers specify how a product will be fabricated, and Section 2.3 introduces you to the major classes of manufacturing processes.

Focus On PRODUCT ARCHAEOLOGY

Perhaps you have heard someone say that engineers *discover* new technologies, much like how archaeologists discover past technologies. Although the notion of discovery drives both archaeologists and engineers, archaeologists discover what has already existed, whereas engineers discover what has never existed. However, engineers can learn a tremendous amount about design by studying existing technologies using product archaeology.

Product archaeology is the process of reconstructing the life cycle of a product—the customer requirements, design specifications, and manufacturing processes used to produce it—in order to understand the decisions that led to its development. Product archaeology was first introduced in 1998 as a way to measure the design attributes that drive cost through the analysis of the physical products themselves.[2] More recently, product archaeology has been broadened to study not only the manufacturing cost of a product, but also the global and societal context that influences its development. Product archaeology also enables engineers to study the environmental impact of a product by considering the energy and material usage throughout its life cycle. When implemented in an engineering classroom, product archaeology allows students to place themselves in the minds of designers and in the time frame during which a specific product was developed in order to try to recreate the global and local conditions that led to its development.

For example, in a number of mechanical engineering programs around the world, students are engaged in various types of product archaeology projects and exercises by mimicking the process archaeologists use:

1. *Preparation*: Background research about a product, including market research,

patent searches, and benchmarking existing products

2. *Excavation:* Physical dissection of the product, cataloging of the materials, manufacturing processes used, and primary function of each component

3. *Evaluation:* Conducting material analysis and testing, comparing performance to other similar products, assessing the economic and environmental impact of the product components

4. *Explanation*: Drawing conclusions about the global, economic, environmental, and societal issues that shaped the design of the product and that currently shape the design of similar products

For example, at Penn State University, students conduct "archaeological digs" of bicycles. As part of their research, dissection, and product analysis, the students unearthed the following information about bicycles that can help shape the future design of bicycles for a wide range of global markets.

Bicycles in a global context:

- Bicycles are used as ambulances in sub-Saharan Africa.

- Japan has so many bicycles that they have bicycle parking structures.

- In countries such as Holland, there are entire transportation infrastructures just for bicycles, including lanes, traffic signals, parking lots, road signs, and tunnels.

- Many bicycles in China are electric.

Bicycles in a societal context:

- A number of bicycle cafes serve organic foods and loan bikes to people to travel around town.

---

[2]K. T. Ulrich and S. Pearson, "Assessing the Importance of Design Through Product Archaeology," *Management Science,* 1998, 44(3), 352–369.

- Henry Ford was a bike mechanic, and the Wright brothers used bicycle tubing for their first flight.
- The bicycle served as a catalyst for so-called rational dress reform among women as part of their emancipation movement.

Environmental impact of bicycles:

- There are many bike-sharing programs in European countries.
- There are a wide variety of programs to encourage people to bike to work to save on carbon emissions.
- Students found many statistics on bicycle commuters in U.S. cities, including information on bicycle efficiencies.

Economic issues in bicycle design:

- The relative costs of bicycles versus cars including the costs to make, operate, and maintain each.
- There are cost savings between plastic bicycles versus bicycles made with traditional materials.
- When bicycles are a prominent mode of transportation they lead to reduced healthcare costs.

One of the benefits of learning design principles using product archaeology is that doing so clearly demonstrates the multidisciplinary nature of product development and how essential design principles are to mechanical engineering.

Archaeological-inspired product "digs" can also help mechanical engineers develop their design aptitude. For instance, an archaeologist at a dig site may find stone artifacts, textiles, pieces of pottery, and some ancient utensils and then reconstruct the culture and technology of the time. Mechanical engineers can likewise study engineered components and then use their knowledge of matter, energy, and information to reconstruct the product. See if you can determine what product these "artifacts" come from (answer at the end of the chapter):

- Infrared Sensor
- Motor
- Gears
- Storage compartment
- Window
- Tube
- Microswitch
- Spout

▶ 2.2 The Design Process

In the broad view, mechanical design is the systematic process for devising a product or system that meets one of the global society's technical needs. As illustrated by the Grand Challenges (Section 2.1) and the top ten list of the mechanical engineering profession's achievements (Section 1.3), the need could lie in the area of healthcare, transportation, technology, communication, energy, or security. Engineers conceive solutions to those problems and turn their conceptions into functioning hardware.

Although a mechanical engineer might specialize in a field such as materials selection or fluids engineering, the day-to-day activities often focus on design. In some instances, a designer starts from scratch and has the freedom to develop an original product from the concept stage onward. The technology that is developed could be so revolutionary as to create entirely new markets and business opportunities. Smart phones and hybrid vehicles are examples of how technology is changing the way people think about communication and transportation. In other cases, an engineer's design work is incremental and focuses on improving an existing product. Examples are the addition of a high-definition video camera to a cell phone and the small yearly variations made to an automobile model.

Where does the life of a new product begin? First, a company identifies new business opportunities and defines requirements for a new product, system, or service. Past, current, and potential customers are surveyed, online product reviews and feedback forums are studied, and related products are researched. Marketing, management, and engineering all provide additional input to help develop a comprehensive set of system requirements.

In the next phase, engineers exercise their creativity and develop potential concepts, select the top concept using the requirements as decision criteria, develop details (such as layout, material choice, and component sizing), and bring the hardware to production. Does the product meet the initial requirements, and can it be produced economically and safely? To answer such questions, engineers make many approximations, trade-offs, and decisions along the way. Mechanical engineers are mindful that the level of precision needed in any calculation gradually grows as the design matures from concept to final production. Resolving specific details (Is a grade 1020 steel strong enough? What must the viscosity of the oil be? Should ball or tapered roller bearings be used?) doesn't make much sense until the design gets reasonably close to its final form. After all, in the early stages of a design, the specifications for the product's size, weight, power, or performance could change. Design engineers learn to be comfortable with order-of-magnitude calculations (Section 3.6), and they are able to develop products even in the presence of ambiguity and requirements that might change over time.

Focus On

INNOVATION

Many people think that people are simply born innovative or not. Although some people are stronger right-brain thinkers, everyone can learn to become more innovative. Innovation, a familiar concept to industrial designers, artists, and marketers, is becoming a critical topic in the development of strategies around the world to solve complex social, environmental, civic, economic, and technical challenges. Initiatives centered on technology and scientific innovation are underway across the globe.

- The U.S. federal government developed "A Strategy for American Innovation," which includes an Office of Innovation and Entrepreneurship and the formation of a

National Advisory Council on Innovation and Entrepreneurship.

- For the first time in history, the United States government has a chief technology officer.

- The China Standards and Innovation Policy initiative aims to analyze relationships between standards and innovation in order to better inform global leaders.

- In Australia, the minister for Innovation, Industry, Science and Research developed, for the entire nation, a Framework of Principles for Innovation Initiatives.

- The Department of Science and Technology within the government of India has developed the India Innovation Initiative (i3) to create an innovation network, encouraging and promoting innovators and commercialization across the country.

- The Agricultural Innovation in Africa project, funded by the Bill and Melinda Gates Foundation, is supporting efforts that contribute to agricultural scientific innovations and to technology policy improvement through Africa's Regional Economic Communities.

Not only are innovation initiatives underway at the national level, but many companies have developed innovation centers to drive new product, process, and service development. Companies like Microsoft, Procter & Gamble, Accenture, IBM, AT&T, Computer Sciences Corporation, Qualcomm, and Verizon have all opened innovation centers focused on developing key scientific and technological innovations.

Mechanical engineers play significant roles in these corporate and national innovation initiatives. Recognizing and understanding how mechanical engineering design impacts the success of innovative technologies is vital to solving the Grand Challenges. You will encounter design again in your curriculum, but you now must understand how innovation can develop a wide range of technologies to provide better engineered solutions. Figure 2.2 shows a 2 × 2 chart with Style (low/high) on the vertical axis and Technology (low/high) on the horizontal axis. This chart provides a framework to strategically develop innovative products for a wide range of customers.

Each quadrant contains a different digital music player. In the lower left, the Low-Style/Low-Tech version is a standard, affordable player, designed for customers who just want to play music. The player, while not the most stylish or high-tech type, provides solid, expected playback of digital music. In the lower right, the Low-Style/High-Tech version is the SwiMP3 player from FINIS. This player integrates waterproof technologies with revolutionary bone conduction of sound to provide swimmers with clear sound underwater. While the player is functionally effective, it does not need to be high in style for its intended market. In the upper left, the

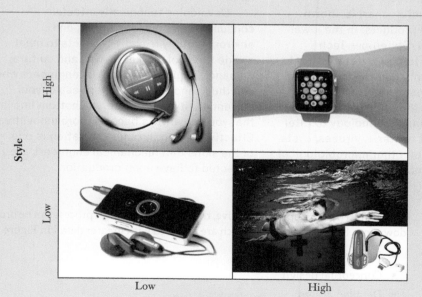

Figure 2.2

Style versus technology chart for digital music players.

Istvan Csak/Shutterstock.com; Allies Interactive/Shutterstock.com; charnsitr/Shutterstock.com; Courtesy of FINIS, Inc.

Figure 2.3

Style versus technology chart for water purification systems.

Steven Coling/Shutterstock.com; Courtesy of Applica Waters Products; Courtesy of Hague Quality Water International; Courtesy of Lifesaver Systems Ltd.

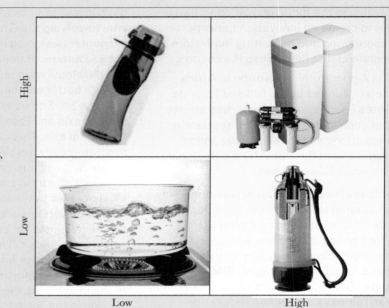

High-Style/Low-Tech version is a standard player with a stylish shape and control interface. In the upper right, the High-Style/High-Tech version is an Apple™ Watch, for customers who want the latest technologies along with stylish features.

In Figure 2.3, a similar chart is shown for water purification products. In the lower left quadrant, the Low-Style/Low-Tech product is a pot used to boil water in order to eliminate micro-organisms using basic heat technology. In the lower right, the Low-Style/High-Tech product is the LIFESAVER® bottle that uses advanced nano technology to filter out even the smallest bacteria, viruses, cysts, parasites, fungi, and all other microbiological waterborne pathogens. In the upper left, the

High-Style/Low-Tech product is a Fashion Bottle filtration system from Clear2O®. In the upper right, the High-Style/High-Tech product is the Hague WaterMax® system, a custom-designed water treatment system for an entire house.

Developing technically effective, consumer-safe, globally aware, and environmentally friendly products to meet a wide range of market, social, and cultural demands requires mechanical engineers who can think innovatively. Regardless of your current ability to innovate, the bottom line is that you can always become more innovative. Effective design through innovation is one of the skills that mechanical engineers are expected to have upon graduation.

From a macroscopic perspective, the mechanical design process can be broken down into four major stages, which are outlined with greater detail in Figure 2.4:

- Requirements development
- Conceptual design
- Detailed design
- Production

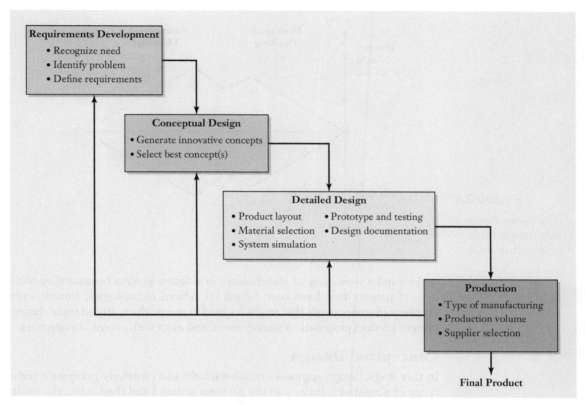

Requirements Development
- Recognize need
- Identify problem
- Define requirements

Conceptual Design
- Generate innovative concepts
- Select best concept(s)

Detailed Design
- Product layout
- Material selection
- System simulation
- Prototype and testing
- Design documentation

Production
- Type of manufacturing
- Production volume
- Supplier selection

Final Product

Requirements Development

Engineering design begins when a basic need has been identified. This could be a technical need from a certain market or a basic human need like clean water, renewable energy, or protection from natural disasters. Initially, a design engineer develops a comprehensive set of system requirements considering the following issues:

- *Functional performance*: What the product must accomplish
- *Environmental impact*: Effects during all phases of production, use, and retirement
- *Manufacturing*: Resource and material limitations
- *Economic issues*: Budget, cost, price, profit
- *Ergonomic concerns*: Human factors, aesthetics, ease of use
- *Global issues*: International markets, needs, and opportunities
- *Life cycle issues*: Use, maintenance, planned obsolescence
- *Social factors*: Civic, urban, cultural issues

The system requirements essentially represent the constraints that the design must eventually satisfy. To develop the requirements, engineers conduct extensive research and gather background information from a diverse set of sources. As mentioned in Section 1.4, engineers need to be able to communicate with and

Figure 2.4

Flowchart of the prototypical mechanical design process.

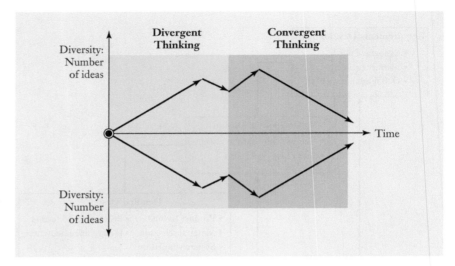

Figure 2.5

The generation and selection of ideas in conceptual design.

understand a wide array of stakeholders in a design process because they need to read patents that have been issued for related technologies, consult with suppliers of components that might be used in the product, attend trade shows, present product proposals to management, and meet with potential customers.

Conceptual Design

In this stage, design engineers collaboratively and creatively generate a wide range of potential solutions to the problem at hand and then select the most promising one(s) to develop. Initially, as shown in Figure 2.5, the process is guided by *divergent thinking*—a diverse set of creative ideas is developed. Some people think that creativity is reserved for artists, who are born with the ability to be innovative, and that engineers need to be practical, leaving the creative tasks to others. The reality is that being creative is a critical part of an engineer's job; product design requires engineers who are part rational scientists and part innovative artists. Engineers can learn to be more creative, giving themselves a necessary skill set to survive in their academic and workplace careers. Many times the most creative solutions come from a collaborative innovation session where people can discuss ideas with others from varied backgrounds—different professions, industries, ages, educations, cultures, and nationalities.

Once a rich set of concepts has been generated, the process is guided by *convergent thinking*, as engineers begin to eliminate ideas and converge on the best few concepts. The requirements list from the first stage is used to eliminate infeasible or inferior designs and to identify the concepts with the most potential to satisfy the requirements. These evaluations can be performed using a list of pros and cons or a matrix with ratings of the concepts using preliminary calculations to compare key requirements. Computer models and prototype hardware might also be produced at this stage to help with the selection process. At this stage, the design remains relatively fluid and changes can be made inexpensively, but the further along a product is in the development process, the more difficult and expensive changes become. This stage culminates in identifying the most promising design concept.

Detailed Design

At this point in the design process, the team has defined, innovated, analyzed, and converged its way to the best concept. However, many design and manufacturing details remain open, and each must be resolved before the product hardware can be produced. In the detailed design of the product, a number of issues must be determined:

- Developing product layout and configuration
- Selecting materials for each component
- Addressing design-for-X issues (such as design for reliability, manufacturing, assembly, variation, costing, recycling)
- Optimizing the final geometry, including appropriate tolerances
- Developing completed digital models of all components and assemblies
- Simulating the system using digital and mathematical models
- Prototyping and testing critical components and modules
- Developing the production plans

An important general principle in the detailed design stage is *simplicity*. A simpler design concept is better than a complex one, because fewer things can go wrong. Think of the most successfully engineered products, and many times they are characterized by an effective integration of design innovation, sound engineering, and functional simplicity. Keeping things as simple as possible has a well-earned reputation among engineers.

Simplicity

In addition, engineers need to be comfortable with the concept of iteration in a design process. *Iteration* is the process of making repeated changes and modifications to a design to improve and perfect it. For instance, if none of the generated concepts satisfactorily meet the requirements, then engineers must either revisit the requirements list or return to the concept ideation stage. Similarly, if the production plan of the final design is not feasible, then engineers must revisit the design details and choose different materials, new configurations, or some other design detail. With each iteration, the design gradually improves—performing better, more efficiently, and more elegantly. Iteration enables you to turn hardware that works into hardware that works well.

Iteration

Although engineers clearly are concerned with a design's technical aspects (forces, materials, fluids, energy, and motion), they also recognize the importance of a product's appearance, ergonomics, and aesthetics. Whether it is a consumer electronics product, the control room of an electrical power plant, or the flight deck of a commercial jetliner, the interface between the user and the hardware should be comfortable, simple, and intuitive. The *usability* of a product can become particularly problematic as its technology becomes more sophisticated. No matter how impressive the technology may be, if it is difficult to operate, customers will not embrace it as enthusiastically as they may have otherwise. In this regard, engineers often collaborate with industrial designers and psychologists to improve the appeal and usability of their products. In the end, engineering is a business venture that meets the needs of its customers.

Usability

Documentation

Engineers must be very diligent in documenting engineering drawings from the design process, meeting minutes, and written reports so that others can understand the reasons behind each of the decisions. Such *documentation* is also useful for future design teams who want to learn from and build on the present team's experiences. A design notebook (see Section 3.7) is an effective way to capture the information and knowledge created during a design process.

Patents

Design notebooks—preferably bound, numbered, dated, and even witnessed—also support the patenting of important new technology that a company wants to prevent others from using. Drawings, calculations, photographs, test data, and a listing of the dates on which important milestones were reached are important to capture accurately how, when, and by whom the invention was developed. *Patents* are a key aspect of the business side of engineering because they provide legal protection for the inventors of new technology. Patents are one aspect of intellectual property (a field that also encompasses copyrights, trademarks, and trade secrets), and they are a right to property, analogous to the deed for a building or a parcel of land.

Patents are granted for a new and useful process, machine, article of manufacture, or composition of matter or for an improvement to them. Patents are agreements between an inventor and a national government. An inventor is granted the legal right to exclude others from making, using, offering for sale, selling, or importing an invention. In exchange, the inventor agrees to disclose and explain the invention to the general public in the written document called a *patent*. A patent is a monopoly on the new technology that expires after a certain number of years, whose duration depends on the type of patent issued and the nation issuing it. It can be argued that the patent system has formed the economic foundation on which our society has made its technological progress. Patents stimulate corporate research and product development because they provide a financial incentive (a limited monopoly) for innovation. By being creative, an inventor can use the protection offered by a patent to obtain an advantage over business competitors.

The United States Constitution provides the Congress with the authority to enact patent laws. Interestingly, this authority is mentioned before other (perhaps more well-known) powers of the Congress, such as declaring war and maintaining an army.

There are three primary types of patents in the United States: plant, design, and utility. As the name implies, a *plant patent* is issued for certain types of asexually reproduced plants, and it is not commonly encountered by mechanical engineers.

Design patent

A *design patent* is directed at a new, original, and ornamental design. A design patent is intended to protect an aesthetically appealing product that is the result of artistic skill; it does not protect the product's functional characteristics. For instance, a design patent could protect the shape of an automobile's body if it is attractive, pleasing to look at, or gives the vehicle a sporty appearance. However, the design patent would not protect the functional characteristic of the body, such as reducing wind drag or offering improved crash protection.

As a result, companies with the largest design patent portfolios typically have products that are functionally indistinguishable from their competitors. For instance, companies developing products such as athletic shoes, cleaning supplies, consumer electronics, furniture, watches, and personal hygiene products often have significant design patent portfolios.

More commonly encountered in mechanical engineering, the *utility patent* protects the function of an apparatus, process, product, or composition of matter. The utility patent generally contains three main components:

Utility patent

- The *specification* is a written description of the purpose, construction, and operation of the invention.
- The *drawings* show one or more versions of the invention.
- The *claims* explain in precise language the specific features that the patent protects.

The description provided in the patent must be detailed enough to show someone else how to use the invention. As an example, the cover page from a United States utility patent is shown in Figure 2.6 (see page 44). The page includes the patent's title and number, the date the patent was granted, the names of the inventors, a bibliography of related patents, and a brief summary of the invention. Utility patents become valid on the date the patent is granted, and recently issued ones expire 20 years after the date of the application, which must be filed within one year of the inventor having publicly disclosed or used the invention (e.g., by selling or offering to sell it to others, by demonstrating it at an industrial trade show, or by publishing an article on it).

In 2013, IBM received the most utility patents in the United States with 6,788 patents. Samsung was second on the list with 4,652 patents. The other top 15 companies in 2013 include Canon, Sony, LG Electronics, Microsoft, Toshiba, Panasonic, Hitachi, Google, Qualcomm, General Electric, Siemens, Fujitsu, and Apple.

To apply for a patent, engineers normally work with patent attorneys who conduct a search of existing patents, prepare the application, and interact with a national trademark and patent office. In 2013, the United States alone granted over 277,000 utility patents and over 23,000 design patents. Approximately half of all 2013 patents have U.S. origins. The following table shows the top ten countries ranked by the number of patents granted in the United States in 2013 (data from the United States Patent and Trademark Office).

Country	Number of Patents Granted in the United States	Percentage Increase from 2000
Japan	54,170	65%
Germany	16,605	53%
South Korea	15,745	353%
Taiwan	12,118	108%
Canada	7272	85%
China (PRC)	6597	3947%
France	6555	57%
United Kingdom	6551	60%
Israel	3152	277%
Italy	2930	49%

United States Patent [19]

Wickert et al.

[11] **Patent Number:** 5,855,257

[45] **Date of Patent:** Jan. 5, 1999

---

[54] **DAMPER FOR BRAKE NOISE REDUCTION**

[75] Inventors: **Jonathan A. Wickert**, Allison Park; **Adnan Akay**, Sewickley, both of Pa.

[73] Assignee: **Chrysler Corporation**, Auburn Hills, Mich.

[21] Appl. No.: **761,879**

[22] Filed: **Dec. 9, 1996**

[51] Int. Cl.^{6} **F16D 65/10**

[52] U.S. Cl. .... **188/218 XL**; 188/218 A

[58] Field of Search .... 188/18 A, 218 A, 188/218 R, 218 XL; 74/574

[56] **References Cited**

U.S. PATENT DOCUMENTS

1,745,301	1/1930	Johnston .
1,791,495	2/1931	Frey .
1,927,305	9/1933	Campbell .
1,946,101	2/1934	Norton .
2,012,838	8/1935	Tilden .
2,081,605	5/1937	Sinclair .
2,197,583	4/1940	Koeppen et al. .
2,410,195	10/1946	Baselt et al. .
2,506,823	5/1950	Wyant .
2,639,195	5/1953	Bock .
2,702,613	2/1955	Walther, Sr. .
2,764,260	9/1956	Fleischman .
2,897,925	8/1959	Strohm .
2,941,631	6/1960	Fosberry et al. .... 188/218 A
3,250,349	5/1966	Byrnes et al. .
3,286,799	11/1966	Shilton .... 188/218 A
3,292,746	12/1966	Robiette .
3,368,654	2/1968	Wegh et al. .

3,435,925	4/1969	Harrison .
3,934,686	1/1976	Stimson et al. .
4,043,431	8/1977	Ellege .... 188/218 A
4,656,899	4/1987	Contoyonis .
5,004,078	4/1991	Oono et al. .
5,383,539	1/1995	Bair et al. .

FOREIGN PATENT DOCUMENTS

123707	7/1931	Australia .	
2275692	1/1976	France .... 188/218 A
58-72735	4/1983	Japan .	
63-308234	12/1988	Japan .	
141236	9/1984	Rep. of Korea .... 18/218 A
254561	9/1925	United Kingdom .... 188/218 A
708083	10/1952	United Kingdom .	
857043	12/1960	United Kingdom .... 188/218 A
934096	8/1963	United Kingdom .... 188/218 A
2181199	4/1987	United Kingdom .	
2181802	4/1987	United Kingdom .	

Primary Examiner—Robert J. Oberleitner
Assistant Examiner—Chris Schwartz
Attorney, Agent, or Firm—Roland A. Fuller, III

[57] **ABSTRACT**

An apparatus for reducing unwanted brake noise has a ring damper affixed around a periphery of a brake rotor in a disk brake system in a manner that permits relative motion and slippage between the ring damper and the rotor when the rotor vibrates during braking. In a preferred embodiment, the ring damper is disposed in a groove formed in the periphery of the disk and is pre-loaded against the rotor both radially and transversely. The ring damper is held in place by the groove itself and by the interference pre-load or pre-tension between the ring damper and the disk brake rotor.

30 Claims, 4 Drawing Sheets

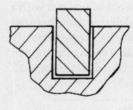
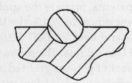
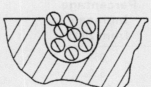
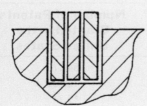

Figure 2.6

Cover page from a United States patent.

Source: United States Patent, Wickert et al., Patent Number: 5,855,257, Jan. 5, 1999.

Compare that to the following table showing the top ten countries ranked by the percentage increase of patents granted in the United States between 2013 and 2000 (including a minimum of 100 patents issued in 2013).

Country	Percentage Increase from 2000
China	3947%
India	2900%
Saudi Arabia	1158%
Poland	769%
Malaysia	389%
South Korea	353%
Czech Republic	329%
Israel	277%
Hungary	271%
Singapore	254%

While obtaining a patent from a particular country protects an individual or company in that country, sometimes international patent protection is preferred. The World Intellectual Property Organization (WIPO—www.wipo. int) offers individual and corporate patent applicants a way to obtain patent protection internationally. In 2013, the number of international patent applications filed with WIPO surpassed 200,000 for the first time with the United States being the most active country, followed by Japan, China, and Germany.

The year 2013 was an important one in patent law, as the United States switched from a "first-to-invent" patent system to a "first-to-file" system. This means that the actual date of a product's invention is no longer meaningful; rather, the owner of the patent is the first person to file the patent regardless of when the invention was conceived. This significant change in patent law has caused many companies around the world to re-think how they go about developing, disclosing, and protecting their new product ideas and design changes.

Sometimes engineers want a quick prototype to finalize some product features in preparation for a patent application, for product documentation, or to communicate product details to others. A picture may be worth a thousand words, but a physical prototype is often useful for engineers to visualize complex machine components. Many times these prototypes can be physically tested so that trade-off decisions are made based on the results of measurements and analysis. Methods for producing such components are called *rapid prototyping*,   Rapid prototyping
3-D printing, or additive manufacturing. The key capability of these processes is that complex three-dimensional objects are fabricated directly from a computer-generated drawing, often in a matter of hours.

Some rapid prototyping systems use lasers to fuse layers of a liquid polymer together (a process known as *stereolithography*) or to fuse raw material in powder form. Another prototyping technique moves a printhead (similar to one used in an ink-jet printer) to spray a liquid adhesive onto a powder and "glue-up" a prototype bit by bit. In essence, the rapid prototyping system is a

(a)

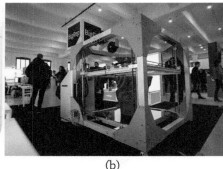
(b)

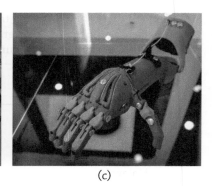
(c)

Figure 2.7

Three-dimensional
printing systems and
a multi-material 3-D
printed prosthetic
hand.

Chesky_W/iStock/Thinkstock; Anadolu
Agency/Getty Images; Anadolu Agency/
Getty Images

three-dimensional printer capable of transforming an electronic representation of the component into plastic, ceramic, and/or metallic parts. Figure 2.7 depicts two 3-D printing rapid prototyping systems [(a) and (b)] and a multimaterial 3-D printed prosthetic hand [(c)]. These rapid prototyping technologies can produce durable and fully functional prototypes that are fabricated from polymers and other materials. The components can be assembled, tested, and increasingly used as production parts.

Production

The engineer's involvement is not over once the working prototype has been delivered and the finishing touches have been placed on the drawings. Mechanical engineers work in a broad environment, and their designs are viewed with a critical eye beyond the criterion of simply whether the solution functions as intended. After all, if the product is technically superb but requires expensive materials and manufacturing operations, customers might avoid the product and select one that is more balanced in its cost and performance.

Therefore, even at the requirements development stage, engineers must take into account manufacturing requirements for the production stage. After all, if you're going to take the time to design something, it had better be something that actually can be built, preferably at a low cost. The materials selected for a product influence how it can be manufactured. A part that is machined from metal might be best for one design concept, but a plastic component produced by injection molding might be the better choice for another. In the end, the design's function, shape, materials, cost, and manner of production are tightly interconnected and balanced throughout the design process.

Once the detailed design is completed, a designer becomes involved with the fabrication and production of the product. In part, the fabrication techniques that an engineer selects depend on the time and expense of setting up the tooling and machines necessary for production. Some systems—for instance, automobiles, air conditioners, microprocessors, hydraulic valves, and computer hard disk drives—are mass-produced, a term that denotes the widespread use of mechanical automation. As an example, Figure 2.8 shows an assembly line where robots perform various assembly tasks in

Figure 2.8

Robots perform various assembly tasks on a mass-production line for vehicles.

© imageBROKER/Alamy

an automotive manufacturing plant. Historically, these kinds of assembly lines comprised custom tools and specialized fixtures that were capable of efficiently producing only certain types of components for certain types of vehicles. But now flexible manufacturing systems allow a production line to quickly reconfigure for different components of different vehicles. Because by *mass production* finished products can be produced relatively quickly, a company can cost-effectively allocate a large amount of factory floor space and many expensive machine tools, even though any one of them might perform only simple tasks, like drilling a few holes or polishing a single surface.

Mass production

Aside from hardware produced by means of mass manufacturing, other products (such as commercial jetliners) are made in relatively small quantities, or they are unique (such as the Hubble Space Telescope). Some one-of-a-kind products can even be produced directly from a computer-generated drawing by using what is, in essence, a three-dimensional printer (Figure 2.7). The best production method for a product depends on the quantity to be produced, the allowable cost, and the necessary level of precision. The next section reviews the most prominent production and manufacturing methods.

Focus On   GLOBAL DESIGN TEAMS

Technological advances in simulation and virtual prototyping are making geographic separation between product design teams around the world increasingly irrelevant. As a result, products and systems can now be developed by collaborative global design teams. Boeing recently unveiled not only one of the most technically advanced commercial jets ever in the 787 Dreamliner, but they also undertook one of the most ambitious engineering projects ever, working closely with almost 50 top-tier system suppliers at 135 sites across four continents. Some of Boeing's major suppliers from around the world are shown in Figure 2.9.

Boeing also used sophisticated database management software that digitally linked all of the work sites. This ensured that all of their engineering staff worked on the same set of solid models and computer-aided design (CAD) representations. So while manufacturing and rapid prototyping technologies continue to advance, Boeing and many other companies are taking advantage of virtual prototyping as an effective tool in engineering design. Virtual prototyping takes advantage of advanced visualization and simulation technologies available in the fields of virtual reality, scientific visualization, and computer-aided

Figure 2.9

The major global suppliers for the Boeing 787 Dreamliner.

REUTERS/LANDOV

787 Dreamliner structure suppliers

Selected component and system suppliers.

Part name
Company (country)

Wingtips
KAA (Korea)

Fixed & movable
leading edge
Spirit (U.S.)

Wing
Mitsubishi
(Japan)

Centre fuselage
Alenia (Italy)

Forward fuselage
Spirit (U.S.)
Kawasaki (Japan)

Centre wing box
Fuji (Japan)

Landing gear structure
Messier-Dowty (France)

Lithium-ion batteries
GS Yuasa (Japan)

Movable
trailing edge
(U.S., Canada,
Australia)

Rear fuselage
Boeing (U.S.)

Wing-to-body
fairing
Boeing (U.S.)

Engine
nacelles
Goodrich (U.S.)

Engine
Rolls-Royce (U.K.)
General
Electric (U.S.)

Horizontal
stabilizer
Alenia
(Italy)

Tail fin
Boeing
(U.S.)

Passenger
entry doors
Latecoere (France)

Lithium-ion batteries
GS Yuasa (Japan)

Main landing gear
wheel well
Kawasaki (Japan)

Fixed trailing edge
Kawasaki (Japan)

OTHERS

Wing/body fairing
Boeing (Canada)

Cargo access doors
Saab (Sweden)

design to provide realistic digital representations of components, modules, and products, as shown in Figure 2.10. Using these tools, design engineers can simulate and test many different scenarios at a fraction of the cost, giving them confidence when actual parts are manufactured.

When used in design processes, virtual prototyping facilitates the implementation of many iterations because design changes can be made rapidly on the digital model. Also, 3-D scanners are allowing digital models to be created quickly and accurately from physical models.

(a)

(b)

Figure 2.10

(a) A wireframe structural model of the 787 Dreamliner. (b) The Boeing virtual reality lab in Seattle where engineers can simulate various design configurations and operational scenarios.

Boeing Images; Bob Ferguson/Boeing Images

▶ 2.3 **Manufacturing Processes**

Manufacturing technologies are economically important because they are the means for adding value to raw materials by converting them into useful products. Each of the many different manufacturing processes is well suited to a particular need based on environmental impact, dimensional accuracy, material properties, and the mechanical component's shape. Engineers select processes, identify the machines and tools, and monitor production to ensure that the final product meets its specifications. The main classes of manufacturing processes are as follows:

- *Casting*   is the process whereby liquid metal, such as gray iron, aluminum, or bronze, is poured into a mold, cooled, and solidified.
- *Forming*   encompasses a family of techniques whereby a raw material is shaped by stretching, bending, or compression. Large

forces are applied to plastically deform a material into its new permanent shape.

- *Machining*   refers to processes where a sharp metal tool removes material by cutting it. The most common machining methods are drilling, sawing, milling, and turning.

- *Joining*   operations are used to assemble subcomponents into a final product by welding, soldering, riveting, bolting, or adhesively bonding them. Many bicycle frames, for instance, are welded together from individual pieces of metal tubing.

- *Finishing*   steps are taken to treat a component's surface to make it harder, improve its appearance, or protect it from the environment. Some processes include polishing, electroplating, anodizing, and painting.

In the remainder of this section, we describe the processes of casting, forming, and machining in additional detail.

Casting   In *casting*, liquid metal is poured into the cavity of a mold, which can be expendable or reusable. The liquid then cools into a solid object with the same shape as the mold. An attractive feature of casting is that complex shapes can be produced as solid objects without the need to join any pieces. Casting is an efficient process for creating many copies of a three-dimensional object, and, for that reason, cast components are relatively inexpensive. On the other hand, defects can arise if the metal solidifies too soon and prevents the mold from filling completely. The surface finish of cast components generally has a rough texture, and they might require additional machining operations to produce smooth and flat surfaces. Some examples of cast components include automotive engine blocks, cylinder heads, and brake rotors and drums (Figure 2.11).

Rolling   One kind of a forming operation is called *rolling,* which is the process of reducing the thickness of a flat sheet of material by compressing it between rollers, not unlike making cookie or pizza dough. Sheet metal that is produced in this manner is used to make aircraft wings and fuselages, beverage containers, Forging   and the body panels of automobiles. *Forging* is another forming process, which

Figure 2.11

Examples of hardware produced by casting include a disk-brake rotor, automotive-oil pump, piston, bearing mount, V-belt sheave, model-airplane engine block, and a two-stroke engine cylinder.

Image courtesy of the authors.

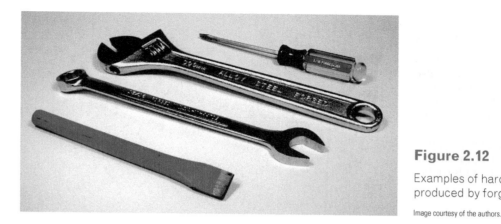

Figure 2.12

Examples of hardware produced by forging.

Image courtesy of the authors.

is based on the principle of heating, impacting, and plastically deforming metal into a final shape. Industrial-scale forging is the modern version of the blacksmith's art of working metal by hitting it with a hammer against an anvil. Components that are produced by forging include some crankshafts and connecting rods in internal combustion engines. Compared to castings, a forged component is strong and hard, and for that reason, many hand tools are produced this way (Figure 2.12).

The forming process known as *extrusion* is used to create long straight metal parts whose cross sections may be round, rectangular, L-, T-, or C-shaped, for instance. In extrusion, a mechanical or hydraulic press is used to force heated metal through a tool (called a die) that has a tapered hole ending in the shape of the finished part's cross section. The die used to shape the raw material is made from a metal that is much harder than what is being formed. Conceptually, the process of extrusion is not unlike the familiar experience of squeezing toothpaste out of a tube. Figure 2.13 shows examples of aluminum extrusions with a variety of cross sections.

Extrusion

Machining refers to processes whereby material is gradually removed from a workpiece in the form of small chips. The most common machining methods

Machining

Figure 2.13

Examples of aluminum extrusions.

Image courtesy of the authors.

Figure 2.14

This body for a hydraulic valve assembly was first cast from aluminum (left) and then machined in order to produce holes, flatten surfaces, and cut threads (right).

Image courtesy of the authors.

are called drilling, sawing, milling, and turning. Machining operations are capable of producing mechanical components with dimensions and shapes that are far more precise than their cast or forged counterparts. One drawback of machining is that (by its very nature) the removed material is wasted. In a production line, machining operations are often combined with casting and forging when cast or forged components require additional operations to flatten surfaces, make holes, and cut threads (Figure 2.14).

Machining tools include drill presses, bandsaws, lathes, and milling machines. Each tool is based on the principle of removing unwanted material from a workpiece by means of the cutting action of sharpened blades. The *drill press* shown in Figure 2.15 is used to bore round holes into a workpiece. A drill bit is held in the rotating chuck, and, as a machinist turns the pilot wheel, the

Drill press

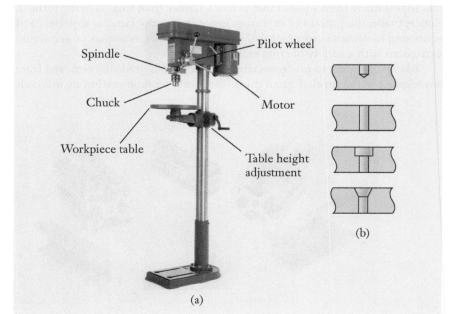

Spindle

Pilot wheel

Chuck

Motor

Workpiece table

Table height adjustment

(a)

(b)

Figure 2.15

(a) Major components of a drill press.

© David J. Green-engineering themes /Alamy.

(b) Different types of holes that can be produced.

bit is lowered into the workpiece's surface. As should be the case whenever metal is machined, the point where the bit cuts into the workpiece is lubricated. The oil reduces friction and helps remove heat from the cutting region. For safety reasons, vises and clamps are used to hold the workpiece securely and to prevent material from shifting unexpectedly.

A machinist uses a *band saw* to make rough cuts through metal (Figure 2.16). **Band saw**
The blade is a long, continuous loop with sharp teeth on one edge, and it rides on drive and idler wheels. A variable-speed motor enables the operator to adjust the blade's speed depending on the type and thickness of the material being cut. The workpiece is supported on a table that is capable of being tilted for cuts that are made at an angle. The machinist feeds the workpiece into the blade and guides it by hand to make straight or slightly curved cuts. When the blade becomes dull and needs to be replaced or if it breaks, the bandsaw's internal blade grinder and welder are used to clean up the blade's ends, connect them, and reform a loop.

A *milling machine (or mill)* is used for machining the rough surfaces of a **Milling machine**
workpiece to be flat and smooth and for cutting slots, grooves, and holes (Figure 2.17, see page 54). The milling machine is a versatile machine tool in which the workpiece is moved slowly relative to a rotating cutting tool. The workpiece is held by a vise on an adjustable table so that the part can be accurately moved in three directions (along the plane of the table and perpendicular to it) to locate the workpiece precisely beneath the cutting bit. A piece of metal plate

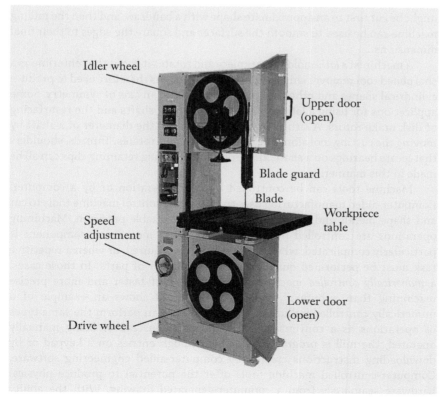

Figure 2.16

Major elements of a bandsaw.

Image courtesy of the authors.

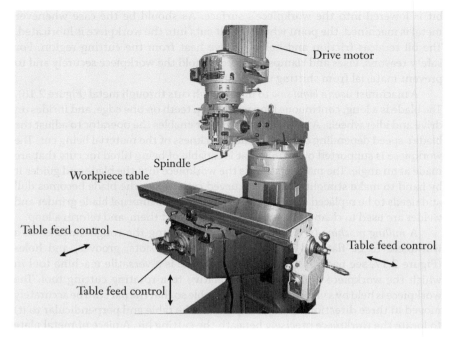

Drive motor

Spindle

Workpiece table

Table feed control

Table feed control

Table feed control

Figure 2.17

Major elements of a
milling machine.

Image courtesy of the authors.

might be cut first to an approximate shape with a bandsaw, and then the milling machine can be used to smooth the surfaces and square the edges to their final dimensions.

Lathe

A machinist's *lathe* holds a workpiece and rotates it about the centerline as a sharpened tool removes chips of material. The lathe is therefore used to produce cylindrical shapes and other components that have an axis of symmetry. Some applications for using a lathe are the production of shafts and the resurfacing of disk brake rotors. A lathe can be used to reduce the diameter of a shaft by moving the cutting tool along the shaft's length as it rotates. Threads, shoulders that locate bearings on a shaft, and grooves for holding retaining clips can all be made in this manner.

Machine tools can be controlled by hand operation or by a computer. Computer-aided manufacturing uses computers to control machine tools to cut and shape metals and other materials with remarkable precision. Machining operations are controlled by a computer when a mechanical component is particularly complicated, when high precision is required, or when a repetitive task must be performed quickly on a large number of parts. In those cases,

Numerical control

a *numerically controlled* machine tool is capable of faster and more precise machining than a human operator. Figure 2.18 shows an example of a numerically controlled milling machine. This mill can perform the same types of operations as a conventional one. However, instead of being manually operated, the mill is programmed either through entries on a keypad or by downloading instructions created by computer-aided engineering software. Computer-controlled machine tools offer the potential to produce physical hardware seamlessly from a computer-generated drawing. With the ability

Figure 2.18

A numerically controlled milling machine can produce hardware directly from instructions created by 3-D CAD software packages.

© David J. Green/Alamy

to quickly reprogram machine tools, even a small general-purpose shop can produce high-quality machined components.

Some of the same technologies used to create rapid product prototypes for design evaluation are beginning to be used for *custom production*. Rapid or direct digital manufacturing is a class of additive fabrication techniques to produce custom or replacement parts. A mass-manufacturing line takes advantage of mechanical automation, but those systems are intended to produce many identical parts. Rapid manufacturing systems take precisely the opposite viewpoint: one-of-a-kind components are produced directly from a computer-generated digital file. Digital models can be produced by using computer-aided design software or by scanning a physical object. This capability offers the potential for creating complex customized products at a reasonable cost. While rapid prototyping typically uses thermoplastics, photopolymers, or ceramics to create the parts, rapid manufacturing technologies can now also use a variety of metals and alloys. This allows engineers to create fully functional parts extremely quickly. For instance, an electron beam melts metal powder in a vacuum chamber creating very strong parts that can withstand high temperatures. Customized production is giving engineers the ability to manufacture a product as soon as someone orders it as well as being able to produce one-of-a-kind specifications by taking advantage of rapid manufacturing technologies.

Customized
production

Summary

The creative process behind mechanical design cannot be set forth fully in one chapter—or even in one textbook for that matter. With this material as a starting point, you can continue to develop design skills and hands-on experience throughout your professional career. Even the most seasoned engineer grapples with all the decisions and trade-offs in a design process that transforms an idea into manufactured hardware that can be sold at a reasonable cost. The subject of mechanical design has many facets. In this chapter we introduced you to a basic design process and some of the issues that guide how a new product is designed, manufactured, and ultimately protected in the business environment through patents. We hope that you recognize that design principles can be used to develop and produce a diverse set of products, systems, and services to meet the always complex technical, global, societal, and environmental challenges that our world faces.

As we described in Chapter 1, engineers apply their skills in mathematics, science, and computer-aided engineering for the purpose of making things that work safely and transform lives. At the highest level, engineers apply the design process described in Section 2.2 to reduce an open-ended problem to a sequence of manageable steps: definition of system requirements, conceptual design where concepts are generated and narrowed down, and detailed design where all the geometric, functional, and production details of a product are developed. Engineering is ultimately a business venture, and you should be aware of that broader context in which mechanical engineering is practiced. When developing a new product, an engineer, a team of engineers, or a company often wants to protect the new technology with a patent. Patents provide the inventor with a limited monopoly on the product in exchange for the invention being explained to others.

In the end, successful design is a function of creativity, elegance, usability, and cost. During a design process, engineers use their judgment and make order-of-magnitude calculations to move ideas to concepts and concepts to detailed designs. Engineers also specify the methods to be used to produce hardware, and those decisions are based on the quantity to be produced, the allowable cost, and the level of necessary precision. Although rapid prototyping is becoming an increasingly viable way to quickly manufacture custom products, the primary mass production manufacturing processes still include casting, forming, machining, joining, and finishing. Each technique is well suited for a specific application based on the shape of the mechanical component and the materials used. Machining operations are conducted using bandsaws, drill presses, milling machines, and lathes, whereby each of these machine tools uses a sharpened tool to remove material from a workpiece. Machine tools can be numerically controlled to fabricate high-precision components based on designs that are developed through computer-aided engineering software packages.

Self-Study and Review

2.1. What are the major stages of a mechanical design process?

2.2. Discuss the importance of innovation in the design process.

2.3. What are the categories of system requirements that design engineers must consider when they start a design process?

2.4. To what extent should detailed decisions be made early in the design process regarding dimensions, materials, and other factors?

2.5. Discuss some of the interpersonal and communication issues that arise when engineers work together in a cross-disciplinary team on a design project that has global impact.

2.6. Explain how simplicity, iteration, and documentation play significant roles in a design process.

2.7. What are the major classes of manufacturing processes?

2.8. Give examples of hardware produced by casting, rolling, forging, extrusion, and machining.

2.9. Explain how a bandsaw, drill press, and milling machine are used.

2.10. What is rapid prototyping technology, and when could the technology best be used?

2.11. What are some differences between design and utility patents?

2.12. How long is a recently issued utility patent valid in your country?

Problems

For P2.1–P2.6, the product doesn't happen to be a certain shape or color; it *must*, by regulation and/or fundamental function, be a certain shape or color.

P2.1
Give three examples of engineered products that must be circular in shape and explain why. Any ball is not allowed as an answer!

P2.2
Give three examples of engineered products that must be triangular in shape and explain why.

P2.3
Give three examples of engineered products that must be rectangular in shape and explain why.

P2.4

Give three examples of engineered products that must be green in color.

P2.5

Give three examples of engineered products that must be black in color.

P2.6

Give three examples of engineered products that must be transparent.

P2.7

Give three examples of engineered products that have a specific minimum weight but no specified maximum weight, and specify the approximate minimum weight.

P2.8

Give three examples of engineered products that have to be precisely a certain weight, and provide the weight.

P2.9

Give three examples of engineered products that fulfill their designed purpose by failing or breaking.

P2.10

Give three examples of engineered products that are designed to work well over a million times.

P2.11

List three products that can be used equally well by people with and without visual impairments, and explain why.

P2.12

Select a type of product that can have versions in all four quadrants in the style-versus-technology design chart (see *Focus On Innovation*). Show the four versions of the product and clearly explain why you think they fall into their specific quadrants.

P2.13

Imagine you are tasked with designing a coffee maker that can be marketable to cafés around the globe. Conduct research on coffee makers to determine a set of global, social, environmental, and economic issues you need to consider as you begin the design process. (This is essentially part of the preparation phase in product archaeology; see *Focus on Product Archaeology*.)

P2.14

Imagine you are tasked with designing a single dishwasher for both the European and American markets. Determine a set of global, social, environmental, and economic issues you need to consider in the design process. (This is essentially part of the preparation phase in product archaeology; see *Focus on Product Archaeology*.)

P2.15

Find a product specification sheet for a consumer product such as an automobile, appliance, TV, motor, or something similar, and determine whether the specifications are easy to interpret. For example, as an engineering student, do you understand what all the specifications mean? Why or why not? Would the typical nonengineer, nontechnical customer be able to understand them? Why or why not? Also explain how the specifications on the sheet address both environmental and economic issues, either directly or indirectly. Include your specification sheet with your submission.

P2.16

Develop 15 ways to determine which direction is north. Provide descriptions and/or sketches of each idea.

P2.17

Generate 15 ideas on how to improve the classroom for this course. Provide descriptions and/or sketches of each idea.

P2.18

Develop 10 ideas for a packaging system that can prevent the shell of a raw egg from breaking if it is dropped down a flight of stairs. Provide descriptions and/or sketches of each idea.

P2.19

Generate 10 ideas for a system to assist disabled persons as they enter and exit a swimming pool. The device is intended to be installed into either new or existing pools. Provide descriptions and/or sketches of each idea.

P2.20

Develop 10 concepts for a new safety feature that can be incorporated into household ladders to prevent falling accidents. Provide descriptions and/or sketches of each idea.

P2.21

In 2010, a giant 100-square-mile iceberg (four times the size of Manhattan) broke off of the northwestern tip of Greenland. The iceberg could float into the waters between Greenland and Canada disrupting critical shipping lanes. Develop 10 ideas to keep the giant iceberg in place. Provide descriptions and/or sketches of each idea.

P2.22

In August 2010, massive downpours caused widespread flooding across Pakistan, killing at least 1500 people. Many times flood waters breached a riverbank and cut a new course for hundreds of miles, destroying villages days later that were not warned of the impending flood waters. Develop 10 ideas that could have prevented the deaths in these villages. Provide descriptions and/or sketches of each idea.

P2.23

The water table under the North China Plain has been dropping steadily by about 4 feet a year while demand for water has been increasing. Projections have the region running out of water by 2035. Develop 10 solutions to solve this issue. Provide descriptions and/or sketches of each idea.

P2.24

On thousands of college campuses around the world, millions of students generate kinetic energy walking around campus every day. Develop 10 ideas for capturing and storing the kinetic energy from students walking around campus. Provide descriptions and/or sketches of each idea.

P2.25

A team of design students working on a mousetrap-powered vehicle decides to use computer compact discs as wheels. The discs are lightweight and readily available. However, they must also be aligned and securely attached to the axles. Develop five design concepts for attaching wheels to a 5-mm-diameter axle. Compact discs are 1.25 mm thick, and they have inner and outer diameters of 15 mm and 120 mm.

P2.26

Go to Google Patents and find a patent for an innovative mechanical device that you think would *not* be a success in the marketplace. Since it is patented, it is considered innovative. Therefore, explain the device and why you think it would not be a success.

P2.27

For the magnesium camera body pieces shown, provide an explanation for which processes you think were used in its manufacture and why. See Figure P2.27.

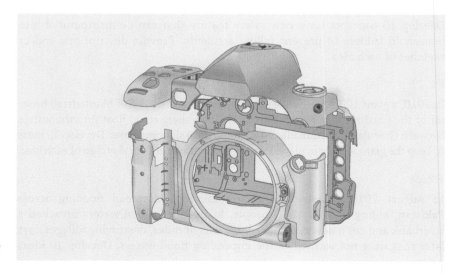

Figure P2.27

P2.28

For the aluminum structural member shown, provide an explanation for the processes you think were used in its manufacture and why. See Figure P2.28.

Figure P2.28

P2.29

In 2010, the Eyjafjallajökull volcano in Iceland erupted, impacting millions of lives by disrupting air travel throughout Europe for weeks. Generate five ideas for a system that could prevent future ash clouds from this volcano from impacting European air traffic.

P2.30

For the ideas in P2.29, create a table that lists the advantages and disadvantages of each idea and make a recommendation about which idea would be the best, considering technical effectiveness, cost, environmental issues, and societal impact.

P2.31

Given the following components and clues, determine what product(s) they describe.

- Diesel engine
- Pulley
- Heat from friction limits the engine power that can be used
- Most components are cast iron or steel
- Most manufacturers are in Europe, Asia, and Africa
- Applied pressure can be manually adjusted

- Auger
- Africa is one place this is commonly used
- Sieve
- Operates using shear forces

P2.32

Using a product currently in your possession or near you, develop ideas for how it could be re-designed to improve its function or decrease its cost. Come up with as many ideas as possible.

P2.33*

As a group, identify a product that is at least one decade old and research the global, social, environmental, and economic factors that may have impacted its design (e.g., shape, configuration, materials, manufacturing) given its intended market, price, and function. Prepare a technical report that describes each set of factors using appropriate evidence from your research sources (e.g., the product itself, specification sheets, user manual, company website, user reviews).

P2.34*

As a group, develop a list of "bad" designs that you think are inefficient, ineffective, inelegant, or provide solutions to problems that are not worth solving. These can be products, processes, systems, or services. Prepare a two minute presentation on these designs.

References

Cagan, J., and Vogel, C.M., *Creating Breakthrough Products.* Upper Saddle River, NJ: FT Press, 2012.

Dieter, G., and Schmidt, L., *Engineering Design,* 4th ed. New York: McGraw-Hill, 2009.

Haik, Y., and Shahin, T. M., *Engineering Design Process.* Stamford, CT: Cengage Learning, 2011.

Juvinall, R. C., and Marshek, K. M., *Fundamentals of Machine Component Design,* 2nd ed. Hoboken, NJ: Wiley, 1991.

Kalpakjian, S., and Schmid, S. R., *Manufacturing Processes for Engineering Materials,* 4th ed. Upper Saddle River, NJ: Prentice-Hall, 2003.

Selinger, C., "The Creative Engineer," *IEEE Spectrum,* August 2004, pp. 48–49.

Shigley, J. E., Mitchell, L. D., and Budynas, R. G., *Mechanical Engineering Design,* 7th ed. New York: McGraw-Hill, 2004.

Ullman, D., *The Mechanical Design Process,* 4th ed. New York: McGraw-Hill, 2009.

Ulrich, K. T., and Eppinger, S. D., *Product Design and Development.* New York: McGraw-Hill, 1998.

Answer to product archaeology dig from page 35: *Automatic soap dispenser*

CHAPTER 3

Technical Problem-Solving and Communication Skills

- Understand the fundamental process to analyze and solve engineering problems.
- Report both a numerical value and its unit in each calculation performed.
- List the base units in the United States Customary System and the Système International d'Unités, and state some of the derived units used in mechanical engineering.
- Understand the need for proper handling of units when making engineering calculations and the implications of not doing so.

- Convert numerical quantities between the United States Customary System and the Système International d'Unités.
- Check your calculations to verify that they are dimensionally consistent.
- Understand how to perform order-of-magnitude approximations.
- Recognize why communication skills are important to engineers, and be able to clearly present solutions in written, oral, and graphical forms.

▶ 3.1 Overview

In this chapter, we will outline the fundamental steps that engineers follow when they solve technical problems and perform calculations in their daily work. These problems arise frequently as part of any engineering design process, and, in order to support their design decisions, mechanical engineers must obtain numerical answers to questions that involve a remarkable breadth of variables and physical properties. In the first portion of this chapter, we will study a fundamental process used by mechanical engineers to analyze technical problems, generating solutions that they understand and can communicate to others. Some of the quantities that you will encounter as you solve problems in your study of mechanical engineering are force, torque, thermal conductivity, shear stress, fluid viscosity, elastic modulus, kinetic energy, Reynolds number, and specific heat. The list is very long indeed. The only way that you can make sense of so many quantities is to be very clear about them in calculations and when you explain results to others. Each quantity that is encountered in mechanical engineering has two components: a numerical value and a dimension. One is meaningless without the other. Practicing engineers pay as close and careful attention to the units in a calculation as they do to the numbers. In the second portion of this chapter, we will discuss fundamental concepts for systems of units, conversions between them, and a procedure for checking dimensional consistency that will serve you well.

Many times in a design process, engineers are asked to estimate quantities rather than finding an exact value. They must answer a number of questions often in the face of uncertainty and incomplete information: About how strong? Approximately how heavy? Roughly how much power? Around what temperature? Also, exact values for material properties are never known; so there always will be some variation between samples of materials.

For these reasons, mechanical engineers need to be comfortable making approximations in order to assign numerical values to quantities that are otherwise unknown. They use their common sense, experience, intuition, and knowledge of physical laws to find answers through a process called *order-of-magnitude approximation*. In Section 3.3, we will illustrate how the fundamental problem-solving process can be used to make order-of-magnitude approximations.

Lastly, the ability to effectively communicate the results of a calculation to others is a critical skill mechanical engineers must possess. Obtaining the answer to a technical question is only half of an engineer's task; the other half involves describing the result to others in a clear, accurate, and convincing manner. Other engineers must be able to understand your calculations and what you did. They must be able to respect your work and have confidence that you solved the problem correctly. Thus, in the final portion of this chapter, we will discuss how to effectively organize and present engineering calculations in a manner that others can follow.

Element 2:
Professional
practice

The topics covered in this chapter fall under the category of *Professional practice* in the hierarchy of mechanical engineering topics (Figure 3.1), and the skills learned in this chapter will support your activities in every other category in the curriculum. These skills include proficiency with technical problem-solving, dimensions, unit systems, conversions, significant digits, approximation, and communication.

The importance of communication and problem-solving skills for an engineering professional cannot be underestimated. Effectively designed systems can be destroyed during operation by the smallest error in analysis, units, or dimension. We can learn much from past examples of these kinds of errors, including the total loss of the *Mars Climate Orbiter* spacecraft in 1999. Shown in Figure 3.2 (see page 66), this spacecraft weighed some 1387 pounds and was part of a $125-million interplanetary exploration program. The spacecraft was designed to be the first orbiting weather satellite for the planet Mars.

The *Mars Climate Orbiter* (MCO) was launched aboard a Delta II rocket from Cape Canaveral, Florida. As the MCO approached the northern hemisphere of Mars, the spacecraft was to fire its main engine for 16 minutes and 23 seconds at a thrust level of 640 N. The engine burn would slow down the spacecraft and place it into an elliptical orbit.

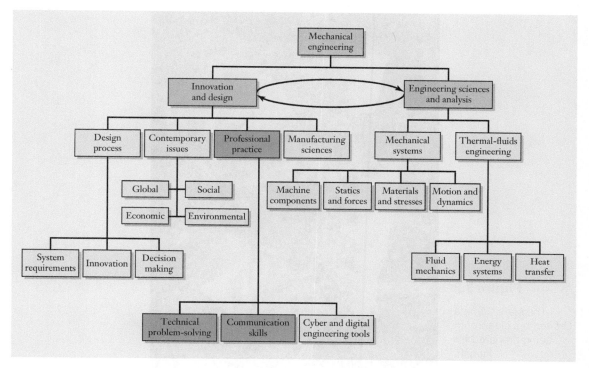

However, following the first main engine burn, the National Aeronautics and Space Administration (NASA) suddenly issued the following statement:

Mars Climate Orbiter is believed to be lost due to a suspected navigation error. Early this morning at about 2 A.M. Pacific Daylight Time the orbiter fired its main engine to go into orbit around the planet. All the information coming from the spacecraft leading up to that point looked normal. The engine burn began as planned five minutes before the spacecraft passed behind the planet as seen from Earth. Flight controllers did not detect a signal when the spacecraft was expected to come out from behind the planet.

The following day, it was announced that

Flight controllers for NASA's Mars Climate Orbiter are planning to abandon the search for the spacecraft at 3 P.M. Pacific Daylight Time today. The team has been using the 70-meter-diameter (230-foot) antennas of the Deep Space Network in an attempt to regain contact with the spacecraft.

What went wrong? A close look at the spacecraft's flight trajectory revealed that during its final approach to the planet, the *Mars Climate Orbiter* apparently passed only 60 km above the Martian surface, rather than the planned closest approach of between 140 km and 150 km. The implication of the unexpectedly low altitude as the spacecraft approached Mars was that either the spacecraft burned up and crashed, or it skipped off the thin Martian atmosphere like a stone on the surface of a lake and began to orbit the sun. Either way, the spacecraft was lost.

Figure 3.1

Relationship of the topics emphasized in this chapter (shaded blue boxes) relative to an overall program of study in mechanical engineering.

Figure 3.2

The *Mars Climate Orbiter* in a cleanroom being prepared for launch.

Courtesy of NASA.

The *Mars Climate Orbiter* Mishap Investigation Board found that the primary problem was a units error that occurred when information was transferred between two teams collaborating on the spacecraft's operation and navigation. To steer the spacecraft and make changes to its velocity, one team of scientists and engineers needed to know the engine's impulse, or the net effect of the rocket engine's thrust over the time that the engine burns. Impulse has the dimensions of (force) × (time), and mission specifications called for it to have been given in the units of newton-seconds, which are the standard units in the International System of Units. However, another team reported numerical values for the impulse without indicating the dimensions, and the data were mistakenly interpreted as being given in the units of pound-seconds, which are the standard units in the United States Customary System. This error resulted in the main engine's effect on the spacecraft's trajectory being underestimated by a factor of 4.45, which is precisely the conversion factor between newtons and pounds.

In another example of the potentially disastrous consequences of unit errors, in July of 1983 Air Canada Flight 143 was en route from Montreal to Edmonton. The Boeing 767 had three fuel tanks, one in each wing and one in the fuselage, which supplied the plane's two jet engines. Flying through clear sky on a summer day, remarkably, one fuel pump after another stopped as each tank on the jetliner ran completely dry. The engine on the left wing was the

Figure 3.3

The landing of Air
Canada Flight 143

© Winniper Free Press. Reprinted with
permission.

first to stop, and 3 minutes later, as the plane descended, the second engine
stopped. Except for small auxiliary backup systems, this sophisticated aircraft
was without power.

The flight crew and air traffic controllers decided to make an emergency
landing at an old airfield. Through their training and skill, the flight crew
was fortunately able to land the plane safely, narrowly missing race cars and
spectators on the runway who had gathered that day for amateur auto racing
(Figure 3.3). Despite the collapse of the front landing gear and the subsequent
damage that occurred to the plane's nose, the flight crew and passengers
suffered no serious injuries.

After a thorough investigation, a review board determined that one of
the important factors behind the accident was a refueling error in which the
quantity of fuel that should have been added to the tanks was incorrectly
calculated. Before takeoff, it was determined that 7682 liters (L) of fuel were
already in the plane's tanks. However, fuel consumption of the new 767s was
being calculated in kilograms, and the airplane needed 22,300 kilograms (kg)
of fuel to fly from Montreal to Edmonton. In addition, the airline had been
expressing the amount of fuel needed for each flight in the units of pounds (lb).
As a result, fuel was being measured by volume (L), weight (lb), and mass (kg)
in two different systems of units.

In the refueling calculations, the conversion factor of 1.77 was used to
convert the volume of fuel (L) to mass (kg), but the units associated with that
number were not explicitly stated or checked. As a result, 1.77 was assumed
to mean 1.77 kg/L (the density of jet fuel). However, the correct density
of jet fuel is 1.77 lb/L, not 1.77 kg/L. As a result of the miscalculation,
roughly 9,000 L instead of 16,000 L of fuel were added to the plane. As
Flight 143 took off for western Canada, it was well short of the fuel required
for the trip.

The proper accounting of units does not have to be difficult, but it
is critical to professional engineering practice. In the following sections,
we begin developing good practices for analyzing engineering problems,
keeping track of units in calculations, and understanding what the solutions
mean.

▶ 3.2 General Technical Problem-Solving Approach

Over the years and across such industries as those described in Chapter 1, engineers have developed a reputation for paying attention to details and getting their answers right. The public gives its trust to the products that engineers design and build. That respect is based in part on confidence that the engineering has been done properly.

Engineers expect one another to present technical work in a manner that is well documented and convincing. Among other tasks, engineers perform calculations that are used to support decisions about a product's design encompassing forces, pressures, temperatures, materials, power requirements, and other factors. The results of those calculations are used to make decisions—sometimes having substantial financial implications for a company—about what form a design should take or how a product will be manufactured. Decisions of that nature can quite literally cost or save a company millions of dollars, and so it is important that the decisions are made for the right reason. When an engineer offers a recommendation, people depend on it being correct.

With that perspective in mind, while your calculations must clearly make sense to you, they must also make sense to others who want to read, understand, and learn from your work—but not necessarily decipher it. If another engineer is unable to follow your work, it could be ignored and viewed as being confusing, incomplete, or even wrong. Good problem-solving skills—writing clearly and documenting each step of a calculation—include not only getting the answer correct, but also communicating it to others convincingly.

Problem-solving process

One of the important communication skills that you can begin developing in the context of this textbook is following a consistent *problem-solving process* for engineering analysis problems. You should view your solutions as a type of technical report that documents your approach and explains your results in a format that can be followed and understood by others. By developing and presenting a systematic solution, you will reduce the chance of having common—but preventable—mistakes creep into your work. Errors involving algebra, dimensions, units, conversion factors, and misinterpretation of a problem statement can be kept to a minimum by paying attention to the details.

For the purpose of solving the end-of-chapter problems in this textbook, you should try to organize and present your work according to the following three steps, which also define the structure in which the examples are presented:

1. **Approach.** The intent of this step is to make sure that you have a plan of attack in mind for solving the problem. This is an opportunity to think about the problem up front before you start crunching numbers and putting pencil to paper. Write a short summary of the problem and explain the general approach that you plan to take, and list the major concepts, *assumptions*, equations, and conversion factors that you expect to use. Making the proper set of assumptions is critical to solving the problem accurately. For instance, if gravity is assumed to be present, then the weight of all the components in the problem may need to be accounted for. Similarly, if friction is assumed present, then the

Assumptions

equations must account for it. In most analysis problems, engineers have to make important assumptions about many key parameters including gravity, friction, distribution of applied forces, stress concentrations, material inconsistencies, and operating uncertainties. By stating these assumptions, identifying the given information, and summarizing the knowns and unknowns, the engineer fully defines the scope of the problem. By being clear about the objective, you are able to disregard extraneous information and focus on solving the problem efficiently.

2. **Solution.** Your solution to an engineering analysis problem will generally include text and diagrams along with your calculations to explain the major steps that you are taking. If appropriate, you should include a simplified drawing of the physical system being analyzed, label the major components, and list numerical values for relevant dimensions. In the course of your solution, and as you manipulate equations and perform calculations, it is good practice to solve for the unknown variable symbolically before inserting numerical values and units. In that manner, you can verify the dimensional consistency of the equation. When you substitute a numerical value into an equation, be sure to include the units as well. At each point in the calculation, you should explicitly show the units associated with each numerical value, and the manner in which the dimensions are canceled or combined. A number without a unit is meaningless, just as a unit is meaningless without a numerical value assigned to it. At the end of the calculation, present your answer using the appropriate number of significant digits, but keep more digits in the intermediate calculations to prevent round-off errors from accumulating.

3. **Discussion.** This final step must always be addressed because it demonstrates an understanding of the assumptions, equations, and solutions. First, you must use your intuition to determine whether the answer's order of magnitude seems reasonable. Second, you must evaluate your assumptions to make sure they are reasonable. Third, identify the major conclusion that you are able to draw from the solution, and explain what your answer means from a physical standpoint. Of course, you should always double-check the calculations and make sure that they are dimensionally consistent. Finally, underline, circle, or box your final result so that there is no ambiguity about the answer that you are reporting.

Putting this process into practice with repeated success requires an understanding of dimensions, units, and the two primary unit systems.

▶ 3.3 Unit Systems and Conversions

Engineers specify physical quantities in two different—but conventional—systems of units: the *United States Customary System (USCS)* and the *International System of Units* (Système International d'Unités or *SI*). Practicing mechanical engineers must be conversant with both systems. They need to convert quantities from one system to the other, and they must be able to perform calculations equally well in either system. In this textbook,

United States Customary System (USCS) and International System of Units (SI)

examples and problems are formulated in both systems so that you can learn to work effectively with the USCS and the SI. As we introduce new physical quantities in the following chapters, the corresponding USCS and SI units for each will be described, along with their conversion factors.

Base and Derived Units

Given some perspective from the *Mars Climate Orbiter's* loss and the emergency landing of the Air Canada flight on the importance of units and their bookkeeping, we now turn to the specifics of the USCS and SI. A unit is defined as an arbitrary division of a physical quantity, which has a magnitude that is agreed on by mutual consent. Both the USCS and SI are made up of base units

Base units

and derived units. A *base unit* is a fundamental quantity that cannot be broken down further or expressed in terms of any simpler elements. Base units are independent of one another, and they form the core building blocks of any unit system. As an example, the base unit for length is the meter (m) in the SI and the foot (ft) in the USCS.

Derived units

Derived units, as their name implies, are combinations or groupings of several base units. An example of a derived unit is velocity (length/time), which is a combination of the base units for length and time. The liter (which is equivalent to 0.001 m^3) is a derived unit for volume in the SI. Likewise, the mile (which is equivalent to 5280 ft) is a derived unit for length in the USCS. Unit systems generally have relatively few base units and a much larger set of derived units. We next discuss the specifics of base and derived units in both the USCS and SI and conversions between them.

International System of Units

In an attempt to standardize the different systems of measurement around the world, in 1960 the *International System of Units* was named as the measurement standard structured around the seven base units in Table 3.1. In addition to the mechanical quantities of meters, kilograms, and seconds, the SI includes base units for measuring electric current, temperature, the amount of substance, and light intensity. The SI is colloquially referred to as the metric system, and it conveniently uses powers of ten for multiples and divisions of units.

The base units in the SI are today defined by detailed international agreements. However, the units' definitions have evolved and changed slightly

Quantity	SI Base Unit	Abbreviation
Length	meter	m
Mass	kilogram	kg
Time	second	s
Electric current	ampere	A
Temperature	kelvin	K
Amount of substance	mole	mol
Light intensity	candela	cd

Table 3.1

Base Units in the SI

as measurement technologies have become more precise. The origins of the meter, for instance, trace back to the eighteenth century. The meter was originally intended to be equivalent to one ten-millionth of the length of the meridian passing from the northern pole, through Paris, and ending at the equator (namely, one-quarter of the Earth's circumference). Later, the meter was defined as the length of a bar that was made from a platinum-iridium metal alloy. Copies of the bar, which are called *prototype meters*, were distributed to governments and laboratories around the world, and the bar's length was always measured at the temperature of melting water ice. The meter's definition has been updated periodically to make the SI's length standard more robust and repeatable, all the while changing the actual length by as little as possible. As of October 20, 1983, the meter is defined as the length of the path traveled by light in vacuum during a time interval of 1/299,792,458 of a second, which in turn is measured to high accuracy by an atomic clock.

Prototype meter

In a similar vein, at the end of the eighteenth century, the kilogram was defined as the mass of 1000 cm^3 of water. Today, the kilogram is determined by the mass of a physical sample that is called the *standard kilogram*, and like the previously used prototype meter, it is also made of platinum and iridium. The standard kilogram is kept in a vault in Sèvres, France, by the International Bureau of Weights and Measures, and duplicate copies are maintained in other laboratories throughout the world. Although the meter is today based on a reproducible measurement involving the speed of light and time, the kilogram is not. Scientists are researching alternative means to define the kilogram in terms of an equivalent electromagnetic force or in terms of the number of atoms in a carefully machined silicon sphere, but for the time being the kilogram is the only base unit in the SI that continues to be defined by a human-made artifact.

Standard kilogram

With respect to the other base units in the SI, the second is defined in terms of the time required for a certain quantum transition to occur in a cesium-133 atom. The kelvin (abbreviated K without the degree (°) symbol) is based on the triple point of pure water, which is a special combination of pressure and temperature where water can exist as a solid, liquid, or gas. Similar fundamental definitions have been established for the ampere, mole, and candela.

A few of the derived units used in the SI are listed in Table 3.2 (see on page 72). The newton (N) is a derived unit for force, and it is named after the British physicist Sir Isaac Newton. While his classical laws of motion are presented in more detail in Chapter 4, his *second law of motion*, $F = ma$, states that the force F acting on an object is equivalent to the product of its mass m and acceleration a. The newton is therefore defined as the force that imparts an acceleration of one meter per second per second to an object having a mass of one kilogram:

Second law of motion

$$1\text{N} = (1\text{kg})\left(1\frac{\text{m}}{\text{s}^2}\right) = 1\frac{\text{kg} \cdot \text{m}}{\text{s}^2} \qquad (3.1)$$

The base unit kelvin (K), and the derived units joule (J), pascal (Pa), watt (W), and other that are named after individuals are not capitalized, although their abbreviations are.

Quantity	SI Derived Unit	Abbreviation	Definition
Length	micrometer or micron	μm	$1\ \mu\text{m} = 10^{-6}$ m
Volume	liter	L	$1\ \text{L} = 0.001\ \text{m}^3$
Force	newton	N	$1\ \text{N} = 1\ (\text{kg} \cdot \text{m})/\text{s}^2$
Torque, or moment of a force	newton-meter	N $\cdot$ m	—
Pressure or stress	pascal	Pa	$1\ \text{Pa} = 1\ \text{N/m}^2$
Energy, work, or heat	joule	J	$1\ \text{J} = 1\ \text{N} \cdot \text{m}$
Power	watt	W	$1\ \text{W} = 1\ \text{J/s}$
Temperature	degree Celsius	°C	$°\text{C} = \text{K} - 273.15$
Although a change in temperature of 1 kelvin equals a change of 1 degree Celsius, numerical values are converted using the formula.			

Table 3.2

Certain Derived Units in the SI

Prefix

Base and derived units in the SI are often combined with a *prefix* so that a physical quantity's numerical value does not have a power-of-ten exponent that is either too large or too small. Use a prefix to shorten the representation of a numerical value and to reduce an otherwise excessive number of trailing zero digits in your calculations. The standard prefixes in the SI are listed in Table 3.3. For example, modern wind turbines are now producing over 7,000,000 W of power. Because it is cumbersome to write so many trailing zeroes, engineers prefer to condense the powers of ten by using a prefix. In this case, we describe a turbine's output as being over 7 MW (megawatt), where the prefix "mega" denotes a multiplicative factor of 10^6.

SI conventions

Good practice is not to use a prefix for any numerical value that falls between 0.1 and 1000. Thus, the "deci," "deca," and "hecto" prefixes in Table 3.3 are rarely used in mechanical engineering. Other *conventions* for manipulating dimensions in the SI include the following:

1. If a physical quantity involves dimensions appearing in a fraction, a prefix should be applied to the units appearing in the numerator rather than the denominator. It is preferable to write kN/m in place of N/mm. An exception to this convention is that the base unit kg can appear in a dimension's denominator.

2. Placing a dot or hyphen between units that are adjacent in an expression is a good way to keep them visually separated. For instance, in expanding a newton into its base units, engineers write $(\text{kg} \cdot \text{m})/\text{s}^2$ instead of kgm/s^2. An even worse practice would be to write mkg/s^2, which is particularly confusing because the numerator could be misinterpreted as a millikilogram!

Name	Symbol	Multiplicative Factor
tera	T	$1,000,000,000,000 = 10^{12}$
giga	G	$1,000,000,000 = 10^{9}$
mega	M	$1,000,000 = 10^{6}$
kilo	k	$1000 = 10^{3}$
hecto	h	$100 = 10^{2}$
deca	da	$10 = 10^{1}$
deci	d	$0.1 = 10^{-1}$
centi	c	$0.01 = 10^{-2}$
milli	m	$0.001 = 10^{-3}$
micro	μ	$0.000,001 = 10^{-6}$
nano	n	$0.000,000,001 = 10^{-9}$
pico	p	$0.000,000,000,001 = 10^{-12}$

Table 3.3

Order-of-Magnitude Prefixes in the SI

3. Dimensions in plural form are not written with an "s" suffix. Engineers write 7 kg rather than 7 kgs because the trailing "s" could be misinterpreted to mean seconds.

4. Except for abbreviations of derived units that are named after individuals and the abbreviation for liter, dimensions in the SI are written in lowercase.

United States Customary System of Units

The use of the SI in the United States was legalized for commerce by the Congress in 1866. The Metric Conversion Act of 1975 later outlined voluntary conversion of the United States to the SI:

> It is therefore the declared policy of the United States to designate the metric system of measurement as the preferred system of weights and measures for United States trade and commerce.

That policy notwithstanding, the so-called process of metrification in the United States has been a slow one, and, at least for the time being, the United States continues to employ two systems of units: the SI and the United States Customary System (USCS). The USCS includes such measures as pounds, tons, feet, inches, miles, seconds, and gallons. Sometimes referred to as the English/British System, or the *Foot-Pound-Second System*, the USCS is a historical representation of units with its origin tracing back to the ancient Roman Empire. In fact, the abbreviation for pound (lb) is taken from the Roman unit of weight, *libra*, and the word "pound" itself comes from the Latin *pendere*, meaning "to weigh." The USCS was originally used in Great Britain, but it is today primarily used in the United States. Most other industrialized countries have adopted the SI as their uniform standard of measurement for business

Foot-Pound-Second System

and commerce. Engineers practicing in the United States or in companies with U.S. affiliations need to be skillful with both the USCS and SI.

Why does the United States stand out in retaining the USCS? The reasons are complex and involve economics, logistics, and culture. The vast continent-sized infrastructure already existing in the United States is based on the USCS, and immediate conversion away from the current system would involve significant expense. Countless structures, factories, machines, and spare parts already have been built using USCS dimensions. Furthermore, while most American consumers have a sense for, say, how much a gallon of gasoline or a pound of apples costs, they are not as familiar with the SI counterparts. That being noted, standardization to the SI in the United States is proceeding because of the need for companies to interact with, and compete against, their counterparts in the international business community. Until such time as the United States has made a full transition to the SI (and don't hold your breath), it will be necessary—and indeed essential—for you to be proficient in both unit systems.

As shown in Table 3.4, the seven base units in the USCS are the foot, pound, second, ampere, degree Rankine, mole, and candela. One of the major distinctions between the SI and USCS is that mass is a base unit in the SI (kg), whereas force is a base unit in the USCS (lb). It is also acceptable to refer to the pound as a pound-force with the abbreviation lbf. In the branch of mechanical engineering dealing with forces, materials, and structures, the shorter terminology "pound" and the abbreviation "lb" are more common, and that convention is used throughout this textbook.

Another distinction between the USCS and SI is that the USCS employs two different dimensions for mass: the *pound-mass* and the *slug*. The pound-mass unit is abbreviated lbm. There is no abbreviation for the slug, and so the full name is written out adjacent to a numerical value. It is also conventional to use the plural "slugs." This dimension's name appears to have been chosen historically to refer to a chunk or lump of material, and it is unrelated to the small land mollusk of the same name. In mechanical engineering, the slug is the preferred unit for calculations involving such quantities as gravitation, motion, momentum, kinetic energy, and acceleration. However, the pound-mass is a more convenient dimension for engineering calculations involving the material, thermal, or combustion properties of liquids, gases, and fuels. Both dimensions for mass will be used in their conventional mechanical engineering contexts throughout this textbook.

Pound-mass
Slug

Quantity	USCS Base Unit	Abbreviation
Length	foot	ft
Force	pound	lb
Time	second	s
Electric current	ampere	A
Temperature	degree Rankine	°R
Amount of substance	mole	mol
Light intensity	candela	cd

Table 3.4

Base Units in the USCS

In the final analysis, however, the slug and pound-mass are simply two different derived units for mass. Because they are measures of the same physical quantity, they are also closely related to one another. In terms of the USCS's base units of pounds, seconds, and feet, the slug is defined as follows:

$$1 \text{ slug} = 1\frac{\text{lb} \cdot \text{s}^2}{\text{ft}} \qquad (3.2)$$

Referring to the second law of motion, one pound of force will accelerate a one-slug object at the rate of one foot per second per second:

$$1 \text{ lb} = (1 \text{ slug})\left(1\frac{\text{ft}}{\text{s}^2}\right) = 1\frac{\text{slug} \cdot \text{ft}}{\text{s}^2} \qquad (3.3)$$

On the other hand, the pound-mass is defined as the quantity of mass that weighs one pound. One pound-mass will accelerate at the rate of 32.174 ft/s^2 when one pound of force acts on it:

$$1 \text{ lbm} = \frac{1 \text{ lb}}{32.174 \text{ ft/s}^2} = 3.1081 \times 10^{-2}\frac{\text{lb} \cdot \text{s}^2}{\text{ft}} \qquad (3.4)$$

The numerical value of 32.174 ft/s^2 is taken as the reference acceleration because it is the Earth's gravitational acceleration constant. By comparing Equations (3.2) and (3.4), we see that the slug and pound-mass are related by

$$1 \text{ slug} = 32.174 \text{ lbm} \qquad 1 \text{ lbm} = 3.1081 \times 10^{-2} \text{ slugs} \qquad (3.5)$$

In short, the slug and pound-mass are both defined in terms of the action of a one-pound force, but the reference acceleration for the slug is 1 ft/s^2, and the reference acceleration is 32.174 ft/s^2 for the pound-mass. By agreement among the measurement standards laboratories of English-speaking countries, 1 lbm is also equivalent to 0.45359237 kg.

Despite the fact that the pound-mass and pound denote different physical quantities (mass and force), they are often improperly interchanged. One reason for the confusion is the similarity of their names. Another reason is related to the definition itself of the pound-mass: a quantity of matter having a mass of 1 lbm also weighs 1 lb, assuming Earth's gravity. By contrast, an object having a mass of 1 slug weighs 32.174 lb on Earth. You should realize, however, that the USCS is not alone with the dubious potential of confusing mass and weight. One will sometimes see the SI's kilogram used improperly to denote force. Some tires and pressure gages, for instance, are labeled with inflation pressures having the dimensions of kg/m^2. Some scales that are used in commerce have weights tabulated in kilograms or in terms of a defunct unit called the kilogram-force that is not even part of the SI.

Aside from mass, other derived units can be formed as combinations of the USCS's base units. Some that arise in mechanical engineering are listed in Table 3.5 (see on page 76), which includes the *mil* (1000th of an inch, or 1/12,000th of a foot), the *foot-pound* (for energy, work, or heat), and the *horsepower* (550 (ft · lb)/s). Also note that the abbreviation for inch typically includes a period to distinguish it from the word "in" within technical documents.

Quantity	Derived Unit	Abbreviation	Definition
Length	mil	mil	1 mil = 0.001 in.
	inch	in.	1 in. = 0.0833 ft
	mile	mi	1 mi = 5280 ft
Volume	gallon	gal	1 gal = 0.1337 ft^3
Mass	slug	slug	1 slug = 1(lb · s^2)/ft
	pound-mass	lbm	1 lbm = 3.1081×10^{-2} (lb· s^2)/ft
Force	ounce	oz	1 oz = 0.0625 lb
	ton	ton	1 ton = 2000 lb
Torque, or moment of a force	foot-pound	ft · lb	—
Pressure or stress	pound/inch^2	psi	1 psi = 1 lb/in^2
Energy, work, or heat	foot-pound	ft · lb	—
	British thermal unit	Btu	1 Btu = 778.2 ft · lb
Power	horsepower	hp	1 hp = 550 (ft · lb)/s
Temperature	degree Fahrenheit	°F	°F = °R − 459.67

Although a change in temperature of 1° Rankine also equals a change of 1° F, numerical values are converted using the formula.

Table 3.5

Certain Derived Units in the USCS

Focus On   MASS AND WEIGHT

Mass is an intrinsic property of an object based on the amount and density of material from which it is made. Mass m measures the quantity of matter that is contained in the object, and, as such, it does not vary with position, motion, or changes in the object's shape. Weight, on the other hand, is the force that is needed to support the object against gravitational attraction, and it is calculated as

$$w = mg$$

based on the gravitational acceleration

$$g = 32.174 \frac{ft}{s^2} \approx 32.2 \frac{ft}{s^2} \quad \text{(USCS)}$$

$$g = 9.8067 \frac{m}{s^2} \approx 9.81 \frac{m}{s^2} \quad \text{(SI)}$$

By international agreement, these accelerations are standard values at sea level and a latitude of 45°. The gravitational acceleration at a specific location on the Earth's surface, however, does vary with latitude, the slightly irregular shape of the Earth, the density of the Earth's crust, and the size of any nearby land masses. Although an object's weight depends on gravitational acceleration, its mass does not. For most mechanical engineering calculations, it is sufficient to approximate g to three significant digits.

Converting Between the SI and USCS

A numerical value in one unit system can be transformed into an equivalent value in the other system by using unit conversion factors. Conversion factors between some of the USCS and the SI quantities that arise in mechanical engineering are listed in Table 3.6. The conversion process requires changes made both to

Quantity	Conversion		
Length	1 in.	=	25.4 mm
	1 in.	=	0.0254 m
	1 ft	=	0.3048 m
	1 mi	=	1.609 km
	1 mm	=	3.9370×10^{-2} in.
	1 m	=	39.37 in.
	1 m	=	3.2808 ft
	1 km	=	0.6214 mi
Area	1 in^2	=	645.16 mm^2
	1 ft^2	=	9.2903×10^{-2} m^2
	1 mm^2	=	1.5500×10^{-3} in^2
	1 m^2	=	10.7639 ft^2
Volume	1 ft^3	=	2.832×10^{-2} m^3
	1 ft^3	=	28.32 L
	1 gal	=	3.7854×10^{-3} m^3
	1 gal	=	3.7854 L
	1 m^3	=	35.32 ft^3
	1 L	=	3.532×10^{-2} ft^3
	1 m^3	=	264.2 gal
	1 L	=	0.2642 gal
Mass	1 slug	=	14.5939 kg
	1 lbm	=	0.45359 kg
	1 kg	=	6.8522×10^{-2} slugs
	1 kg	=	2.2046 lbm
Force	1 lb	=	4.4482 N
	1 N	=	0.22481 lb
Pressure or stress	1 psi	=	6895 Pa
	1 psi	=	6.895 kPa
	1 Pa	=	1.450×10^{-4} psi
	1 kPa	=	0.1450 psi

Table 3.6

Conversion Factors between Certain Quantities in the USCS and SI

Work, energy, or heat	1 ft · lb	=	1.356 J
	1 Btu	=	1055 J
	1 J	=	0.7376 ft · lb
	1 J	=	9.478×10^{-4} Btu
Power	1 (ft · lb)/s	=	1.356 W
	1 hp	=	0.7457 kW
	1 W	=	0.7376 (ft · lb)/s
	1 kW	=	1.341 hp

Table 3.6

Continued

the numerical value and to the dimensions associated with it. Regardless, the physical quantity remains unchanged, being made neither larger nor smaller, since the numerical value and units are transformed together. In general terms, the procedure for converting between the two systems is as follows:

1. Write the given quantity as a number followed by its dimensions, which could involve a fractional expression such as kg/s or N/m.
2. Identify the units desired in the final result.
3. If derived units such as L, Pa, N, lbm, or mi are present in the quantity, you may find it necessary to expand them in terms of their definitions and base units. In the case of the pascal, for instance, we would write

$$\text{Pa} = \frac{\text{N}}{\text{m}^2} = \left(\frac{1}{\text{m}^2}\right)\left(\frac{\text{kg} \cdot \text{m}}{\text{s}^2}\right) = \frac{\text{kg}}{\text{m} \cdot \text{s}^2}$$

where we canceled the meter dimension algebraically.

4. Likewise, if the given quantity includes a prefix that is not incorporated in the conversion factors, expand the quantity according to the prefix definitions listed in Table 3.3. The kilonewton, for instance, would be expanded as 1 kN = 1000 N.
5. Look up the appropriate conversion factor from Table 3.6, and multiply or divide, as necessary, the given quantity by it.
6. Apply the rules of algebra to cancel dimensions in the calculation and to reduce the units to the ones that you want in the final result.

You cannot escape conversion between the USCS and SI, and you will not find your way through the maze of mechanical engineering without being proficient with both sets of units. The decision as to whether the USCS or SI should be used when solving a problem will depend on how the information in the statement is specified. If the information is given in the USCS, then you should solve the problem and apply formulas using the USCS alone. Conversely, if the information is given in the SI, then formulas should be applied using the SI alone. It is bad practice for you to take data given in the USCS, convert to SI, perform calculations in SI, and then convert back to USCS (or vice versa). The reason for this recommendation is twofold. First, from the practical day-to-day matter of being a competent engineer, you will need to be fluent in both the USCS and SI. Furthermore, the additional steps involved when quantities are converted from one system to another and back again are just other opportunities for errors to creep into your solution.

Example 3.1 | *Engine Power Rating*

A gasoline-powered engine produces a peak output of 10 hp. Express the given power P in the SI.

Approach
As listed in Table 3.5 for derived units in the USCS, the abbreviation "hp" refers to horsepower. The SI unit for power is the watt (W). The final row of Table 3.6 lists the conversions for power, and we read off the factor 1 hp = 0.7457 kW. Here the "kilo" prefix in Table 3.3 denotes a multiplier of 1000.

Solution
Applying the conversion factor to the engine's power rating, we have

$$P = (10 \text{ hp})\left(0.7457\frac{\text{kW}}{\text{hp}}\right)$$

$$= 7.457(\text{hp})\left(\frac{\text{kW}}{\text{hp}}\right)$$

$$= 7.457 \text{ kW}$$

Discussion
The conversion process involves two steps: algebraically combining the numerical values and the dimensions. In our calculation, we canceled the hp dimension, and we explicitly showed that step in our solution. In terms of the derived unit watt, the motor produces an output of 7457 W. However, since this numerical value is greater than 1000, the prefix "kilo" is used.

$$P = 7.457 \text{ kW}$$

Example 3.2 | *Fire Spinkler*

The specification for a certain residential fire-suppression system is that water should be sprayed at the rate q of 10 gal/min. For the revision of a technical manual intended for customers outside the United States, express the flow rate in the SI based on a time interval of 1 s.

Example 3.2 | *continued*

Approach

To complete this problem, we need to apply conversion factors for both volume and time. Referring to Table 3.6, we could express volume in the SI using the dimensions of either m^3 or L. We assume that a cubic meter is much larger than the amount of water we expect to be sprayed each second; so we initially decide to convert volume to liters with the conversion factor 1 gal = 3.785 L. We will double-check that assumption after we complete the calculation.

Solution

Converting dimensions for both volume and time,

$$q = \left(10 \frac{\text{gal}}{\text{min}}\right)\left(\frac{1}{60} \frac{\text{min}}{\text{s}}\right)\left(3.785 \frac{\text{L}}{\text{gal}}\right)$$

$$= 0.6308 \left(\frac{\cancel{\text{gal}}}{\cancel{\text{min}}}\right)\left(\frac{\cancel{\text{min}}}{\text{s}}\right)\left(\frac{\text{L}}{\cancel{\text{gal}}}\right)$$

$$= 0.6308 \frac{\text{L}}{\text{s}}$$

Discussion

We first combined the numerical values and then algebraically canceled the dimensions. It is good practice (and a good double check) to show how the units clearly cancel in the conversion process. We could have alternatively expressed the flow rate in the units of m^3/s. However, since a cubic meter is 1000 times larger than a liter, the dimension L/s turns out to be better suited for the problem at hand since the numerical value of 0.6308 does not involve a power-of-ten exponent.

$$q = 0.6308 \frac{\text{L}}{\text{s}}$$

Example 3.3 | *Laser*

Helium-neon lasers are used in engineering laboratories, in robot vision systems, and even in the barcode readers found in supermarket checkout counters. A certain laser has a power output of 3 mW and produces light of wavelength $\lambda = 632.8$ nm. The lowercase Greek character lambda (λ) is a

Example 3.3 | *continued*

conventional symbol used for wavelength; Appendix A summarizes the names and symbols of other Greek letters. (a) Convert the power rating to horsepower. (b) Convert the wavelength to inches.

Approach

Referring to Table 3.3, the power dimension mW refers to a milliwatt or 10^{-3} W, and nm denotes a billionth of a meter (10^{-9} m). The conversion factors for power and length are listed in Table 3.6 as 1 kW = 1.341 hp and 1 m = 39.37 in.

Solution

(a) We first convert the power's SI prefix from milli to kilo in order to apply the conversion factor as listed in the table. The laser produces 3×10^{-3} W = 3×10^{-6} kW. We convert this small quantity into the USCS as follows:

$$P = (3 \times 10^{-6}\,\text{kW})\left(1.341\,\frac{\text{hp}}{\text{kW}}\right)$$

$$= 4.023 \times 10^{-6}\,(\text{kW})\left(\frac{\text{hp}}{\text{kW}}\right)$$

$$= 4.023 \times 10^{-6}\,\text{hp}$$

(b) Expressed in scientific notation, the laser's wavelength is 632.8×10^{-9} m = 6.328×10^{-7} m, and the length conversion becomes

$$\lambda = (6.328 \times 10^{-7}\,\text{m})\left(39.37\,\frac{\text{in.}}{\text{m}}\right)$$

$$= 2.491 \times 10^{-5}\,(\text{m})\left(\frac{\text{in.}}{\text{m}}\right)$$

$$= 2.491 \times 10^{-5}\,\text{in.}$$

Discussion

Because the dimensions of horsepower and inch are so much larger than the laser's power and wavelength, they are not very convenient for describing its characteristics.

$$P = 4.023 \times 10^{-6}\,\text{hp}$$

$$\lambda = 2.491 \times 10^{-5}\,\text{in.}$$

▶ 3.4 Significant Digits

A *significant digit* is one that is known to be correct and reliable in the light of inaccuracy that is present in the supplied information, any approximations that have been made along the way, and the mechanics of the calculation itself. As a general rule, the last significant digit that you report in the answer to a problem should have the same order of magnitude as the last significant digit in the given data. It would be inappropriate to report more significant digits in the answer than were given in the supplied information, since that implies that the output of a calculation is somehow more accurate than the input to it.

The *precision* of a number is half as large as the place of the last significant digit present in the number. The factor is one-half because the last digit of a number represents the rounding off of the trailing digits, either higher or lower. For instance, suppose an engineer records in a design notebook that the force acting on the bearing of a hard disk drive's motor caused by rotational imbalance is 43.01 mN. That statement means that the force is closer to 43.01 mN than it is to either 43.00 mN or 43.02 mN. The reported value of 43.01 mN and its number of significant digits mean that the actual physical value of the force could lie anywhere between 43.005 and 43.015 mN (Figure 3.4(a)). The precision of the numerical value is ±0.005 mN, the variation that could be present in the force reading and still result in a rounded-off value of 43.01 mN. Even when we write 43.00 mN, a numerical value that has two trailing zeros, four significant digits are present, and the implied precision remains ±0.005 mN.

Alternatively, suppose that the engineer had written the force as being 43.010 mN. That statement implies that the value is known quite accurately indeed, and means that the force lies closer to 43.010 mN than to either 43.009 mN or 43.011 mN. The precision is now ±0.0005 mN (Figure 3.4(b)).

Likewise, you can see how some ambiguity arises when a quantity such as 200 lb is reported. On the one hand, the value could mean that the measurement was made only within the nearest 100 pounds, meaning that the

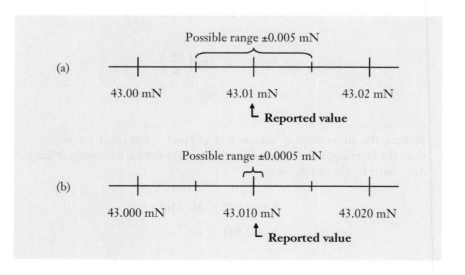

Figure 3.4

The precision of a numerical value for force depends on the number of significant digits that are reported. The actual physical value lies in a range centered around the reported value. (a) Two trailing digits. (b) Three trailing digits.

Possible range ±0.005 mN

(a)

43.00 mN   43.01 mN   43.02 mN

↳ Reported value

Possible range ±0.0005 mN

(b)

43.000 mN   43.010 mN   43.020 mN

↳ Reported value

actual magnitude of the force is closer to 200 lb than to either 100 lb or 300 lb. On the other hand, the value could mean that the force is indeed 200 lb, and not 199 lb or 201 lb, and that the real value lies somewhere between 199.5 lb and 200.5 lb. Either way, it's vague to write a quantity such as 200 lb without giving some other indication as to the precision of the number.

As a general rule, during the intermediate steps of a calculation, retain several more significant digits than you expect to report in the final answer. In that manner, rounding errors will not creep into your solution, compound along the way, and distort the final answer. When the calculation is complete, you can always truncate the numerical value to a reasonable number of significant digits. These considerations lead us to the following rule-of-thumb guideline:

> For the purpose of examples and problems in this textbook, treat the supplied information as being exact. Recognizing engineering approximations and limits on measurements, though, report your answers to only four significant digits.

You should be aware of the misleading sense of accuracy that is offered by the use of calculators and computers. While a calculation certainly can be performed to eight or more significant digits, nearly all dimensions, material properties, and other physical parameters encountered in mechanical engineering are known to far fewer digits. Although the computation itself might be very accurate, the input data that is supplied to calculations will rarely have the same level of accuracy.

▶ 3.5 Dimensional Consistency

When you apply equations of mathematics, science, or engineering, the calculations must be dimensionally consistent, or they are wrong. *Dimensional consistency* means that the units associated with the numerical values on each side of an equality sign match. Likewise, if two terms are combined in an equation by summation, or if they are subtracted from one another, the two quantities must have the same dimensions. This principle is a straightforward means to double-check your algebraic and numerical work.

In paper-and-pencil calculations, keep the units adjacent to each numerical quantity in an equation so that they can be combined or canceled at each step in the solution. You can manipulate dimensions just as you would any other algebraic quantity. By using the principle of dimensional consistency, you can double-check your calculation and develop greater confidence in its accuracy. Of course, the result could be incorrect for a reason other than dimensions. Nevertheless, performing a double check on the units in an equation is always a good idea.

The principle of dimensional consistency can be particularly useful when you perform calculations involving mass and force in the USCS. The definitions of the slug and pound-mass in terms of different reference accelerations is often a point of confusion when mass quantities are converted between the USCS and SI. In those cases, the principle of dimensional consistency can be applied to confirm that the units in the calculation are correct. Dimensional consistency can be illustrated by as simple a calculation as finding the weights of two objects, the first having a mass of 1 slug and the second having a mass of 1 lbm. In the first case, the weight of a 1 slug object is

$$w = (1 \text{ slug})\left(32.174\,\frac{\text{ft}}{\text{s}^2}\right)$$

$$= 32.174\,\frac{\text{slug} \cdot \text{ft}}{\text{s}^2}$$

$$= 32.174 \text{ lb}$$

In the final step of this calculation, we used the definition of the slug from Table 3.5. This object, having a mass of one slug, weighs 32.174 lb. On the other hand, for the object having a mass of 1 lbm, direct substitution in the equation $w = mg$ would give the dimensions of lbm · ft/s², which is neither the same as a pound nor a conventional unit for force in the USCS. For the calculation to be dimensionally consistent, an intermediate step is necessary to convert m to the units of slug using Equation (3.5):

$$m = (1 \text{ lbm})\left(3.1081 \times 10^{-2}\,\frac{\text{slug}}{\text{lbm}}\right)$$

$$= 3.1081 \times 10^{-2} \text{ slugs}$$

In the second case, the weight of the 1 lbm object becomes

$$w = (3.1081 \times 10^{-2} \text{ slugs})\left(3.174\,\frac{\text{ft}}{\text{s}^2}\right)$$

$$= 1\,\frac{\text{slug} \cdot \text{ft}}{\text{s}^2}$$

$$= 1 \text{ lb}$$

The principle of dimensional consistency will help you to make the proper choice of mass units in the USCS. Generally speaking, the slug is the preferred unit for calculations involving Newton's second law ($f = ma$), kinetic energy ($\frac{1}{2}mv^2$), momentum (mv), gravitational potential energy (mgh), and other mechanical quantities. The process of verifying the dimensional consistency of an equation by keeping track of units is illustrated in the following examples.

Example 3.4 | *Aerial Refueling*

The KC-10 Extender tanker aircraft of the United States Air Force is used to refuel other planes in flight. The Extender can carry 365,000 lb of jet fuel, which can be transferred to another aircraft through a boom that temporarily connects the two planes. (a) Express the mass of the fuel in the units of slugs and lbm. (b) Express the mass and weight of the fuel in SI units.

Example 3.4 | *continued*

Approach
We will calculate the mass m in terms of the fuel's weight w and the gravitational acceleration $g = 32.2$ ft/s^2. With w expressed in pounds and g having the units of ft/s^2, the expression $w = mg$ is dimensionally consistent when mass has the units of slugs. We will then convert from slugs to lbm using the conversion factor 1 slug $= 32.174$ lbm from Equation (3.5). In part (b), we convert the fuel's mass in the USCS to the SI using the conversion factor 1 slug $= 14.59$ kg from Table 3.6.

Solution
(a) We first determine the mass of the fuel in the units of slugs:

$$m = \frac{3.65 \times 10^5 \text{ lb}}{32.2 \text{ ft/s}^2} \quad \leftarrow \left[m = \frac{w}{g} \right]$$

$$= 1.134 \times 10^4 \frac{\text{lb} \cdot \text{s}^2}{\text{ft}}$$

$$= 1.134 \times 10^4 \text{ slugs}$$

In the last step, we used the definition 1 slug $= 1$ (lb · s^2)/ft. Since an object that weighs 1 lb has a mass of 1 lbm, the mass of the fuel can alternatively be expressed as 365,000 lbm.

(b) We convert the mass quantity 1.134×10^4 slugs into the SI unit kg:

$$m = (1.134 \times 10^4 \text{ slugs})\left(14.59 \frac{\text{kg}}{\text{slug}} \right)$$

$$= 1.655 \times 10^5 (\text{slugs})\left(\frac{\text{kg}}{\text{slug}} \right)$$

$$= 1.655 \times 10^5 \text{ kg}$$

Since the numerical value for mass has a large power-of-ten exponent, an SI prefix from Table 3.3 should be applied. We first write $m = 165.5 \times 10^3$ kg so that the exponent is a multiple of 3. Since the "kilo" prefix already implies a factor of 10^3 g, $m = 165.5 \times 10^6$ g or 165.5 Mg, where "M" denotes the prefix "mega." The fuel's weight in the SI is

$$w = (1.655 \times 10^5 \text{ kg})\left(9.81 \frac{\text{m}}{\text{s}^2} \right) \quad \leftarrow [w = mg]$$

$$= 1.62 \times 10^6 \frac{\text{kg} \cdot \text{m}}{\text{s}^2}$$

$$= 1.62 \times 10^6 \text{ N}$$

Example 3.4 | *continued*

Because this quantity also has a large power-of-ten exponent, we use the SI prefix "M" to condense the factor of 1 million. The fuel weighs 1.62 MN.

Discussion

To double-check the calculation of weight in the SI, we note that we can convert the fuel's 365,000 lb weight directly to the dimensions of newtons. By using the conversion factor 1 lb = 4.448 N from Table 3.6, *w* becomes

$$w = (3.65 \times 10^5 \, \text{lb}) \left(4.448 \, \frac{\text{N}}{\text{lb}} \right)$$

$$= 1.62 \times 10^6 \, (\text{lb}) \left(\frac{\text{N}}{\text{lb}} \right)$$

$$= 1.62 \times 10^6 \, \text{N}$$

or 1.62 MN, confirming our previous answer.

$$m = 1.134 \times 10^4 \text{ slugs}$$
$$m = 365{,}000 \text{ lbm}$$
$$m = 165.5 \text{ Mg}$$
$$w = 1.62 \text{ MN}$$

Example 3.5 | *Orbital Debris Collision*

The International Space Station has hundreds of shields made of aluminum and bulletproof composite materials that are intended to offer protection against impact with debris present in low Earth orbit (Figure 3.5). With sufficient advance warning, the station's orbit can even be adjusted slightly to avoid close approaches of larger objects. Over 13,000 pieces of debris have been identified by the United States Space Command, including paint chips, spent booster casings, and even an astronaut's glove. (a) Calculate the kinetic energy $U_k = \frac{1}{2} mv^2$ of an $m = 1$ g particle of debris traveling at $v = 8$ km/s, which is a typical velocity in low Earth orbit. (b) How fast would a 0.31 lb baseball have to be thrown to have the same kinetic energy?

Approach

We first convert the debris particle's mass and velocity to the dimensionally consistent units of kg and m/s, respectively, using the definition of the "kilo" prefix (Table 3.3). The conventional SI unit for energy in Table 3.2 is the joule, defined as $1 \text{ N} \cdot \text{m}$. In part (b), we will convert the kinetic energy to

Example 3.5 | *continued*

0

Figure 3.5

The International Space Station.

Image courtesy Johnson Space Center Office of Earth Sciences/NASA

the USCS using the factor $1\,\text{J} = 0.7376\ \text{ft} \cdot \text{lb}$ from Table 3.6. Since the baseball's weight is specified in the problem statement, we will make an intermediate calculation for its mass.

Solution

(a) With $m = 0.001$ kg and $v = 8000$ m/s, the kinetic energy of the debris particle is

$$U_k = \frac{1}{2}(0.001\ \text{kg})\left(8000\ \frac{\text{m}}{\text{s}}\right)^2 \quad \leftarrow \left[U_k = \frac{1}{2}mv^2\right]$$

$$= 32{,}000\ (\text{kg})\left(\frac{\text{m}^2}{\text{s}^2}\right)$$

$$= 32{,}000\ \left(\frac{\text{kg} \cdot \text{m}}{\text{s}^2}\right)(\text{m})$$

$$= 32{,}000\ \text{N} \cdot \text{m}$$

$$= 32{,}000\ \text{J}$$

Applying an SI prefix to suppress the trailing zeroes, the kinetic energy of the particle is 32 kJ.

(b) Expressed in the USCS, the particle's kinetic energy is

$$U_k = (32{,}000\ \text{J})\left(0.7376\ \frac{\text{ft} \cdot \text{lb}}{\text{J}}\right)$$

$$= 23{,}603\ (\cancel{\text{J}})\left(\frac{\text{ft} \cdot \text{lb}}{\cancel{\text{J}}}\right)$$

$$= 23{,}603\ \text{ft} \cdot \text{lb}$$

Example 3.5 | *continued*

For a dimensionally consistent calculation of kinetic energy in the USCS, we will determine the baseball's mass in the units of slugs:

$$m = \frac{0.31 \text{ lb}}{32.2 \text{ ft/s}^2} \qquad \leftarrow \left[m = \frac{w}{g} \right]$$

$$= 9.627 \times 10^{-3} \frac{\text{lb} \cdot \text{s}^2}{\text{ft}}$$

$$= 9.627 \times 10^{-3} \text{ slugs}$$

since 1 slug = 1(lb · s²)/ft in Equation (3.2). To have the same kinetic energy as the debris particle, the baseball must be thrown with velocity

$$v = \sqrt{\frac{2(23{,}603 \text{ ft} \cdot \text{lb})}{9.627 \times 10^{-3} \text{ slugs}}} \qquad \leftarrow \left[v = \sqrt{\frac{2U_k}{m}} \right]$$

$$= 2214 \sqrt{\frac{\text{ft} \cdot \text{lb}}{\text{slug}}}$$

$$= 2214 \sqrt{\frac{\text{ft} \cdot (\text{slug} \cdot \text{ft/s}^2)}{\text{slug}}}$$

$$= 2214 \frac{\text{ft}}{\text{s}}$$

Discussion

Even though orbiting dust and debris particles may be small in size, they can convey large amounts of kinetic energy because their velocities are so great. The equivalent speed of a baseball is some 1500 mph, or about fifteen times the speed of a major league fastball pitch or the fastest professional cricket bowl.

$$U_k = 32 \text{ kJ}$$

$$v = 2214 \frac{\text{ft}}{\text{s}}$$

Example 3.6 | *Bending of a Drill Bit*

This example illustrates the full problem-solving process from
Section 3.2, which incorporates the principles of dimensional analysis
from Sections 3.3–3.5.

   A drill press holds sharpened bits in a rotating chuck and is used to
bore holes in a workpiece. The steel drill bit has a diameter $d = 6$ mm and
length $L = 65$ mm. The bit is accidentally bent as the workpiece shifts
during a drilling operation, and it is subjected to the side force of
$F = 50$ N. As derived in mechanical engineering courses on stress analysis,
the sideways deflection of the tip is calculated by using the equation

$$\Delta x = \frac{64\,FL^3}{3\pi\,Ed^4}$$

where the terms have the following units:

Δx (length) the deflection of the tip
F (force) the magnitude of the force applied at the tip
L (length) the drill bit's length
E (force/length^2) a property of the drill bit's material, called the elastic
     modulus
d (length) the drill bit's diameter

By using the numerical value $E = 200 \times 10^9$ Pa for steel, calculate the
amount Δx that the tip deflects. (See Figure 3.6.)

Approach

We are tasked to solve for the deflection at the tip of the steel drill bit given the
applied force. We first make a number of assumptions about the system:

- The curved flutes on the bit are small and can be neglected in the
  analysis

Figure 3.6

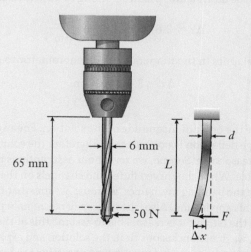

Example 3.6 | *continued*

- The force is perpendicular to the primary bending axis of the bit
- The channels that spiral along the bit have minimal impact on the bending and can be ignored

We will first combine the units of each quantity in the given equation according to the rules of algebra and verify that the units appearing on each side of the equation are identical. Then we will insert the known quantities, including the bit length, diameter, elastic modulus, and applied force, to solve for the deflection.

Solution

The quantity $64/3\pi$ is a dimensionless scalar, and it therefore has no units to influence dimensional consistency. The units of each quantity in the given equation are cancelled:

$$(\text{length}) = \frac{(\text{force})(\text{length})^3}{((\text{force})(\text{length})^2)(\text{length})^4}$$

$$= (\text{length})$$

The equation is indeed dimensionally consistent. The tip moves sideways by the amount

$$\Delta x = \frac{64(50 \text{ N})(0.065 \text{ m})^3}{3\pi(200 \times 10^9 \text{ Pa})(6 \times 10^{-3} \text{ m})^4} \quad \leftarrow \left[\Delta x = \frac{64 \, FL^3}{3\pi \, Ed^4} \right]$$

Next, we combine the numerical values and dimensions

$$\Delta x = 3.6 \times 10^{-4} \frac{\text{N} \cdot \text{m}^3}{\text{Pa} \cdot \text{m}^4}$$

and then expand the derived unit pascal according to its definition in Table 3.2:

$$\Delta x = 3.6 \times 10^{-4} \frac{\text{N} \cdot \text{m}^3}{(\text{N/m}^2)(\text{m}^4)}$$

Finally, we cancel units in the numerator and denominator to obtain

$$\Delta x = 3.6 \times 10^{-4} \text{ m}$$

Discussion

First, we evaluate the order of magnitude of the solution. For a steel bit of this length, a large deflection is not expected. Therefore the solution's order of magnitude is reasonable. Second, we revisit our assumptions to make sure they are reasonable. While the curved flutes and channels on the bit may slightly influence the bending mechanics, we must assume that their impact is negligible for this application. Also, the force may not remain perfectly perpendicular to the bit, but it is reasonable to assume this at the moment of deflection. Third, we draw conclusions from the solution and explain its physical meaning. Drill bits undergo many forces during operation, and it makes sense

Example 3.6 | *continued*

that most bits would be made of steel to minimize deflection. Because the numerical value has a large negative exponent, we convert it to standard form by using the SI prefix "milli" to represent a factor of 10^{-3}. The tip moves by the amount $\Delta x = 0.36$ mm, just over one-third of a millimeter.

$$\Delta x = 0.36 \text{ mm}$$

Example 3.7 | *Elevator acceleration*

This example also illustrates the full problem-solving process from Section 3.2, which incorporates the principles of dimensional analysis from Sections 3.3–3.5.

A person with a mass of 70 kg is standing on a scale in an elevator that reads 140 lb at a given instant. Determine which way the elevator is moving and whether it is accelerating. From Newton's second law, if a body is accelerating, then the sum of the forces equals the body's mass m times its acceleration a through the equation

$$\Sigma F = ma$$

If the sum of the forces in any direction is equal to zero, then the body is not accelerating in that direction. (See Figure 3.7.)

Figure 3.7

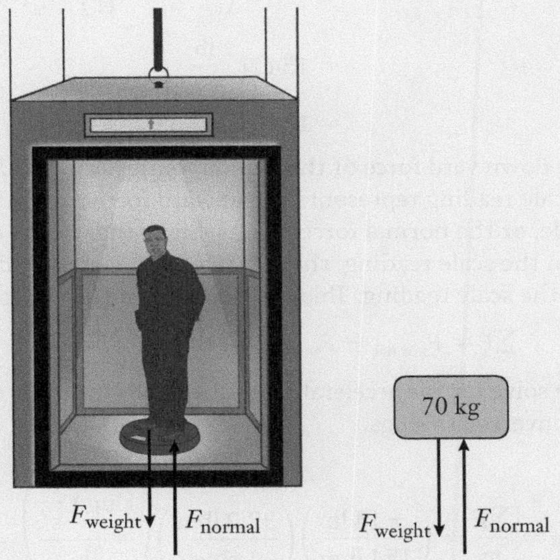

F_{weight} F_{normal} F_{weight} F_{normal}

70 kg

Example 3.7 | *continued*

Approach

We are tasked to determine which direction the elevator is moving and whether the elevator is accelerating. We first make a number of assumptions about the system:

- The person and elevator are moving in tandem; so we only have to analyze the forces on the person
- The only movement is in the vertical or y-direction
- Our analysis is done on Earth, and therefore gravity is 9.81 m/s² or 32.2 ft/s²
- The scale does not move relative to the elevator floor or the person

First convert the mass of the person in kilograms to the equivalent weight in newtons. Then, convert their weight in newtons to pounds. Compare the person's weight to the reading on the scale to determine which direction the elevator is moving. Then use the difference in weight to determine the acceleration of the elevator. If there is no difference in weight, then we know that the elevator is not accelerating.

Solution

The weight W of the person can be found as follows:

$$W = (\text{mass})(\text{gravity}) = (70 \text{ kg})(9.81 \text{ m/s}^2) = 687 \text{ N}$$

Then, the weight can be converted into pounds using the factor from Table 3.6 as follows,

$$W = 687 \text{ N}\left(0.22481 \frac{\text{lb}}{\text{N}}\right)$$

$$= 154 \text{ N} \cdot \frac{\text{lb}}{\text{N}}$$

$$= 154 \text{ lb}$$

This is the downward force of the person, represented by F_{weight} in Figure 3.7. The scale reading represents the upward force exerted on the person by the scale, or the normal force, F_{normal}. Since the weight of the person is larger than the scale reading, the elevator is accelerating downward which decreases the scale reading. This is illustrated in the equation

$$\Sigma F = F_{\text{normal}} - F_{\text{weight}} = 140 \text{ lb} - 154 \text{ lb} = -14 \text{ lb}$$

Finally, we solve for the acceleration, noting that the mass of the person must be converted to slugs.

$$a = \frac{\Sigma F}{m} = \left(\frac{-14 \text{ lb}}{154 \text{ lbm}}\right)\left(\frac{32.2 \text{ lbm}}{1 \text{ slug}}\right)\left(\frac{\text{slug} \cdot \frac{\text{ft}}{\text{s}^2}}{\text{lb}}\right) = -2.9 \frac{\text{ft}}{\text{s}^2}$$

Example 3.7 | *continued*

Discussion

First, we evaluate the order of magnitude of the solution. The acceleration is not large, which is not expected since the scale reading is not significantly different from the person's weight. Second, we revisit our assumptions to make sure they are reasonable. All the assumptions are very logical. The person, scale, and elevator may undergo some motion relative to one another in reality, but the impact on the analysis would be minimal. Third, we draw conclusions from the solution and explain its physical meaning. The acceleration is negative, indicating a downward acceleration, which aligns with the scale reading. When an elevator begins to accelerate downward, the passengers temporarily feel lighter. Their mass does not change since gravity did not change. However, their perceived weight has changed, which is what the scale measures.

Note that the same analysis can be done using the SI. First, we convert the scale reading to newtons

$$W = (140 \text{ lb}) \left(\frac{4.45 \text{ N}}{1 \text{ lb}} \right)$$

$$= 623 \text{ lb} \cdot \frac{\text{N}}{\text{lb}}$$

$$= 623 \text{ N}$$

Since this is less than the person's actual weight of 687 N, we conclude that the elevator is accelerating downward. Solving for acceleration gives

$$a = \frac{\Sigma F}{m} = \left(\frac{623 \text{ N} - 687 \text{ N}}{70 \text{ kg}} \right)$$

$$= \frac{-64 \text{ N}}{70 \text{ kg}} = -0.91 \left(\frac{\text{kg} \cdot \frac{\text{m}}{\text{s}^2}}{\text{kg}} \right) = -0.91 \frac{\text{m}}{\text{s}^2}$$

Using the conversion 1 ft = 0.3048 m from Table 3.6, this solution can be converted to −2.9 ft/s², which matches our previous analysis.

$$a = -2.9 \frac{\text{ft}}{\text{s}^2}$$

▶ 3.6 Estimation in Engineering

I n the later stages of a design process, engineers certainly make precise calculations as they solve technical problems. However, in the earlier stages of design, engineers nearly always make approximations when they solve technical problems. Those estimates are made to reduce a real system, as imperfect and nonideal as it may be, into its most basic and essential elements. Approximations are also made to remove extraneous factors that complicate the problem but that otherwise have little influence on a final result. Engineers are comfortable making reasonable approximations so that their mathematical models are as simple as possible, while leading to a result that is accurate enough for the task at hand. If the accuracy needs to be increased later on, for instance, as a design becomes finalized, then they need to incorporate more physical phenomena or details about the geometry, and the equations to be solved would likewise become more complicated.

Order-of-magnitude estimates

Given that some imperfections and uncertainty are always present in real hardware, engineers often make *order-of-magnitude estimates*. Early in the design process, for instance, order-of-magnitude approximations are used to evaluate potential design options for their feasibility. Some examples are estimating the weight of a structure or the amount of power that a machine produces or consumes. Those estimates, made quickly, are helpful to focus ideas and narrow down the options available for a design before significant effort has been put into figuring out details.

Engineers make order-of-magnitude estimates while fully aware of the approximations involved and recognizing that reasonable approximations are going to be necessary to reach an answer. In fact, the term "order of magnitude" implies that the quantities considered in the calculation (and the ultimate answer) are accurate to perhaps a factor of 10. A calculation of this type might estimate the force carried by a certain bolted connection to be 1000 lb, implying that the force probably isn't as low as 100 lb or as great as 10,000 lb, but it certainly could be 800 lb or 3000 lb. At first glance, that range might appear to be quite wide, but the estimate is nevertheless useful because it places a bound on how large the force could be. The estimate also provides a starting point for any subsequent, and presumably more detailed, calculations that a mechanical engineer would need to make. Calculations of this type are educated estimates, admittedly imperfect and imprecise, but better than nothing. These calculations are sometimes described as being made on the *back of an envelope* because they can be performed quickly and informally.

Back-of-an-envelope calculations

Order-of-magnitude estimates are made when engineers in a design process begin assigning numerical values to dimensions, weights, material properties, temperatures, pressures, and other parameters. You should recognize that those values will be refined as information is gathered, the analysis improves, and the design becomes better defined. The following examples show some applications of order-of-magnitude calculations and the thought processes behind making estimates.

Focus On

IMPORTANCE OF ESTIMATIONS

On April 20, 2010, an explosion destroyed Transocean's Deepwater Horizon oil drilling rig in the Gulf of Mexico, killing 11 people, injuring 17 others, and creating the largest accidental marine oil spill in history. It was not until July 15 that the leak was stopped, but only after 120–180 million gallons of oil had spilled into the gulf and British Petroleum had spent over U.S. $10 billion on the cleanup. During the initial release, multiple fluids flowed simultaneously out of the well, including seawater, mud, oil, and gas. Engineers from around the world quickly started creating various analytical models to estimate future flow rates of these fluids, including the flow shown in Figure 3.8. The approaches varied in complexity and included the following fundamental methods:

- Engineers used video from remotely operated vehicles (ROVs) to estimate flow rates and consistencies

- Engineers used a flow visualization and measurement technique called particle image velocimetry (PIV) to estimate the velocity of the outer surface of the leaking oil plumes

- Engineers collected acoustic Doppler data over periods of minutes to estimate the cross-sectional area of the oil plume

- Engineers used pressure measurements during the closing of the well on July 15 to estimate the flow rate before its closing

- Engineers created models of the oil reservoir to approximate how much oil was left, allowing them to estimate how much oil had been released

- Engineers used data from the actual surface collection to estimate the trends in the gas to oil ratio, allowing for an approximation of the oil release

Despite the different estimation approaches and sets of assumptions, most of the approaches approximated that the flow rates were between 50,000 and 70,000 barrels of oil per day. This is equivalent to between 2 and 3 million gallons, and between 8 and 11 million liters of oil per day. These estimates were not only critical to the cleanup and rescue efforts, but were also pivotal in closing the failed well, to understanding why it failed, to predicting the impact of future spills, and to designing safer machinery in the future.

Many times, technical problem solving is an inexact science, and a rough estimate is the best engineers can do. Models that seem to match actual data sometimes do not effectively predict future performance. In this example, with such a dynamic flow environment being modeled by the engineers using a range of techniques, the range of uncertainty is quite high, and the models at best can provide only an estimation of what happened.

Figure 3.8

The escaping oil is shown underwater two weeks after the accident occurred.

REUTERS/BP/Landov

Example 3.8 | *Aircraft's Cabin Door*

Commercial jet aircraft have pressurized cabins because they travel at high altitude where the atmosphere is thin. At the cruising altitude of 30,000 ft, the outside atmospheric pressure is only about 30% of the sea-level value. The cabin is pressurized to the equivalent of a mountaintop where the air pressure is about 70% that at sea level. Estimate the force that is applied to the door of the aircraft's main cabin by this pressure imbalance. Treat the following information as "given" when making the order-of-magnitude estimate: (1) the air pressure at sea level is 14.7 psi, and (2) the force F on the door is the product of the door's area A and the pressure difference Δp according to the expression $F = A\Delta p$.

Approach

We are tasked with approximating the amount of force exerted on the interior of an aircraft door during flight. The pressure information is given, but we have to make some assumptions about the door and cabin surroundings. We assume that:

- The size of the aircraft's door is approximately 6 × 3 ft, or 18 ft^2
- We can neglect the fact that the door is not precisely rectangular
- We can neglect the fact that the door is curved to blend with the shape of the aircraft's fuselage
- We do not have to account for small changes in pressure due to the movement of passengers inside the cabin during flight

We will first calculate the pressure difference and then calculate the area of the door to find the total force.

Solution

The net pressure acting on the door is the difference between air pressures inside and outside the aircraft.

$$\Delta p = (0.7 - 0.3)(14.7 \text{ psi})$$
$$= 5.88 \text{ psi}$$

Because Δp has the units of pounds per square inch (Table 3.5), in order for the equation $F = A\Delta p$ to be dimensionally consistent, the area must be converted to the units of square inches:

$$A = (18 \text{ ft}^2)\left(12\frac{\text{in.}}{\text{ft}}\right)^2$$

$$= 2592 \; (\cancel{\text{ft}^2})\left(\frac{\text{in}^2}{\cancel{\text{ft}^2}}\right)$$

$$= 2592 \text{ in}^2$$

Example 3.8 | *continued*

The total force acting on the door becomes

$$F = (2592 \text{ in}^2)(5.88 \text{ psi}) \qquad \leftarrow [F = A\Delta p]$$

$$= 15{,}240 \ (\text{in}^2)\left(\frac{\text{lb}}{\text{in}^2}\right)$$

$$= 15{,}240 \text{ lb}$$

Discussion

First, we evaluate the order of magnitude of the solution. The forces created by pressure imbalances can be quite large when they act over large surfaces, even for seemingly small pressures. Therefore, our force seems quite reasonable. Second, our assumptions greatly simplified the problem. But since we need to estimate only the force, these assumptions are realistic. Third, recognizing the uncertainty in our estimate of the door's area and the actual value of the pressure differential, we conclude that the pressure is in the range of 10,000–20,000 lb.

Approximately 10,000–20,000 lb

Example 3.9 | *Human Power Generation*

In an analysis of sustainable sources of energy, an engineer wants to estimate the amount of power that a person can produce. In particular, can a person who is riding an exercise bike power a television (or similar appliance or product) during the workout? Treat the following information as given when making the order-of-magnitude estimate: (1) An average LCD television consumes 110 W of electrical power. (2) A generator converts about 80% of the supplied mechanical power into electricity. (3) A mathematical expression for power P is

$$P = \frac{Fd}{\Delta t}$$

where F is the magnitude of a force, d is the distance over which it acts, and Δt is the time interval during which the force is applied.

Approach

We are tasked with estimating whether it is feasible for a person exercising to independently power a product that requires approximately 110 W. We first make some assumptions to make this estimation:

- To estimate a person's power output while exercising, make a comparison with the rate at which a person can climb a flight of stairs with the same level of effort

Example 3.9 | *continued*

- Assume that a flight of stairs has a 3 m rise and that it can be climbed by a 700 N person in under 10 s

Calculate the power generated by a person climbing stairs and then compare it to the power required.

Solution

The stair climbing analogy provides the estimate of power output as

$$P = \frac{(700 \text{ N})(3 \text{ m})}{10 \text{ s}} \quad \leftarrow \left[P = \frac{Fd}{\Delta t} \right]$$

$$= 210 \frac{\text{N} \cdot \text{m}}{\text{s}}$$

$$= 210 \text{ W}$$

where we have used the definition of the watt from Table 3.2. The useful work that a person can produce is therefore approximately 200 W. However, the generator that would be used to convert mechanical power into electricity is not perfectly efficient. With the stated efficiency of 80%, about $(210 \text{ W})(0.80) = 168 \text{ W}$ of electricity can be produced.

Discussion

First, we evaluate the order of magnitude of the solution. This amount of power seems reasonable, because some nonelectric exercise bikes generate enough power to run their own displays and resistance. Second, we revisit our assumptions to make sure they are reasonable. While the stair-climbing analogy may not be perfect, it provides an effective estimate of the amount of power someone could generate over a sustained time. Third, given the uncertainty in our estimates and the range of exertion while exercising, we conclude that a person can generate 100–200 W over an extended period of time. This would be sufficient for a wide range of LCD televisions.

> Approximately 100–200 W

▶ 3.7 Communication Skills in Engineering

We began this chapter with a case study involving NASA's loss of an interplanetary weather satellite. While the *Mars Climate Orbiter* Mishap Investigation Board found that the main cause of the spacecraft's loss was the improper conversion of the engine's impulse between the USCS and SI, they also concluded that another contributing factor was inadequate communication between the individuals and teams who were cooperating on the spacecraft's mission:

> *It was clear that the operations navigation team did not communicate their trajectory concerns effectively to the spacecraft operations team, or project management. In addition, the spacecraft operations team did not understand the concerns of the operations navigation team.*

The board concluded that, even for such seemingly straightforward engineering concepts as the units of pounds and newtons, "communication is critical," and one of the final recommendations was that NASA take steps to improve communications between project elements. The control of a large, expensive, and complex spacecraft broke down not because of faulty design or technology, but because information was misunderstood by and not exchanged clearly between people working together on the ground.

Stereotypes notwithstanding, engineering is a social endeavor. It is not carried out by people working alone in offices and laboratories. A quick search for mechanical engineering jobs on any major employment website will reveal that most positions specifically require engineers to be able to communicate effectively in a wide range of ways. By the nature of their work, engineers interact daily with other engineers, customers, business managers, marketing staff, and members of the public. Presumably, a person who has earned a degree from an accredited engineering program will have solid technical skills in mathematics, science, and engineering. However, the abilities to work with others, collaborate on a team, and convey technical information in written and verbal formats can distinguish one employee from another. Those factors generally play an important role in one's career advancement. For instance, a survey of over 1000 chief financial officers from United States companies found that interpersonal skills, communication, and listening are critical measures of an employee's ultimate professional success. Also, a study by the American Management Association ranked written and verbal communication skills as the single most important factor in determining one's ultimate professional success.

The most effective engineers can relate their ideas, results, and solutions to others through calculations, one-on-one discussions, written technical reports, formal presentations, letters, and digital communications such as e-mails. In beginning your study of mechanical engineering, you should start to develop some of the problem-solving and technical communication skills that meet the standards of what other engineers and the public will expect from you.

Written Communication

While marketing professionals communicate product information to their customers through advertisements, Facebook, and Twitter, engineers do much of their daily communication of product information through a variety of written documents, including notebooks, reports, letters, memoranda, user's manuals, installation instructions, trade publications, and e-mails. In some large projects, engineers who are physically located in different time zones— and even in different countries—will routinely collaborate on a design. Written documentation is therefore a key and practical means for accurately conveying complex technical information.

Design notebook

An effective way to document engineering projects is using a *design notebook*, introduced in Chapter 2. An engineer's design notebook documents the full history of a product's development. The notebook is a written form of communication containing an accurate record of information that can be used to defend patents, to prepare technical reports, to document research and development tests and results, and to assist other engineers who might follow and build on the work. Because they provide a detailed account of a product's development, design notebooks are the property of an engineer's employer and can become important in legal patent disputes. An employer might set additional requirements for the notebook, including:

- All writing must be in ink
- The pages must be bound and sequentially numbered
- Each entry must be dated and signed by the individual performing the work
- All individuals participating in each task must be listed
- Corrections or alterations must be dated and initialed

Requirements of this nature reflect good practice, and they highlight the fact that an engineer's notebook is a legal document that must be unimpeachable in its accuracy. Engineers rely on the technical and historical information in design notebooks to create engineering reports.

Engineering reports

Engineering reports are used to explain technical information to others and also to archive it for future use. The purpose of a report might be to document the concept and evolution of a new product's design or to analyze why a certain piece of hardware broke. Engineering reports can also include the results obtained by testing a product to demonstrate that it functions properly or to verify that it complies with safety standards. For these reasons, engineering reports can become important in litigation should a product cause an injury. Because they are formal documents, engineering reports can indicate whether a product was carefully developed and show whether diligence was paid to potential safety concerns.

Engineering reports generally include text, drawings, photographs, calculations, and graphs or tables of data. These reports can chronicle the history of a product's design, testing, manufacture, and revision.

Although the format of an engineering report varies depending on the business purpose and the specific issue at hand, the general structure includes the following elements:

- A *cover page* indicates the purpose of the report, the product or technical issue involved, the date, and the names of those involved in preparing the report
- An *executive summary* summarizes the full report for readers, providing them with a 1- to 2-page synopsis of the problem, approach, solution and major conclusions
- If appropriate, a *table of contents* gives readers page numbers for major sections, figures, and tables
- The *body of the report* reviews prior work, brings the reader up-to-date, and then describes in detail the design, the supporting decisions,

the results of testing, performance calculations, and other technical information

- A *conclusion* highlights the major findings and brings the report to closure by offering specific recommendations
- *Appendices* contain information that supports the recommendations made in the report but that is too lengthy or detailed to include in the body

You should use certain practices throughout any technical report. These practices will maximize the effectiveness of your report, regardless of the reading audience.

- When you are making a set of recommendations, assumptions, conclusions, or observations, use a bulleted list with short descriptions.
- When you want to emphasize a key point, phrase, or term, use italics or bold. Use italics or bold to emphasize only the most important points; using them too much will decrease their effect.
- Make use of numbered sections and descriptive section headings to provide structure for the reader. Readers can get lost when reading a long report without sections to apportion and organize the information.
- Provide connecting transitions between sections. Although having sections in a report is an effective practice, unless they are connected logically and flow into each other, they can become detached and confusing.
- Make effective use of any references that you use, including a list of references at the end of the report. These references could include research articles, trade publications, books, websites, internal company documents, and other technical reports.

Graphical Communication

Essential elements of any technical report are graphical communication pieces such as drawings, graphs, charts, and tables. Many engineers tend to think and learn visually, and they find that graphical forms of communication are often the best way to convey complex technical information. An important first step in addressing nearly every engineering problem or design is to represent the situation graphically. Modes of graphical communication include hand sketches, dimensioned drawings, three-dimensional computer-generated renderings, graphs, and tables. Each is useful to convey different types of information. A hand drawing might be included in a design or laboratory notebook. Although a quick sketch might not be drawn to scale or address details, it could define the overall shape of a piece of hardware and show some of its major features. A formal engineering drawing, produced later using a computer-aided design package, would be sufficiently detailed that it could be given to a machine shop to have finished parts fabricated.

Tables and graphs are critical forms of communication for engineers who need to present a wide range of data. Tables should include columns and rows with descriptive headings and appropriate units. The data columns should be presented using consistent significant digits and aligned to aid understanding. Graphs or charts should have descriptive axis labels including appropriate units. If more than one set of data is plotted, then the graph needs to include a legend. Engineers need to carefully consider what type of graph or chart to use; the choice depends on the nature of the data and the type of insights that need to be understood by the reader.

Technical Presentations

Although the previous skills have focused on written communication, engineers also convey technical information in verbal presentations. Weekly status reports on a project are given to supervisors and coworkers, designs are discussed and reviewed in group meetings, and formal proposals are made to potential customers. Engineers also deliver technical presentations at professional conferences, such as the ones organized by the American Society of Mechanical Engineers. Learning about engineering and technology is a lifelong endeavor, and engineers attend conferences and other business meetings to stay up-to-date on new techniques and advances in their fields. At such conferences, engineers present their technical work to an audience of other engineers from around the world, and those presentations are intended to be concise, interesting, and accurate, so that the audience can learn from the experiences of their colleagues.

Engineers need to be able to present information not only effectively, but also efficiently. Many times when presenting to management or potential customers, engineers have only a few minutes to communicate their significant findings and insights to support a key decision or recommendation.

Focus On

INEFFECTIVE COMMUNICATIONS

Regardless of what career your mechanical engineering degree prepares you for, the ability to communicate to a wide range of audiences will be essential to your success. A quick survey of employment opportunities in engineering will reveal the importance of being able to communicate in both written and oral form. There have been many examples in engineering where poor communication led to injury or even death, including the 1986 Space Shuttle *Challenger* disaster. With so much at stake, it is critical that you take every opportunity to develop your skills in presenting technical information in written and oral form.

For example, consider the chart in Figure 3.9. This chart shows the relationship between mass and price for a set of similar consumer electronic products (e.g., Blu-ray players). Sometimes, mass is an effective predictor of price, as the more material used in a product, the more expensive it typically is. Also, additional product features many times require more hardware, increasing the mass and as a result, the price too. Figure 3.9 was created to study the relationships and trends of a product's mass to its price as a way to better inform engineers about the impact of their decisions on product price. While the chart may appear effective, it could lead to erroneous technical conclusions and could be improved upon in many ways.

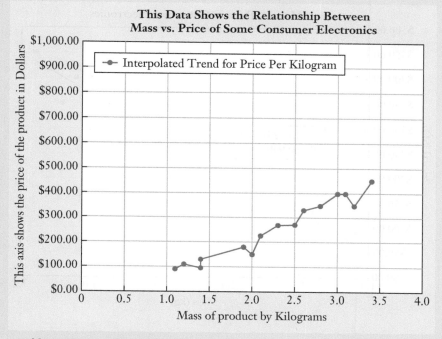

Figure 3.9

Example of a poorly developed technical chart.

Courtesy of Kemper Lewis

- Most importantly, the trend line is not appropriate. The intent is to identify the trend between mass and price and not the exact relationship. Only a sample of products is chosen to estimate the impact of mass on price. Therefore the line that connects the data points is very misleading, as it would predict a series of price increases and decreases as the product mass increases. A simpler trend line would be more appropriate.

- The axis labels are inconsistently capitalized. They are also awkwardly worded and should be shortened.

- The chart has too much space not used by the data. The bounds on the axes need to be tightened.

- The chart title should be a much more succinct statement.

- The legend does not contain any effective information that adds value to the chart.

In Figure 3.10, a revised version of the chart is shown. Note that the underlying data has not changed; only the presentation of the data has been improved.

- A more appropriate linear trend line has been added, clearly demonstrating the overall relationship between mass and price.

- The axis labels are shortened with clear units and concise wording.

- The axis bounds are tightened, allowing for data to be better visualized.

- The chart title has been appropriately revised.

- The legend has been replaced with a summary of the trend line illustrating the estimated mathematical relationship between mass and price ($y = 157.25x - 103.39$), along with statistical support for the underlying relationship ($R^2 = 0.948$).

With the updated chart, more effective conclusions can be drawn, better engineering design decisions can be made, and flaws in the product can be avoided.

Figure 3.10

A revised technical chart supporting better engineering decision making.

Courtesy of Kemper Lewis

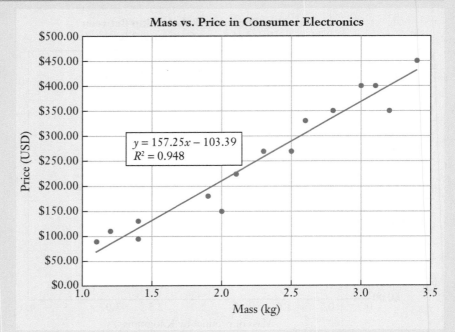

Mass vs. Price in Consumer Electronics

$$y = 157.25x - 103.39$$
$$R^2 = 0.948$$

Price (USD) — Mass (kg)

Example 3.10 | *Written Communication*

A mechanical engineer was running some tests to validate the spring constant of a new spring (part #C134). A mass was placed on a spring, and the resulting compression displacement was measured. Hooke's Law (discussed more in Chapter 5) states that the force exerted on a spring is proportional to the displacement of the spring. This can be expressed by

$$F = kx$$

where F is the applied force, x is the displacement, and k is the spring constant. The data was recorded in the following table in SI units.

Mass	Displacement
0.01	0.0245
0.02	0.046
0.03	0.067
0.04	0.091
0.05	0.114
0.06	0.135
0.07	0.156
0.08	0.1805
0.09	0.207
0.1	0.231

Example 3.10 | *continued*

The engineer is tasked with developing a professional table and graph that communicates the data and explains the Hooke's Law relationship for the spring.

First, the engineer needs to calculate the resulting force from the applied mass using $w = mg$ and construct a table that illustrates the force and displacement data.

Note the following best practices regarding Table 3.7.

- The engineer has added the calculated force values
- Units for each column have been added
- Appropriate borders to separate the data have been added
- The number of significant digits in each column is now consistent
- The headings are capitalized and bolded
- The data is aligned to make each column easy to read

Second, the engineer must communicate the spring rate relationship in the data table. A scatter plot is chosen and created in Figure 3.11. This graph effectively illustrates the relationship between force and displacement and demonstrates how well the data aligns with the linear relationship predicted by Hooke's Law.

Note the best practices regarding Figure 3.11:

- The axes are clearly labeled, including appropriate units
- A descriptive title accompanies the graph
- A trend line clearly demonstrates the linear relationship between the variables
- The number of gridlines is minimal and used only for visual aids
- The data spans the axes, eliminating large areas of empty space in the graph

Mass (kg)	Force (N)	Displacement (m)
0.01	0.098	0.0245
0.02	0.196	0.0460
0.03	0.294	0.0670
0.04	0.392	0.0910
0.05	0.490	0.1140
0.06	0.588	0.1350
0.07	0.686	0.1560
0.08	0.784	0.1805
0.09	0.882	0.2070
0.10	0.980	0.2310

Table 3.7

Results of Spring Test Data

Example 3.10 | *continued*

Figure 3.11

Example of a professional engineering graph.

Courtesy of Kemper Lewis.

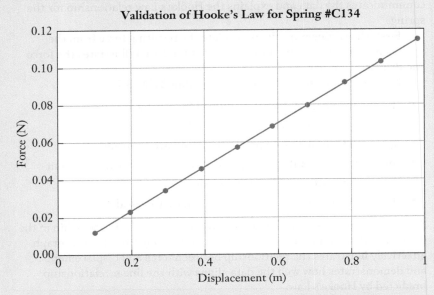

Validation of Hooke's Law for Spring #C134

Using the table and graph, the engineer can quickly estimate and communicate the spring constant for the spring as 4 N/m and validate that against the design requirements.

Summary

Engineers are often described as being proactive people with excellent problem-solving skills. In this chapter, we have outlined some of the fundamental tools and professional skills that mechanical engineers use when they solve technical problems. Numerical values, the USCS and SI systems, unit conversions, dimensional consistency, significant digits, order-of-magnitude approximations, and the ability to communicate technical results effectively are, simply put, everyday issues for engineers. Because each quantity in mechanical engineering has two components—a numerical value and a unit—reporting one without the other is meaningless. Engineers need to be clear about those numerical values and dimensions when they perform calculations and relate their findings to others through written reports and verbal presentations. By following the consistent problem-solving guidelines developed in this chapter, you will be prepared to approach engineering problems in a systematic manner and to be confident of the accuracy of your work.

Self-Study and Review

3.1. Summarize the three major steps that should be followed when solving technical problems in order to present your work clearly.

3.2. What are the base units in the USCS and SI?

3.3. What are examples of derived units in the USCS and SI?

3.4. How are mass and force treated in the USCS and SI?

3.5. What is the major difference in the definitions of the slug and pound-mass in the USCS?

3.6. What is the difference between the pound and pound-mass in the USCS?

3.7. One pound is equivalent to approximately how many newtons?

3.8. One meter is equivalent to approximately how many feet?

3.9. One inch is equivalent to approximately how many millimeters?

3.10. One gallon is equivalent to approximately how many liters?

3.11. How should you decide the number of significant digits to retain in a calculation and to report in your final answer?

3.12. Give an example of when the technical problem-solving process can be used to make order-of-magnitude approximations.

3.13. Give several examples of situations where engineers prepare written documents and deliver verbal presentations.

Problems

P3.1

Express your weight in the units of pounds and newtons, and your mass in the units of slugs and kilograms.

P3.2

Express your height in the units of inches, feet, and meters.

P3.3

For a wind turbine that is used to produce electricity for a utility company, the power output per unit area swept by the blades is 2.4 kW/m^2. Convert this quantity to the dimensions of hp/ft^2.

P3.4

A world-class runner can run half a mile in a time of 1 min and 45 s. What is the runner's average speed in m/s?

P3.5

One U.S. gallon is equivalent to 0.1337 ft^3, 1 ft is equivalent to 0.3048 m, and 1000 L are equivalent to 1 m^3. By using those definitions, determine the conversion factor between gallons and liters.

P3.6

A passenger automobile is advertised as having a fuel economy rating of 29 mi/gal for highway driving. Express the rating in the units of km/L.

P3.7

(a) How much horsepower does a 100 W household lightbulb consume?

(b) How many kW does a 5 hp lawn mower engine produce?

P3.8

The estimates for the amount of oil spilled into the Gulf of Mexico during the 2010 Deepwater Horizon disaster were 120–180 million gal. Express this range in L, m^3, and ft^3.

P3.9

In 1925, the Tri-State Tornado ripped a 219 mi path of destruction through Missouri, Illinois, and Indiana, killing a record 695 people. The maximum winds in the tornado were 318 mph. Express the wind speed in km/h and in ft/s.

P3.10

Uphill water slides are becoming more popular at large water parks. Uphill speeds of riders can reach 19 ft/s. Express this speed in mph.

P3.11

A brand-new engineering hire is late for her first product development team meeting. She gets out of her car and starts running 8 mph. It is exactly 7:58 A.M., and the meeting starts at exactly 8:00 A.M. Her meeting is 500 yd away. Will she make it on time to the meeting? If so, with how much time to spare? If not, how late will she be?

P3.12

A robotic-wheeled vehicle that contains science instruments is used to study the geology of Mars. The rover weighs 408 lb on Earth.

(a) In the USCS dimensions of slugs and lbm, what is the rover's mass?

(b) What is the weight of the rover as it rolls off the lander's platform (the lander is the protective shell that houses the rover during landing)? The gravitational acceleration on the surface of Mars is 12.3 ft/s^2.

P3.13

Calculate various fuel quantities for Flight 143. The plane already had 7682 L of fuel on board prior to the flight, and the tanks were to be filled so that a total of 22,300 kg were present at takeoff.

(a) Using the incorrect conversion factor of 1.77 kg/L, calculate in units of kg the amount of fuel that was added to the plane.

(b) Using the correct factor of 1.77 lb/L, calculate in units of kg the amount of fuel that should have been added.

(c) By what percentage would the plane have been underfueled for its journey? Be sure to distinguish between weight and mass quantities in your calculations.

P3.14

Printed on the side of a tire on an all-wheel-drive sport utility wagon is the warning "Do not inflate above 44 psi," where psi is the abbreviation for the pressure unit pounds per square inch (lb/in^2). Express the tire's maximum pressure rating in (a) the USCS unit of lb/ft^2 (psf) and (b) the SI unit of kPa.

P3.15

The amount of power transmitted by sunlight depends on latitude and the surface area of the solar collector. On a clear day at a certain northern latitude, 0.6 kW/m^2 of solar power strikes the ground. Express that value in the alternative USCS unit of (ft · lb/s)/ft^2.

P3.16

The property of a fluid called *viscosity* is related to its internal friction and resistance to being deformed. The viscosity of water, for instance, is less than that of molasses and honey, just as the viscosity of light motor oil is less than that of grease. A unit used in mechanical engineering to describe viscosity is called the *poise*, named after the physiologist Jean Louis Poiseuille, who performed early experiments in fluid mechanics. The unit is defined by 1 poise = 0.1 (N · s)/m^2. Show that 1 poise is also equivalent to 1 g/(cm · s).

P3.17

Referring to the description in P3.16, and given that the viscosity of a certain engine oil is 0.25 kg/(m · s), determine the value in the units (a) poise and (b) slug/(ft · s).

P3.18

Referring to the description in P3.16, if the viscosity of water is 0.01 poise, determine the value in terms of the units (a) slug/(ft · s) and (b) kg/(m · s).

P3.19

The fuel efficiency of an aircraft's jet engines is described by the *thrust-specific fuel consumption* (TSFC). The TSFC measures the rate of fuel consumption (mass of fuel burned per unit time) relative to the thrust (force) that the engine produces. In that manner, even if an engine consumes more fuel per unit time than a second engine, it is not necessarily more inefficient if it also produces more thrust to power the plane. The TSFC for an early hydrogen-fueled jet engine was 0.082 (kg/h)/N. Express that value in the USCS units of (slug/s)/lb.

P3.20

An automobile engine is advertised as producing a peak power of 118 hp (at an engine speed of 4000 rpm) and a peak torque of 186 ft · lb (at 2500 rpm). Express those performance ratings in the SI units of kW and N · m.

P3.21

From Example 3.6, express the sideways deflection of the tip in the units of mils (defined in Table 3.5) when the various quantities are instead known in the USCS. Use the values $F = 75$ lb, $L = 3$ in., $d = 3/16$ in., and $E = 30 \times 10^6$ psi.

P3.22

Heat Q, which has the SI unit of joule (J), is the quantity in mechanical engineering that describes the transit of energy from one location to another. The equation for the flow of heat during the time interval Δt through an insulated wall is

$$Q = \frac{\kappa A \Delta t}{L}(T_h - T_l)$$

where κ is the thermal conductivity of the material from which the wall is made, A and L are the wall's area and thickness, and $T_h - T_l$ is the difference (in degrees Celsius) between the high- and low-temperature sides of the wall. By using the principle of dimensional consistency, what is the correct dimension for thermal conductivity in the SI? The lowercase Greek character kappa (κ) is a conventional mathematical symbol used for thermal conductivity. Appendix A summarizes the names and symbols of Greek letters.

P3.23

Convection is the process by which warm air rises and cooler air falls. The *Prandtl number* (Pr) is used when mechanical engineers analyze certain heat transfer and convection processes. It is defined by the equation

$$Pr = \frac{\mu c_p}{\kappa}$$

where c_p is a property of the fluid called the specific heat having the SI units kJ/(kg · °C); μ is the viscosity as discussed in Problem P3.16; and κ is the thermal conductivity as discussed in Problem P3.22. Show that Pr is a dimensionless number. The lowercase Greek characters mu (μ) and kappa (κ) are conventional mathematical symbols used for viscosity and thermal conductivity. Appendix A summarizes the names and symbols of Greek letters.

P3.24

When fluid flows over a surface, the Reynolds number will output whether the flow is laminar (smooth), transitional, or turbulent. Verify that the Reynolds number is dimensionless using the SI. The Reynolds number is expressed as

$$R = \frac{\rho V D}{\mu}$$

where ρ is the density of the fluid, V is the free stream fluid velocity, D is the characteristic length of the surface, and μ is the fluid viscosity. The units of fluid viscosity are kg/(m · s).

P3.25

Determine which one of the following equations is dimensionally consistent.

$$F = \frac{1}{2} mx^2, \quad FV = \frac{1}{2} mx^2, \quad Fx = \frac{1}{2} mV^2, \quad Ft = V, \quad FV = 2mt^2$$

where F is force, m is mass, x is distance, V is velocity, and t is time.

P3.26

Referring to Problem P3.23 and Table 3.5, if the units for c_p and μ are Btu/(slug · °F) and slug/(ft · h), respectively, what must be the USCS units of thermal conductivity in the definition of Pr?

P3.27

Some scientists believe that the collision of one or more large asteroids with the Earth was responsible for the extinction of the dinosaurs. The unit of kiloton is used to describe the energy released during large explosions. It was originally defined as the explosive capability of 1000 tons of trinitrotoluene (TNT) high explosive. Because that expression can be imprecise depending on the explosive's exact chemical composition, the kiloton subsequently has been redefined as the equivalent of 4.186×10^{12} J. In the units of kiloton, calculate the kinetic energy of an asteroid that has the size (box-shaped, 13 × 13 × 33 km) and composition (density, 2.4 g/cm^3) of our solar system's asteroid Eros. Kinetic energy is defined by

$$U_k = \frac{1}{2} mv^2$$

where m is the object's mass and v is its speed. Objects passing through the inner solar system generally have speeds in the range of 20 km/s.

P3.28

A structure known as a cantilever beam is clamped at one end but free at the other, analogous to a diving board that supports a swimmer standing on it (Figure P3.28). Using the following procedure, conduct an experiment to measure how the cantilever beam bends. In your answer, report only the significant digits that you know reliably.

(a) Make a small tabletop test stand to measure the deflection of a plastic drinking straw (your cantilever beam) that bends as a force F is applied to the free end. Push one end of the straw over the end of a pencil, and then clamp the pencil to a desk or table. You can also use a ruler, chopstick, or a similar component as the cantilever beam itself. Sketch and describe your apparatus, and measure the length L.

(b) Apply weights to the end of the cantilever beam, and measure the tip's deflection Δy using a ruler. Repeat the measurement with at least a half dozen different weights to fully describe the beam's force–deflection relationship. Coins can be used as weights; one U.S. penny weighs approximately 30 mN. Check the weights of your local currency coins. Make a table to present your data.

(c) Next draw a graph of the data. Show tip deflection on the abscissa and weight on the ordinate, and be sure to label the axes with the units for those variables.

(d) Draw a best-fit line through the data points on your graph. In principle, the deflection of the tip should be proportional to the applied force. Do you find this to be the case? The slope of the line is called the stiffness. Express the stiffness of the cantilever beam either in the units lb/in. or N/m.

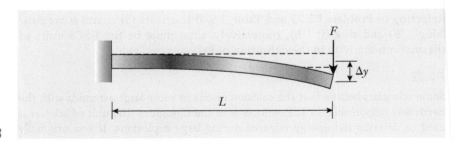

Figure P3.28

P3.29

Perform measurements as described in P3.28 for cantilever beams of several different lengths. Can you show experimentally that, for a given force F, the deflection of the cantilever's tip is proportional to the cube of its length? As in P3.28, present your results in a table and a graph, and report only those significant digits that you know reliably.

P3.30

Using SI units, calculate the change in potential energy of a 150 lb person riding the 15 ft long uphill portion of a water slide (as described in P3.10). The change in potential energy is defined as $mg\Delta h$ where Δh is the change in vertical height. The uphill portion of the slide is set at an angle of 45°.

P3.31

Using the speed given in P3.10, calculate the power required to move the rider in P3.30 up the incline portion of the water slide, where power is the change in energy divided by the time required to traverse the uphill portion.

P3.32

Estimate the force acting on a passenger's window in a commercial jet aircraft due to air pressure differential.

P3.33

Give numerical values for order-of-magnitude estimates for the following quantities. Explain and justify the reasonableness of the assumptions and approximations that you need to make.

(a) The number of cars that pass through an intersection of two busy streets during the evening commute on a typical workday

(b) The number of bricks that form the exterior of a large building on a university campus

(c) The volume of concrete in the sidewalks on a university campus

P3.34

Repeat the exercise of P3.33 for the following systems:

(a) The number of leaves on a mature maple or oak tree

(b) The number of gallons of water in an Olympic-sized swimming pool

(c) The number of blades of grass on a natural turf football field

P3.35

Repeat the exercise of P3.33 for the following systems:

(a) The number of baseballs that can fit into your classroom

(b) The number of people who are born every day

(c) The number of square inches of pizza consumed by students at your university in one semester

P3.36

Make an order-of-magnitude approximation for the volume of gasoline consumed by automobiles each day in the United States by estimating the number of vehicles driven each day, the average distance traveled, and a typical fuel economy rating.

P3.37

It takes the International Space Station about 90 min to complete one trip around the Earth. Estimate the station's orbital velocity in units of mph. Make the approximation that the altitude of the station (approximately 125 mi) is small when compared to the radius of the Earth (approximately 3950 mi).

P3.38

Estimate the size of a square parcel of land that is needed for an airport's 5000-car parking lot. Include space for the access roadways.

P3.39

An automobile assembly plant produces 400 vehicles per day. Make an order-of-magnitude estimate for the weight of the steel needed to make those vehicles. Explain and justify the reasonableness of the assumptions and approximations that you make.

P3.40

Think of some quantity that you encounter in your day-to-day life for which it would be difficult to obtain a highly accurate numerical value but for which an order-of-magnitude approximation can be made. Describe the quantity, make the approximation, and explain and justify the reasonableness of the assumptions and approximations that you need to make.

P3.41

The modulus of elasticity, modulus of rigidity, Poisson's ratio, and the unit weight for various materials are shown below. The data is given as Material; Modulus of Elasticity, E (Mpsi & GPa); Modulus of Rigidity, G (Mpsi & GPa); Poisson's Ratio; and Unit Weight (lb/in^3, lb/ft^3, kN/m^3). Prepare a single table that captures this technical data in a professional and effective manner.

Material	E (Mpsi)	E (GPa)	G (Mpsi)	G (GPa)	Poisson's Ratio	Unit Weight (lb/in³)	Unit Weight (lb/ft³)	Unit Weight (kN/m³)
Aluminum alloys	10.3	71.0	3.8	26.2	0.334	0.098	169	26.6
Beryllium copper	18.0	124.0	7.0	48.3	0.285	0.297	513	80.6
Brass	15.4	106.0	5.82	40.1	0.324	0.309	534	83.8
Carbon steel	30.0	207.0	11.5	79.3	0.292	0.282	487	76.5
Cast iron, grey	14.5	100.0	6.0	41.4	0.211	0.260	450	70.6
Copper	17.2	119.0	6.49	44.7	0.326	0.322	556	87.3
Glass	6.7	46.2	2.7	18.6	0.245	0.094	162	25.4
Lead	5.3	36.5	1.9	13.1	0.425	0.411	710	111.5
Magnesium	6.5	44.8	2.4	16.5	0.350	0.065	112	17.6
Molybdenum	48.0	331.0	17.0	117.0	0.307	0.368	636	100.0
Nickel silver	18.5	127.0	7.0	48.3	0.322	0.316	546	85.8
Nickel steel	30.0	207.0	11.5	79.3	0.291	0.280	484	76.0
Phosphor bronze	16.1	111.0	6.0	41.4	0.349	0.295	510	80.1
Stainless steel	27.6	190.0	10.6	73.1	0.305	0.280	484	76.0

P3.42

For the data in P3.41, prepare a graph that charts the relationship between the modulus of elasticity (y-axis) and unit weight (x-axis) using the USCS unit system data. Explain the resulting trend, including a physical explanation of the trend, noting any deviations from the trend.

P3.43*

If the average person normally takes 12–20 breaths per minute and with an average tidal or normal breath volume of 0.5 liters, what is your group's best estimate for how long it would take the people in your classroom to inhale the entire volume of air in the room you are in currently?

P3.44*

Estimate how many people could fit shoulder to shoulder in your classroom and compare that with the posted maximum capacity per the fire safety code.

P3.45*

Estimate the upper and lower bounds for how many people are in the air flying at any given time around the world.

References

Banks, P., "The Crash of Flight 143," *ChemMatters*, American Chemical Society, October 1996, p. 12.

Burnett, R., *Technical Communication*, 6th ed. Cengage, 2005.

Goldman, D. T., "Measuring Units," In Avallone, E. A., and Baumeister, T., eds., *Marks' Standard Handbook for Mechanical Engineers*, 10th ed. New York: McGraw-Hill Professional, 1996.

Hoffer, W., and Hoffer, M. M., *Freefall: A True Story*. New York: St. Martin's Press, 1989.

Mars Climate Orbiter Mishap Investigation Board Phase I Report. NASA, November 10, 1999.

McNutt, M. K., Camilli, R., Crone, T. J., Guthrie, G. D., Hsieh, P. A., Ryerson, T. B., Savas, O., and Shaffer, F., "Review of Flow Rate Estimates of the Deepwater Horizon Oil Spill," *Proceedings of the National Academy of Sciences*, 109(50), 2011, 20260–20267, doi: 10.1073/pnas.1112139108.

Press Release, "*Mars Climate Orbiter* Failure Board Releases Report, Numerous NASA Actions Underway in Response," NASA, November 10, 1999.

Press Release, "*Mars Climate Orbiter* Mission Status," Media Relations Office, Jet Propulsion Laboratory, September 24, 1999.

Press Release, "NASA's *Mars Climate Orbiter* Believed to Be Lost," Media Relations Office, Jet Propulsion Laboratory, September 23, 1999.

Walker, G., "A Most Unbearable Weight," *Science*, Vol. 304, 2004, pp. 812–813.

CHAPTER 4

Forces in Structures and Machines

- Break a force down into its rectangular and polar components.
- Determine the resultant of a system of forces by using the vector algebra and polygon methods.
- Calculate the moment of a force using the perpendicular lever arm and moment component methods.

- Understand the requirements for equilibrium, and be able to calculate unknown forces in simple structures and machines.
- From the design standpoint, explain the circumstances in which one type of rolling element bearing would be selected for use over another, and calculate the forces acting on them.

▶ 4.1 Overview

When mechanical engineers design products, systems, and hardware, they must apply mathematics and physical principles to model, analyze, and predict system behavior. Successful design is supported by effective engineering analysis; effective engineering analysis relies on an understanding of *forces in structures and machines*. This is the focus of this chapter and the next element of mechanical engineering.

Element 3: Forces in structures and machines

This chapter introduces you to the subject of mechanics, a topic that encompasses forces that act on structures and machines and their tendency either to remain stationary or move. The fundamental principles that form the basis of mechanics are Newton's three laws of motion:

1. Every object remains in a state of rest or uniform motion of constant velocity unless an external unbalanced force acts upon it.

2. An object of mass m, subject to a force F, experiences an acceleration in the same direction as the force with a magnitude directly proportional to the magnitude of the force and inversely proportional to the mass of the object. This relationship can be expressed as $F = ma$.

3. The forces of action and reaction between two objectives are equal, opposite, and collinear.

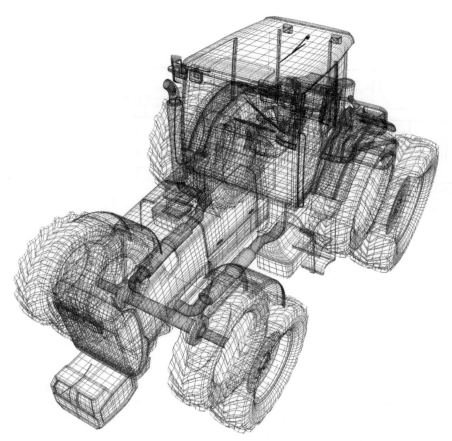

Figure 4.1

Heavy construction
equipment is designed
to support the large
forces developed
during operation.

3DDock/Shutterstock.com

In this and the following chapters, we will explore these principles of forces and the problem-solving skills that are needed to understand their effects on engineering hardware. After developing the concepts of force systems, moments, and static equilibrium, you will see how to calculate the magnitudes and directions of forces acting on and within simple structures and machines. In short, the process of analyzing forces is a first step taken by engineers to see whether a certain piece of hardware will operate reliably (Figure 4.1).

A second objective of this chapter is for you to start understanding the inner workings of mechanical hardware, beginning with rolling element bearings. Just as an electrical engineer might select off-the-shelf resistors, capacitors, and transistors as the elements of a circuit, mechanical engineers have good intuition for specifying bearings, shafts, gears, belts, and other machine components. A working knowledge of hardware and machine components is important for you to develop a technical vocabulary. Mechanical engineering has its own precise language, and, to communicate effectively with other engineers, you will need to learn, adopt, and share that language. That background is also necessary to select the proper component: Should a ball, roller, tapered roller, or thrust roller bearing be used in this design?

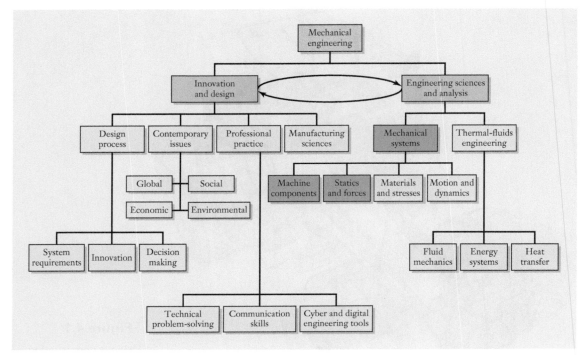

Figure 4.2

Relationship of the topics emphasized in this chapter (shaded blue boxes) relative to an overall program of study in mechanical engineering.

The topics of force systems and machine components discussed in this chapter fit naturally into the hierarchy of mechanical engineering topics outlined in Figure 4.2. The topics fall under the Engineering sciences and analysis branch but provide support for key decisions in the design of innovative systems. Of course, in an introductory textbook, it is not possible to describe every machine and component that embodies mechanical engineering principles, and that is not our intention here or in later chapters. However, by examining just a few machine components, you will develop a growing appreciation for mechanical design issues. It's intellectually healthy for you to be curious about products, wonder how they were made, dissect them, and think about how they could have been made differently or better. In this chapter, we begin that journey by discussing various types of bearings and the forces that act on them. In Chapter 8, we continue that discussion with descriptions of gear, belt, and chain drives.

▶ 4.2 Forces in Rectangular and Polar Forms

Before we can determine the influence of forces on a structure or machine, we first need to describe a force's magnitude and direction. Our analysis will be limited to situations where the forces present all act in the same plane. The corresponding concepts and solution techniques for such two-dimensional problems carry over to the general case of structures and machines

lb	oz	N
1	16	4.448
0.0625	1	0.2780
0.2248	3.597	1

Table 4.1

Conversion Factors Between USCS and SI Units for Force

in three dimensions, but, for our purposes, it's better to avoid the added complexity in algebra and geometry. The properties of forces, equilibrium, and motion in three dimensions are also subjects that you will encounter later in the mechanical engineering curriculum.

Forces are vector quantities since their physical action involves both direction and magnitude. The magnitude of a force is measured by using the dimensions of pounds (lb) or ounces (oz) in the USCS and newtons (N) in the SI. In Chapter 3, the conversion factors between pounds and newtons were listed in Table 3.6, and they are shown in a slightly different format in Table 4.1. This style of listing the conversion factors is a compact way to depict the USCS-to-SI and SI-to-USCS conversion factors. Each row of the table contains equivalent quantities in the units shown at the top of the columns. The three rows of Table 4.1 mean the following:

Row 1: 1 lb = 16 oz = 4.448 N

Row 2: 0.0625 lb = 1 oz = 0.2780 N

Row 3: 0.2248 lb = 3.597 oz = 1 N

In this chapter and the following ones, we will use conversion tables having this type of format for other engineering quantities.

Rectangular Components

Force *vectors* are denoted by using boldface notation, as in **F**. One of the common methods used to represent the influence of a force is in terms of its horizontal and vertical components. Once we set the directions for the x- and y-axes, the force **F** can be broken down into its *rectangular components* along those directions. In Figure 4.3 (see on page 120), the projection of **F** in the horizontal direction (the x-axis) is called F_x, and the vertical projection (y-axis) is called F_y. When you assign numerical values to F_x and F_y, you have described everything about the force **F**. In fact, the pair of numbers (F_x, F_y) is just the coordinates of the force vector's tip.

The *unit vectors* **i** and **j** are used to indicate the directions in which F_x and F_y act. Vector **i** points along the positive x-direction, and **j** is a vector oriented in the positive y-direction. Just as F_x and F_y provide information about the magnitudes of the horizontal and vertical components, the unit vectors give information about the directions of those components. The unit vectors are so named because they have a length of one. By combining the rectangular components and unit vectors, the force is represented in vector algebra notation as

$$\mathbf{F} = F_x\mathbf{i} + F_y\mathbf{j} \qquad (4.1)$$

Vector notation

Rectangular components

Unit vectors

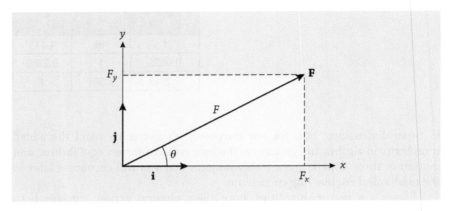

Figure 4.3

Representing a
force vector in terms
of its rectangular
components (F_x, F_y),
and its polar
components (F, θ).

Polar Components

In an alternative view, rather than thinking about a force in terms of how
hard it pulls rightward and upward, you could tell someone how hard the
force pulls, and in which direction it does so. This view is based on polar
coordinates. As shown in Figure 4.3, **F** acts at the angle θ, measured relative
to the horizontal axis. The length of the force vector is a scalar or simple
numerical value, and it is denoted by $F = |\mathbf{F}|$. The $|\ |$ notation designates the
vector's magnitude F, which we write in plain typeface. Instead of specifying
F_x and F_y, we can now view the force vector **F** in terms of the two numbers
Polar components F and θ. This representation of a vector is called the *polar component* or
magnitude-direction form.

Magnitude   The force's *magnitude* and *direction* are related to its horizontal and vertical
Direction   components through

$$F_x = F \cos \theta \qquad \text{(polar to rectangular)}$$
$$F_y = F \sin \theta \qquad\qquad\qquad\qquad (4.2)$$

Trigonometric equations of this sort are reviewed in Appendix B. If we happen
to know the force's magnitude and direction, these equations are used to
determine its horizontal and vertical components. On the other hand, when
we know F_x and F_y, the magnitude and direction of the force are calculated from

$$F = \sqrt{F_x^2 + F_y^2} \qquad \text{(rectangular to polar)}$$

$$\theta = \tan^{-1}\left(\frac{F_y}{F_x}\right) \qquad\qquad\qquad (4.3)$$

Principal value   The inverse tangent operation in Equation (4.3) calculates the *principal
value* of its argument and returns an angle between $-90°$ and $+90°$. Directly
applying the equation $\theta = \tan^{-1}(F_y/F_x)$ will result in a value of θ that lies in
either the first or fourth quadrants of the x-y plane.

Of course, a force could be oriented in any of the plane's four quadrants.
In solving problems, you will need to examine the positive or negative signs
of F_x and F_y and use them to determine the correct quadrant for θ. For
instance, in Figure 4.4(a), where $F_x = 100$ lb and $F_y = 50$ lb, the force's

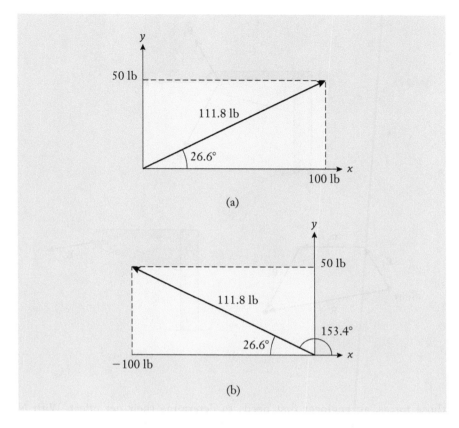

(a)

(b)

Figure 4.4

Determining the angle of action for a force that (a) lies in the first quadrant and (b) lies in the second quadrant.

angle of action is calculated as $\theta = \tan^{-1}(0.5) = 26.6°$. That numerical value correctly lies in the first quadrant because F_x and F_y—the coordinates of the force vector's tip—are both positive. On the other hand, in Figure 4.4(b), when $F_x = -100$ lb and $F_y = 50$ lb, you might be tempted to report $\tan^{-1}(-0.5) = -26.6°$. That angle actually falls in the fourth quadrant, and it is incorrect as a measure of the force's orientation relative to the positive x-axis. As is evident from Figure 4.4(b), **F** forms an angle of 26.6° relative to the negative x-axis. The correct value for the force's angle of action relative to the positive x-axis is $\theta = 180° - 26.6° = 153.4°$.

▶ 4.3 **Resultant of Several Forces**

A *force system* is a collection of several forces that simultaneously act on a structure or machine. Each force is combined with the others to describe their net effect, and the *resultant* **R** measures that cumulative action. As an example, consider the mounting post and bracket of Figure 4.5 (see on page 122). The three forces **F₁**, **F₂**, and **F₃** act in different directions and with different magnitudes. To determine whether the post is capable of supporting

Force system

Resultant

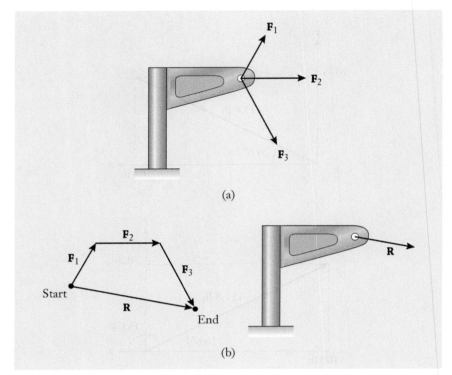

Figure 4.5

(a) A mounting post and bracket that are loaded by three forces. (b) The resultant **R** extends from the start to the end of the chain formed by adding **F**₁, **F**₂, and **F**₃ together.

those forces, an engineer first needs to determine their net effect. With N individual forces $\mathbf{F}_i \, (i = 1, 2, ..., N)$ being present, they are summed according to

$$\mathbf{R} = \mathbf{F}_1 + \mathbf{F}_2 + \cdots + \mathbf{F}_N = \sum_{i=1}^{N} \mathbf{F}_i \qquad (4.4)$$

by using the rules of vector algebra. This summation can be carried out by the vector algebra or vector polygon approaches, which are described next. In different problem-solving situations, you will find that one approach might be simpler than another and that the two approaches can also be used for double-checking your calculations.

Vector Algebra Method

In this technique, each force is broken down into its horizontal and vertical components, which we label as $F_{x,i}$ and $F_{y,i}$ for the ith force. The resultant's horizontal portion R_x is the sum of the horizontal components from all of the individual forces that are present:

$$R_x = \sum_{i=1}^{N} F_{x,i} \qquad (4.5)$$

Likewise, we separately sum the vertical components by using the equation

$$R_y = \sum_{i=1}^{N} F_{y,i} \qquad (4.6)$$

The resultant force is then expressed in vector form as $\mathbf{R} = R_x\mathbf{i} + R_y\mathbf{j}$. Similar to Equation (4.3), we apply the expressions

$$R = \sqrt{R_x^2 + R_y^2}$$

$$\theta = \tan^{-1}\left(\frac{R_y}{R_x}\right) \tag{4.7}$$

to calculate the resultant's magnitude R and direction θ. As before, the actual value for θ is found after considering the positive and negative signs of R_x and R_y, so that θ lies in the correct quadrant.

Vector Polygon Method

An alternative technique for finding the cumulative influence of several forces is the *vector polygon method*. The resultant of a force system can be found by sketching a polygon to represent the addition of the $\mathbf{F}_i$ vectors. The magnitude and direction of the resultant are determined by applying rules of trigonometry to the polygon's geometry. Referring to the mounting post of Figure 4.5(a), the vector polygon for those three forces is drawn by adding the individual $\mathbf{F}_i$'s in a chain according to the *head-to-tail rule*.

Head-to-tail rule

In Figure 4.5(b), the starting point is labeled on the drawing, the three forces are summed in turn, and the endpoint is labeled. The order in which the forces are drawn on the diagram does not matter insofar as the final result is concerned, but diagrams will appear visually different for various addition sequences. The endpoint is located at the tip of the last vector added to the chain. As indicated in Figure 4.5(b), the resultant $\mathbf{R}$ extends from the start of the chain to its end. The action of $\mathbf{R}$ on the bracket is entirely equivalent to the combined effect of the three forces acting together. Finally, the magnitude and direction of the resultant are determined by applying trigonometric identities to the polygon's shape. Some of the relevant equations for right and oblique triangles are reviewed in Appendix B.

We can often obtain reasonably accurate results by summing vectors on a drawing that is made to *scale*, for instance, 1 in. on the drawing corresponds to 100 lb. Such drafting tools as a protractor, scale, and straightedge should be used to construct the polygon and to measure the magnitudes and directions of unknown quantities. It is certainly acceptable to use a purely graphical approach when solving engineering problems, provided that the drawing is large and precise enough that you can determine the answer to a fair number of significant digits.

Scale drawings

Example 4.1 | *Cable Tie-Down*

The eyebolt is fastened to a thick base plate, and it supports three steel cables with tensions 150 lb, 350 lb, and 800 lb. Determine the resultant force that acts on the eyebolt by using the vector algebra approach. The unit vectors $\mathbf{i}$ and $\mathbf{j}$ are oriented with the x-y coordinates as shown. (See Figure 4.6 on page 124.)

Example 4.1 | *continued*

Figure 4.6

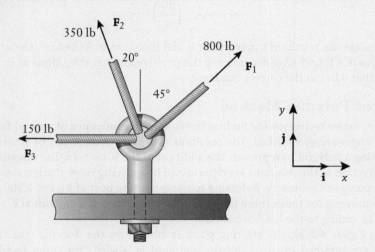

Approach

We are tasked to find the resultant force on the eyebolt. By using Equations (4.1) and (4.2), we will break each force down into its horizontal and vertical components and write them in vector form. Then we will add the respective components of the three forces to find the resultant's components. Given those, the magnitude and angle of action of **R** follow from Equation (4.7).

Solution

The components of the 800 lb force are

$$F_{x,1} = (800 \text{ lb}) \cos 45° \qquad \leftarrow [F_x = F \cos \theta]$$

$$= 565.7 \text{ lb}$$

$$F_{y,1} = (800 \text{ lb}) \sin 45° \qquad \leftarrow [F_y = F \sin \theta]$$

$$= 565.7 \text{ lb}$$

and **F**$_1$ is written in vector form as

$$\mathbf{F}_1 = 565.7\mathbf{i} + 565.7\mathbf{j} \text{ lb} \qquad \leftarrow [\mathbf{F} = F_x\mathbf{i} + F_y\mathbf{j}]$$

By using the same procedure for the other two forces,

$$\mathbf{F}_2 = -(350 \sin 20°)\mathbf{i} + (350 \cos 20°)\mathbf{j} \text{ lb}$$

$$= -119.7\mathbf{i} + 328.9\mathbf{j} \text{ lb}$$

$$\mathbf{F}_3 = -150\mathbf{i} \text{ lb}$$

Example 4.1 | *continued*

To calculate the components of the resultant, the horizontal and vertical components of the three forces are summed separately:

$$R_x = 565.7 - 119.7 - 150 \text{ lb} \quad \leftarrow \left[R_x = \sum_{i=1}^{N} F_{x,i} \right]$$

$$= 296.0 \text{ lb}$$

$$R_y = 565.7 + 328.9 \text{ lb} \quad \leftarrow \left[R_y = \sum_{i=1}^{N} F_{y,i} \right]$$

$$= 894.6 \text{ lb}$$

The magnitude of the resultant force is

$$R = \sqrt{(296.0 \text{ lb})^2 + (894.6 \text{ lb})^2} \quad \leftarrow \left[\sqrt{R_x^2 + R_y^2} \right]$$

$$= 942.3 \text{ lb}$$

and it acts at the angle

$$\theta = \tan^{-1}\left(\frac{894.6 \text{ lb}}{296.0 \text{ lb}} \right) \quad \leftarrow \left[\theta = \tan^{-1}\left(\frac{R_y}{R_x} \right) \right]$$

$$= \tan^{-1}\left(3.022 \, \frac{\text{lb}}{\text{lb}} \right)$$

$$= \tan^{-1}(3.022)$$

$$= 71.69°$$

which is measured counterclockwise from the *x*-axis. (See Figure 4.7.)

Figure 4.7

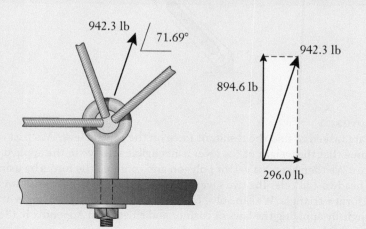

Example 4.1 | *continued*

Discussion

The resultant force is larger than any one force, but less than the sum of the three forces because a portion of $F_{x,3}$ cancels $F_{x,1}$. The resultant force acts upward and rightward on the bolt, which is expected. The three forces acting together place the bolt in tension by pulling up on it, as well as bending it through the sideways loading R_x. As a double check on dimensional consistency when calculating the resultant's angle, we note that the pound units in the argument of the inverse tangent function cancel. Since R_x and R_y are positive, the tip of the resultant vector lies in the first quadrant of the *x-y* plane.

$$R = 942.3 \text{ lb}$$
$$\theta = 71.69° \text{ counterclockwise from the } x\text{-axis}$$

Example 4.2 | *Control Lever*

The 10 lb and 25 lb forces are applied to the control lever of a mechanism. Determine the magnitude and direction of the resultant by using the vector polygon approach. (See Figure 4.8.)

Figure 4.8

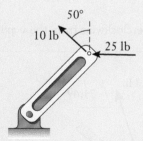

Approach

We are tasked to find the resultant force on the control lever. We first assume that the weight of the lever is negligible relative to the applied forces. We then sketch a vector polygon and combine the forces by using the head-to-tail rule. The two given forces, together with the resultant, will form a triangle. We can solve for the unknown length and angle in the triangle by applying the laws of cosines and sines from Appendix B. (See Figure 4.9.)

Example 4.2 | *continued*

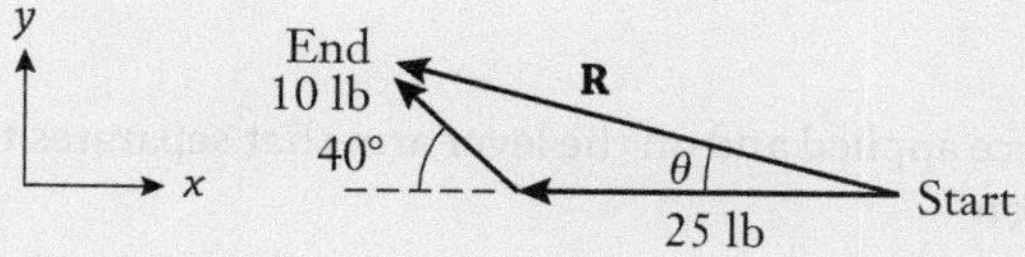

Figure 4.9

Solution
The 25 lb force is sketched first on the diagram, and the 10 lb force is added at the angle of 50° from vertical. The resultant **R** extends from the tail of the 25 lb force vector (which is labeled as the start point) to the head of the 10 lb force vector (the endpoint). The three vectors form a side-angle-side triangle. Applying the law of cosines, we solve for the unknown side length

$$R^2 = (10 \text{ lb})^2 + (25 \text{ lb})^2 \qquad\qquad \leftarrow [c^2 = a^2 + b^2 - 2ab \cos C]$$
$$-2(10 \text{ lb})(25 \text{ lb}) \cos(180° - 40°)$$

from which $R = 33.29$ lb. The angle θ at which **R** acts is determined by applying the law of sines

$$\frac{\sin(180° - 40°)}{33.29 \text{ lb}} = \frac{\sin \theta}{10 \text{ lb}} \qquad \leftarrow \left[\frac{\sin A}{a} = \frac{\sin B}{b} \right]$$

The resultant acts at $\theta = 11.13°$ clockwise from the $-x$-axis.

Discussion
The resultant force is less than the sum of the two forces because the forces are not aligned in the same direction. Also, the direction of the resultant force, at an angle between the two forces, matches the expectation. To double-check the solution, we could break the two forces down into their horizontal and vertical components, as in the technique in Example 4.1. By using the *x-y* coordinate system, the vector expression for the 10 lb force becomes **F** = $-7.660\mathbf{i} + 6.428\mathbf{j}$ lb. See if you can complete the double check for R and θ by using the vector algebra method.

> $R = 33.29$ lb
>
> $\theta = 11.13°$ clockwise from the $-x$-axis

▶ 4.4 Moment of a Force

When you are trying to loosen a bolt, it is more easily turned if you use a wrench with a long handle. In fact, the longer the handle, the less force you need to apply with your hand. The tendency of a force to make an object rotate is called a *moment*. The magnitude of a moment depends

Table 4.2

Conversion Factors
Between USCS and SI
Units for Moment or
Torque

in · lb	ft · lb	N · m
1	0.0833	0.1130
12	1	1.356
8.851	0.7376	1

both on the force applied and on the lever arm that separates the force from the point of rotation.

Perpendicular Lever Arm Method

The magnitude of a moment is defined by

$$M_o = Fd \tag{4.8}$$

Perpendicular lever arm

Torque

where M_o is the moment of the force about point O and F is the magnitude of the perpendicular force. The distance d is called the *perpendicular lever arm*, and it extends from the force's line of action to point O. The term "torque" is sometimes used interchangeably to describe the effect of a force acting across a lever arm. However, mechanical engineers generally reserve *torque* to describe moments that cause rotation of a shaft in a motor, engine, or gearbox. We will discuss those applications in Chapter 8.

Based on Equation (4.8), the dimensions for a moment are the product of force and length. In the USCS, the unit for a moment is either in · lb or ft · lb. In the SI, the unit N · m is used, and various prefixes can be applied when the numerical value is either very large or very small. For instance, 5000 N · m = 5 kN · m and 0.002 N · m = 2 mN · m. Note that the dot symbol (·) is used to show multiplication between the dimensions; so it is clear that mN · m means millinewton-meters. Conversion factors between the two systems of units are listed in Table 4.2. Each row of the table is equivalent, and the first row indicates that

$$1 \text{ in} \cdot \text{lb} = 0.0833 \text{ ft} \cdot \text{lb} = 0.1130 \text{ N} \cdot \text{m}$$

Work and energy are other quantities that arise in mechanical engineering, and they also have dimensions that are the product of force and length. When working in the SI, a joule (J) is defined as one newton-meter, and it is the amount of work that is performed by a 1 N force that acts through a distance of 1 m. However, the physical quantities of work and energy are quite different from moments and torques, and, to be clear about distinguishing between them, only the unit N · m should be used in the SI for moment and torque.

The expression $M_o = Fd$ can be best understood by applying it to a specific structure. In Figure 4.10(a), the force **F** is directed generally downward and to the right on the bracket. One might be interested in the moment of **F** about the base of the support post, which is labeled in the figure as point O. The post could break at that location, and an engineer would design the post's diameter and length to make sure that it can support **F**. The moment is calculated based on both the magnitude of **F** and the perpendicular offset d between the force's line of action and point O. The continuous straight line

Line of action

on which a force vector lies is called its *line of action*. In fact, **F** could be applied to the bracket at any point along its line of action, and the moment about O

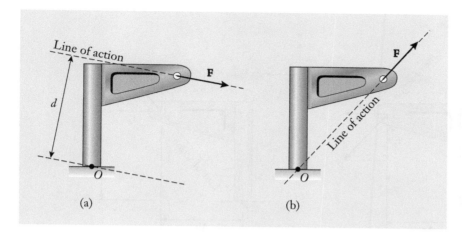

Figure 4.10

Calculating the moment of a force **F**. (a) The line of action of **F** is separated from *O* by the perpendicular lever arm distance *d*. (b) The line of action of **F** passes through *O*, and $M_o = 0$.

would be unchanged. The direction of the moment in Figure 4.10(a) is clockwise because **F** tends to cause the post to rotate that way (even though the rigid mounting would prevent the post from actually moving in this case).

In Figure 4.10(b), the direction of the force has been changed. The force's line of action now passes exactly through point *O*, and so the offset distance is $d = 0$. No moment is produced, and the force tends to pull the post directly out of its base without rotating it. In short, the orientation of a force as well as its magnitude must be taken into account when you calculate a moment.

Moment Components Method

Just as we can break a force down into its rectangular components, it is sometimes useful to calculate a moment in terms of its components. The moment is determined as the sum of portions associated with the two components of the force, rather than the full resultant value of the force. One motivation for calculating the moment in this manner is that the lever arms for individual components are often easier to find than those for the entire resultant force. When applying this technique, we need to use a sign convention and keep track of whether the contribution made by each moment component is clockwise or counterclockwise.

To illustrate this method, let's return to the example of the post and bracket with a force acting on it (Figure 4.11, see on page 130). We will choose the following *sign convention* for rotation directions: a moment that tends to cause counterclockwise rotation is positive, and a clockwise moment is negative. This choice of positive and negative directions is arbitrary; we could just as easily have selected the clockwise direction to be positive. However, once the sign convention is chosen, we need to stick with it and apply it consistently.

The force's components F_x and F_y are shown in the first case (Figure 4.11(a)). Rather than determine the distance from point *O* to the line of action of **F**, which might involve a geometrical construction that we want to avoid, we instead calculate the individual lever arm distances for F_x and F_y, which are more straightforward. Keeping track of the sign convention, the moment about *O* becomes $M_o = -F_x\,\Delta y - F_y\,\Delta x$. The individual contributions to M_o are each negative because both F_x and F_y tend to cause clockwise rotation. Their effects combine constructively to produce the net moment.

Moment sign convention

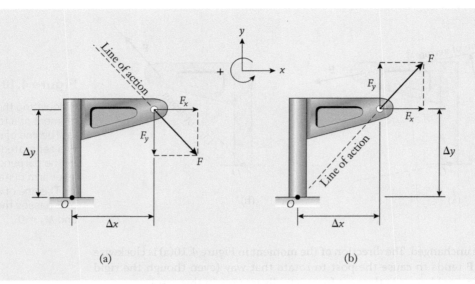

(a)   (b)

Figure 4.11

(a) Both F_x and F_y create clockwise moments about point O. (b) F_x exerts a clockwise moment, but F_y exerts a counterclockwise moment.

The orientation of **F** has been changed in Figure 4.11(b). While F_x continues to exert a negative moment, F_y now tends to cause counterclockwise, or positive, rotation about O. The net moment becomes $M_o = -F_x \Delta y + F_y \Delta x$. Here the two components combine in a destructive manner. In fact, for the special orientation in which $\Delta y/\Delta x = F_y/F_x$, the two terms precisely cancel. The moment in that situation is zero because the line of action for **F** passes through O.

In using the moment components method, we generally write

$$M_o = \pm F_x \Delta y \pm F_y \Delta x \qquad (4.9)$$

where we imply that the numerical values for F_x, Δx, F_y, and Δy are all positive. The positive and negative signs in the equation are assigned while solving a problem depending on whether the moment component tends to cause clockwise or counterclockwise rotation.

Regardless of whether you use the perpendicular lever arm method or the moment components method, when reporting an answer, you should be sure to state (1) the numerical value for the moment's magnitude, (2) the dimensions, and (3) whether the direction is clockwise or counterclockwise. You can indicate the direction by using a plus or minus ($\pm$), provided that you have defined the sign convention on your drawing.

Example 4.3 | *Open-Ended Wrench*

The machinist's wrench is being used to tighten a hexagonal nut. Calculate the moments produced by the 35 lb force about the center of the nut when the force is applied to the wrench in the orientations (a) and (b) as shown. The overall length of the handle, which is inclined slightly upward, is $6\frac{1}{4}$ in. long between centers of the open and closed ends. Report your answer in the units of ft · lb. (See Figure 4.12.)

Example 4.3 | *continued*

Figure 4.12

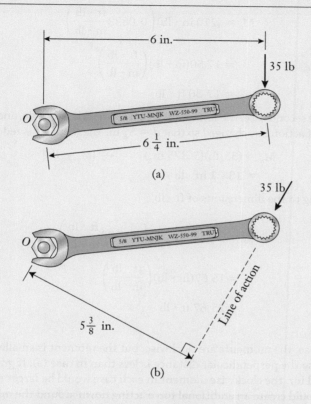

(a)

(b)

Approach

We are tasked with calculating the moments in two orientations. We
first assume that we can neglect the moment created by the weight of
the wrench. Then we will need to use the perpendicular distances in both
cases. When the force acts straight down in case (a), the perpendicular
distance from the center of the nut to the force's line of action is $d = 6$ in.
The inclination and length of the wrench's handle are immaterial insofar as
calculating d is concerned because the handle's length is not necessarily the
same as the perpendicular lever arm's.

Solution

(a) The moment has magnitude

$$M_o = (35 \text{ lb}) (6 \text{ in.}) \qquad \leftarrow [M_o = Fd]$$
$$= 210 \text{ in} \cdot \text{lb}$$

and it is directed clockwise. Apply the conversion factor between
in · lb and ft · lb from Table 4.2 to convert M_o to the desired dimensions:

Example 4.3 | *continued*

$$M_o = (210 \text{ in} \cdot \text{lb})\left(0.0833\frac{\text{ft} \cdot \text{lb}}{\text{in} \cdot \text{lb}}\right)$$

$$= 17.50(\text{in} \cdot \text{lb})\left(\frac{\text{ft} \cdot \text{lb}}{\text{in} \cdot \text{lb}}\right)$$

$$= 17.50 \text{ ft} \cdot \text{lb}$$

(b) In the second case, the force has shifted to an inclined angle, and its line of action has changed so that $d = 5\frac{3}{8}$ in. The moment is reduced to

$$M_o = (35 \text{ lb})(5.375 \text{ in.}) \qquad \leftarrow [M_o = Fd]$$

$$= 188.1 \text{ in} \cdot \text{lb}$$

Converting to the dimensions of ft · lb,

$$M_o = (188.1 \text{ in} \cdot \text{lb})\left(0.0833\frac{\text{ft} \cdot \text{lb}}{\text{in} \cdot \text{lb}}\right)$$

$$= 15.67(\text{in} \cdot \text{lb})\left(\frac{\text{ft} \cdot \text{lb}}{\text{in} \cdot \text{lb}}\right)$$

$$= 15.67 \text{ ft} \cdot \text{lb}$$

Discussion

In each case, the moments are clockwise, but the moment is smaller in case (b) because the perpendicular distance is less than in case (a). If gravity was accounted for, the clockwise moment in each case would be larger since gravity would create an additional force acting down around the middle of the wrench.

When we report the final answer, we indicate the numerical value, dimensions, and direction.

Case (a): $M_o = 17.50 \text{ ft} \cdot \text{lb}$ (clockwise)

Case (b): $M_o = 15.67 \text{ ft} \cdot \text{lb}$ (clockwise)

Example 4.4 | *Adjustable Wrench*

Determine the moment about the center of the nut when the 250 N force is applied to the adjustable wrench. Use (a) the perpendicular lever arm method and (b) the moment components method. (See Figure 4.13.)

Example 4.4 | *continued*

Figure 4.13

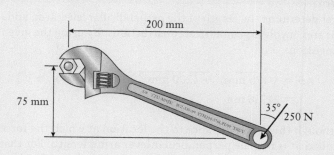

Approach
We are tasked with finding the resulting moment using two methods. We again assume that we can neglect the impact of the wrench's weight. We denote the center of the nut as point A, and the point of application of the force as point B. The moment is calculated by applying Equations (4.8) and (4.9). Use the trigonometric equations from Appendix B to determine the necessary lengths and angles. (See Figure 4.14.)

Figure 4.14

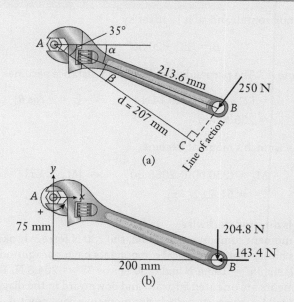

Example 4.4 | *continued*

Solution

(a) First determine the length of the perpendicular lever arm, and this step involves geometrical constructions. By using the given dimensions,

$$AB = \sqrt{(75 \text{ mm})^2 + (200 \text{ mm})^2} \qquad \leftarrow [z^2 = x^2 + y^2]$$

$$= 213.6 \text{ mm}$$

Although this is the distance to the location at which the force is applied, it is not the perpendicular lever arm's length. For that, we need to calculate the length of AC, which is perpendicular to the force's line of action. Because the force is inclined by 35° from vertical, a line perpendicular to it is oriented 35° from horizontal. Line AB lies at the angle

$$\alpha = \tan^{-1}\left(\frac{75 \text{ mm}}{200 \text{ mm}}\right) \qquad \leftarrow \left[\tan\theta = \frac{y}{x}\right]$$

$$= \tan^{-1}\left(0.375\frac{\text{mm}}{\text{mm}}\right)$$

$$= \tan^{-1}(0.375)$$

$$= 20.56°$$

below horizontal, and so it is offset by

$$\beta = 35° - 20.56° = 14.44°$$

from line AC. The perpendicular lever arm distance becomes

$$d = (213.6 \text{ mm})\cos 14.44° \qquad \leftarrow [x = z\cos\theta]$$

$$= 206.8 \text{ mm}$$

and the wrench's moment becomes

$$M_A = (250 \text{ N})(0.2068 \text{ m}) \qquad \leftarrow [M_A = Fd]$$

$$= 51.71 \text{ N} \cdot \text{m}$$

which is directed clockwise.

(b) In the moment components method, the 250 N force is broken down into its horizontal and vertical components having magnitudes (250 N) sin 35° = 143.4 N and (250 N) cos 35° = 204.8 N. Those components are oriented leftward and downward in the diagram. Individually, each exerts a clockwise moment about point A. Referring to our sign convention in the figure, a counterclockwise

Example 4.4 | *continued*

moment is positive. By summing the moment produced by each force component, we have

$$M_A = -(143.4 \text{ N})(0.075 \text{ m}) - (204.8 \text{ N})(0.2 \text{ m}) \leftarrow [M_A = \pm F_x \, \Delta y \pm F_y \, \Delta x]$$
$$= -51.71 \text{ N} \cdot \text{m}$$

Because the numerical value is negative, the net moment is directed clockwise.

Discussion

In this instance, it's probably easier to apply the moment components method because the horizontal and vertical dimensions of the wrench are given in the problem's statement. Also, if gravity was accounted for, the clockwise moment would be larger because gravity would create an additional force acting down around the middle of the wrench.

$$M_A = 51.71 \text{ N} \cdot \text{m (clockwise)}$$

▶ 4.5 Equilibrium of Forces and Moments

With groundwork for the properties of forces and moments now in place, we next turn to the task of calculating (unknown) forces that act on structures and machines in response to other (known) forces that are present. This process involves applying the principles of static equilibrium from Newton's first law to structures and machines that are either stationary or moving at constant velocity. In either case, no acceleration is present, and the resultant force is zero.

Particles and Rigid Bodies

A mechanical system can include either a single object (for instance, an engine's piston) or multiple objects that are connected (the entire engine). When the physical dimensions are unimportant with respect to calculating forces, the object is called a *particle*. This concept idealizes the system as being concentrated at a single point rather than being distributed over an extended area or volume. For the purposes of solving problems, a particle can be treated as having negligible dimensions.

Particle

On the other hand, if the length, width, and breadth of an object are important for the problem at hand, it is called a *rigid body*. As an example, when looking at the motion of a communications satellite as it orbits the Earth, the spacecraft can be regarded as a particle because its dimensions are small compared to the size of the orbit. However, when the satellite is being launched and engineers are interested in the aerodynamics and flight characteristics of

Rigid body

Figure 4.15

A schematic of N forces acting on (a) a single particle and (b) a rigid body. The sets of three dots represent force vectors between F_3 and F_N that are omitted for clarity.

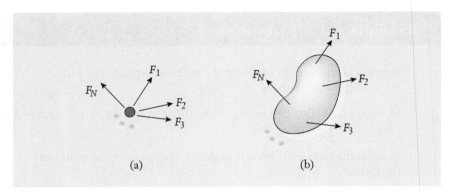

(a)   (b)

the rocket, the launch vehicle would be modeled as a rigid body. Figure 4.15 illustrates the conceptual distinction between forces applied to a particle and to a rigid body. You can see how a force imbalance could cause the rigid body, but not the particle, to rotate.

Force balance

A particle is in equilibrium if the forces acting on it balance with zero resultant. Because forces combine as vectors, the resultant of the N forces that are present must be zero in two perpendicular directions, which we label x and y:

$$\sum_{i=1}^{N} F_{x,i} = 0$$

$$\sum_{i=1}^{N} F_{y,i} = 0 \qquad (4.10)$$

Moment balance

For a rigid body to be in equilibrium, the net moment must also be zero. When those conditions are met, there is no tendency for the object either to move in any direction in response to the forces or to rotate in response to the moments. The requirements for equilibrium of a rigid body comprise Equation (4.10) and

$$\sum_{i=1}^{N} M_{o,i} = 0 \qquad (4.11)$$

The notation $M_{o,i}$ is used to denote the moment of the ith force that is present.

Sign conventions are a good bookkeeping method to distinguish forces that act in opposing directions and moments that are oriented clockwise or counterclockwise. The summations in the equilibrium equations extend over all forces and moments that are present, whether or not their directions and magnitudes are known in advance. Forces that are unknown at the start of the analysis are always included in the summation, algebraic variables are assigned to them, and the equilibrium equations are applied to determine numerical values.

Mathematically speaking, the equilibrium equations for a rigid body comprise a system of three linear equations that involve the unknown forces. One implication of this characteristic is that it is possible to determine at most three unknown quantities when Equations (4.10) and (4.11) are applied to a single rigid body. By contrast, when applying the equilibrium requirements to a particle, the moment equation is not used. In that case, only two *independent equations* exist, and two unknowns at most can be determined. It is not possible to obtain more independent equations of equilibrium by resolving moments about an alternative point or by summing forces in different directions. The additional equations still will be valid, but they simply will be combinations of the other, already derived, ones. As such, they will provide no new information. When you solve equilibrium problems, always check to be sure that you do not have more unknown quantities than available independent equations.

Independent equations

Free Body Diagrams

Free body diagrams are sketches used to analyze the forces and moments that act on structures and machines, and drawing them is an important skill. The free body diagram identifies the mechanical system being examined, and it represents all of the known and unknown forces present. Follow three main steps when drawing a free body diagram:

1. Select an object that will be analyzed by using the equilibrium equations. Imagine that a dotted line is drawn around the object, and note how the line would cut through and expose various forces. Everything within the dotted line is isolated from the surroundings and should appear on the diagram.

2. Next, the coordinate system is drawn to indicate the positive sign conventions for forces and moments. It is meaningless to report an answer of, say, -25 N $\cdot$ m or $+250$ lb without having defined the directions associated with the positive and negative signs.

3. In the final step, all forces and moments are drawn and labeled. These forces might represent weight or contact between the free body and other objects that were removed when the body was isolated. When a force is known, its direction and magnitude should be written on the diagram. At this step in the analysis, forces are included even if their magnitudes and directions are not known. If the direction of a force is unknown (for instance, upward/downward or leftward/rightward), you should just draw it one way or the other on the free body diagram, perhaps using your intuition as a guide. After applying the equilibrium equations and consistently using a sign convention, the correct direction will be determined from your calculations. If you find the quantity to be positive, then you know that the correct direction was chosen at the outset. On the other hand, if the numerical value turns out to be negative, the result simply means that the force acts counter to the assumed direction.

Focus On

ENGINEERING FAILURE ANALYSIS

We have seen how mechanical engineering is based on designing new products, but sometimes it also involves analyzing failures. Many of these structural failures are catastrophic, including the collapses at the Hyatt in Kansas City, Missouri (114 killed), at the Maxima superstore in Riga, Latvia (54 killed), and at the Rana Plaza in Bangladesh (1,129 killed). While engineers are responsible for designing resilient structures and machines using sound force and stress analysis, engineers are also using this kind of analysis to save lives in other ways.

The human body is a biological system that requires a delicate interaction between cellular processes and large-scale organ systems. These interactions produce internal forces and stresses that sometimes cause biological structures to fail. Failures of these systems occur for many reasons, but one cause in particular that demands many resources to mitigate are those failures caused by disease. *Marfan syndrome* is one such disease, causing catastrophic failure to the heart's ascending aorta. It is a genetic connective tissue disorder that deteriorates the structural integrity of the vital organs, which most commonly leads to the eventual rupture of the ascending aorta.

Until recently the only way to correct the effects of Marfan syndrome was to remove the aorta and replace it with an artificial valve and aorta, followed by a lifetime regimen of powerful anticoagulation medications to ensure the proper function of the new parts. One person afflicted with this disease, Tal Golesworthy, was at the point where he needed an aorta transplant to prevent catastrophic failure and death. Unenthusiastic about the prospects of surgery and a lifetime of drug

therapy, he set out to use his knowledge of structural forces to change the entire treatment for aortic dilation.

You see, Tal Golesworthy is an engineer, and after understanding more about the problems caused by Marfan syndrome, he began to see his ailment as a structural design issue. He recognized that the only problem with the aorta was that it lacked sufficient tensile strength. He studied the forces and stresses in similar tubular environments and determined that a feasible solution could be to externally wrap the weakened area allowing the region to remain stable. Mr. Golesworthy then worked with a multidisciplinary team of mechanical engineers, medical device engineers, surgeons, and other medical specialists to solve this structural problem.

Their solution now uses magnetic resonance imaging (MRI) technology to create a model of a patient's aorta, which is then structurally designed based on force and stress analysis. The new mesh wrap is then fabricated using 3-D printing allowing for the wrap to be customized perfectly to the recipient's aortic geometry (see Figure 4.16). Finally, a surgeon implants the mesh in a relatively short procedure, all the while leaving the patient's circulatory system completely intact. This solution also eliminates the need for any ongoing medication.

Mechanical engineers have the opportunity to save lives in many domains by creating innovative medical devices, transportation systems, nanotechnologies, and advanced materials. Applying sound structural analysis to design for a wide range of forces and stresses is critical to the safe use of such systems.

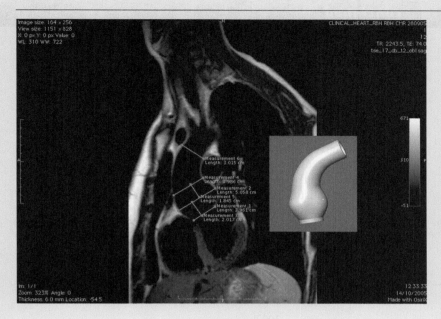

Figure 4.16

Imaging data used to create a customized digital model of the aorta, which can then be fabricated using 3D printing.

John Pepper, Mario Petrou, Filip Rega, Ulrich Rosendahl, Tal Golesworthy, Tom Treasure, "Implantation of an individually computer-designed and manufactured external support for the Marfan aortic root," *Mutimedia Manual of Cardio-Thoracic Surgery*, 2013, by permission of Oxford University Press.

Example 4.5 | *Seat Belt Buckle*

During crash testing of an automobile, the lap and shoulder seat belts develop tensions of 300 lb. Treating the buckle B as a particle, determine the tension T in the anchor strap AB and the angle at which it acts. (See Figure 4.17.)

Figure 4.17

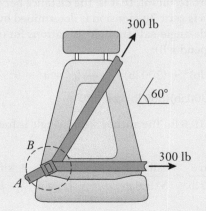

Example 4.5 | *continued*

Approach

We are tasked with finding the magnitude and direction of the tension in the seat belt anchor strap. By treating the buckle as a particle, we can assume that all the force from the shoulder and lap belts act on the anchor strap. The weights of the straps are assumed to be negligible. The free body diagram of the buckle is drawn along with the x-y coordinate system to indicate our sign convention for the positive horizontal and vertical directions. Three forces act on the buckle: the two given 300 lb forces and the unknown force in the anchor strap. For the buckle to be in equilibrium, these three forces must balance. Although both the magnitude T and direction θ of the force in strap AB are unknown, both quantities are shown on the free body diagram for completeness. There are two unknowns, T and θ, and the two expressions in Equation (4.10) are available to solve the problem. (See Figure 4.18.)

Figure 4.18

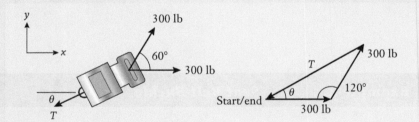

Solution

We will combine the three forces by using the vector polygon approach. The polygon's start and endpoints are the same because the three forces combine to have zero resultant; that is, the distance between the polygon's start and endpoints is zero. The tension is determined by applying the law of cosines to the side-angle-side triangle (equations for oblique triangles are reviewed in Appendix B):

$$T^2 = (300 \text{ lb})^2 + (300 \text{ lb})^2 \qquad \leftarrow [c^2 = a^2 + b^2 - 2ab \cos C]$$
$$- 2(300 \text{ lb})(300 \text{ lb}) \cos 120°$$

We calculate $T = 519.6$ lb. The anchor strap's angle is found from the law of sines:

$$\frac{\sin \theta}{300 \text{ lb}} = \frac{\sin 120°}{519.6 \text{ lb}} \qquad \leftarrow \left[\frac{\sin A}{a} = \frac{\sin B}{b}\right]$$

and $\theta = 30°$.

Example 4.5 | *continued*

Discussion

The three forces in the vector polygon form an isosceles triangle. As a double check, the interior angles sum to 180°, as they should. In reality, the forces in the shoulder and lap belts will act on the buckle, which then acts on the anchor strap. By assuming the buckle to be a particle, we were able to simplify the analysis. In an alternative solution, we could break down the two 300 lb forces into their components along the x- and y-axes and apply Equation (4.10).

$T = 519.6$ lb

$\theta = 30°$ counterclockwise from the $-x$-axis

Example 4.6 | *Wire Cutters*

A technician applies the 70 N gripping force to the handles of the wire cutters. What are the magnitudes of the cutting force on the electrical wire at A and the force carried by the hinge pin at B? (See Figure 4.19.)

Figure 4.19

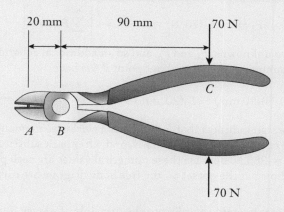

Approach

We are tasked to find the forces at the cutting point and the hinge. We assume that the weight of the wire cutters is negligible compared to the applied force. We then show the coordinate system and positive sign convention for the directions of forces and moments. The free body diagram is drawn for one jaw/handle assembly, which is a rigid body

Example 4.6 | *continued*

because it can rotate and the distances between the forces are significant to the problem. When a blade presses against the wire, the wire in turn pushes back on the blade following the principle of action–reaction. We label the jaw's cutting force as F_A, and the force exerted by the hinge pin on the jaw/handle as F_B. The 70 N gripping force is given, and it is also included on the free body diagram. (See Figure 4.20.)

Figure 4.20

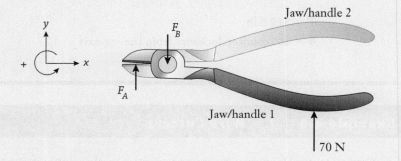

Solution

The cutting force is found by applying the equilibrium equations for a rigid body. The force balance requirement in the vertical direction becomes

$$F_A - F_B + (70 \text{ N}) = 0 \qquad \leftarrow \left[\sum_{i=1}^{N} F_{y,i} = 0 \right]$$

There are two unknowns, F_A and F_B, and so an additional equation is needed. By summing moments about point B, we have

$$(70 \text{ N})(90 \text{ mm}) - F_A(20 \text{ mm}) = 0 \qquad \leftarrow \left[\sum_{i=1}^{N} M_{B,i} = 0 \right]$$

The negative sign indicates that F_A produces a clockwise moment about point B. The cutting force is $F_A = 315$ N, and, after back substitution, we find that $F_B = 385$ N. Because these numerical values are each positive, the directions shown at the outset on the free body diagram are correct.

Discussion

The wire cutters operate according to the principle of a lever. The cutting force is proportional to the force on the handles, and it is also related to the ratio of distances AB and BC. The mechanical advantage for a machine is defined as the ratio of the output and input forces, or in this case, (315 N)/(70 N) = 4.5. The wire cutters magnify the gripping force by 450%.

$$F_A = 315 \text{ N}$$
$$F_B = 385 \text{ N}$$

Example 4.7 | *Forklift Load Capacity*

The forklift weighs 3500 lb and carries an 800 lb shipping container. There are two front wheels and two rear wheels on the forklift. (a) Determine the contact forces between the wheels and ground. (b) How heavy a load can be carried before the forklift will start to tip about its front wheels? (See Figure 4.21.)

Figure 4.21

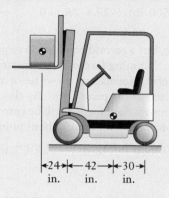

←24→←—42—→←30→
in.   in.   in.

Approach

We are tasked with finding the wheel contact forces and the maximum load the forklift can carry before it tips forward. We assume that the forklift is not moving and then draw the free body diagram along with the positive sign conventions for forces and moments. We label the (known) 3500 lb and 800 lb weights that act through the mass centers of the forklift and container. The (unknown) force between a front wheel and the ground is denoted by F, and the (unknown) force between a rear wheel and the ground is R. On the side view of the free body diagram, the net effects of those forces on the wheel pairs are $2F$ and $2R$. (See Figure 4.22.)

Figure 4.22

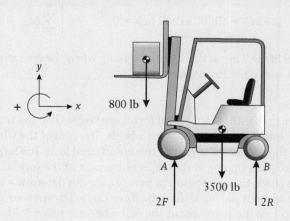

y

$+$

x

800 lb

3500 lb

A B

$2F$ $2R$

Example 4.7 | *continued*

Solution

(a) There are two unknowns (F and R), and therefore two independent equilibrium equations are needed to solve the problem. We first sum forces in the vertical direction

$$-(800 \text{ lb}) - (3500 \text{ lb}) + 2F + 2R = 0 \qquad \leftarrow \left[\sum_{i=1}^{N} F_{y,i} = 0 \right]$$

or $F + R = 2150$ lb, but a second equation is required to determine the two unknowns. Summing forces in the horizontal direction will not provide any useful information, so we use a moment balance. Any location can be chosen as the pivot point. By choosing the point to coincide with the front wheel, force F will be conveniently eliminated from the calculation. Taking moments about point A, we have

$$(800 \text{ lb})(24 \text{ in.}) - (3500 \text{ lb})(42 \text{ in.}) + (2R)(72 \text{ in.}) = 0$$

$$\leftarrow \left[\sum_{i=1}^{N} M_{A,i} = 0 \right]$$

from which $R = 888$ lb. Here, the 800 lb weight and the rear wheel forces exert counterclockwise (positive) moments about A, and the forklift's 3500 lb weight generates a negative moment. Substituting the solution for R into the vertical-force balance

$$F + (888 \text{ lb}) = 2150 \text{ lb}$$

returns $F = 1262$ lb.

(b) When the forklift is on the verge of tipping about the front wheels, the rear wheel has just lost contact with the ground, and $R = 0$. We denote the new weight of the shipping container that causes tipping as w. The moment balance about the front wheels is

$$(w)(24 \text{ in.}) - (3500 \text{ lb})(42 \text{ in.}) = 0 \qquad \leftarrow \left[\sum_{i=1}^{N} M_{A,i} = 0 \right]$$

The forklift will be on the verge of tipping when the operator attempts to lift a $w = 6125$ lb container.

Discussion

It makes sense that the front wheel forces are larger than the rear wheel, because the load is ahead of the front wheels. The sum of the wheel forces also equals the combined weight of the forklift and load. The large weight necessary to cause tipping is also expected as part of the forklift's design. As a double check to our solution of part (a), we can determine the front wheel force without having to find the force carried by the rear wheels. The key is to sum moments about the rear wheels at point B. The unknown

Example 4.7 | *continued*

force R passes through that point, and, by having a perpendicular lever arm with zero length, it will be eliminated in the moment calculation. The moment balance now reads

$$(800 \text{ lb})(96 \text{ in.}) + (3500 \text{ lb})(30 \text{ in.}) - (2F)(72 \text{ in.}) = 0 \quad \longleftarrow \left[\sum_{i=1}^{N} M_{B,i} = 0 \right]$$

We obtain $F = 1262$ lb through only one equation and without any intermediate steps involving R. In short, by carefully choosing the moment's pivot point, we can reduce the amount of algebra by directly eliminating unknown forces.

Wheel forces: $F = 1262$ lb and $R = 888$ lb
Maximum tipping load: $w = 6125$ lb

▶ 4.6 Design Application: Rolling-Element Bearings

n the preceding sections, we discussed the properties of forces and moments and applied the requirements of equilibrium to examine the forces acting on structures and machines. We now consider a specific application to mechanical design and the forces that act on the machine components called *rolling-element bearings*. Bearings are used to hold shafts that rotate relative to fixed supports (for instance, the housing of a motor, gearbox, or transmission). In designing power transmission equipment, mechanical engineers will often perform a force or equilibrium analysis to choose the correct size and type of bearing for a particular application.

Bearings are classified into two broad groups: rolling contact and journal. In this section, rolling contact bearings are the focus, and each comprises the following components:

- An inner race
- An outer race
- Rolling elements in the form of balls, cylinders, or cones
- A separator that prevents the rolling elements from rubbing up against one another

Rolling-element bearings are so common in machine design that they are found in applications as diverse as hard disk drives, bicycle wheels, robotic joints, and automobile transmissions. *Journal bearings*, on the other hand, have no rolling elements. Instead, the shaft simply rotates within a polished sleeve that is lubricated by oil or another fluid. Just as the puck on an air hockey table slides

Journal bearing

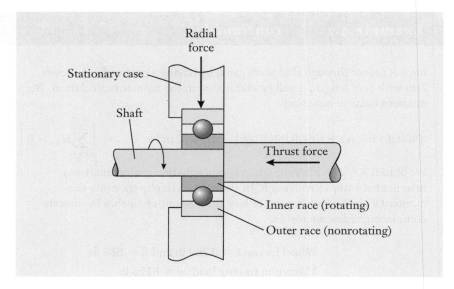

Figure 4.23

An installation of a
ball bearing.

smoothly over a thin film of air, the shaft in a journal bearing slides over and
is supported by a thin film of oil. Although they may be less familiar to you,
journal bearings are also quite common, and they are used to support shafts in
internal-combustion engines, pumps, and compressors.

A sample installation of a rolling contact bearing is shown in Figure 4.23.
The shaft and the bearing's inner race rotate together, while the outer race and
the case are stationary. When a shaft is supported in this way, the bearing's outer
race will fit tightly into a matching circular recess that is formed in the case. As
the shaft turns and transmits power, perhaps in a geartrain or transmission,
the bearing could be subjected to forces oriented either along the shaft (a *thrust
force*) or perpendicular to it (a *radial force*). An engineer will make a decision on
the type of bearing to be used in a machine depending on whether thrust forces,
radial forces, or some combination of the two are going to act on the bearing.

The most common type of rolling-element bearing is the *ball bearing*, which
incorporates hardened, precision-ground steel spheres. Figure 4.24 depicts the
major elements of a standard ball bearing: the inner race, outer race, balls, and
separator. The *inner and outer races* are the bearing's connections to the shaft
and the case. The *separator* (which is sometimes called the *cage* or *retainer*)
keeps the balls evenly spaced around the bearing's perimeter and prevents
them from contacting one another. Otherwise, if the bearing was used at
high speed or subjected to large forces, friction could cause it to overheat and
become damaged. In some cases, the gap between the inner and outer races
is sealed by a rubber or plastic ring to keep grease in the bearing and dirt out
of it.

In principle, the bearing's balls press against the inner and outer races at
single points, similar to the contact between a marble and a floor. The force
that each ball transfers between the inner and outer races therefore becomes
concentrated on those surfaces in an intense and relatively sharp manner, as
indicated in Figure 4.25(a). If those forces instead could be spread over a larger

Thrust force
Radial force

Ball bearings

*Inner and outer
races*

*Separator, cage,
retainer*

Seals

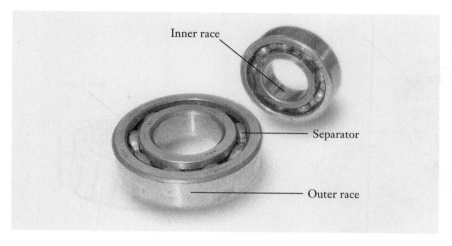

Figure 4.24

Elements of a ball bearing.

Image courtesy of the authors.

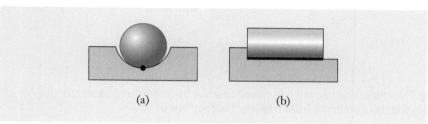

Figure 4.25

Side view of (a) point contact between a spherical ball and the raceway of a bearing, and (b) line contact occurring in a straight roller bearing.

area, we would expect that the bearing would wear less and last longer. With that in mind, rolling-element bearings that incorporate cylindrical rollers or tapered cones in place of spherical balls are one solution for distributing forces more evenly.

Straight (or cylindrical) *roller bearings*, as shown in Figure 4.25(b), can be used to distribute forces more evenly over the bearing's races. If you place a few pens between your hands and then rub your hands together, you have the essence of a straight roller bearing. Figure 4.26(a) (see page 148) illustrates the structure of a straight roller bearing.

Straight roller bearings

While straight roller bearings support forces that are directed mostly radially, *tapered* (or angled) *roller bearings* can support a combination of radial and thrust forces. This is because these bearings are built around rollers shaped like truncated cones (Figure 4.26(b)). One prominent application for tapered roller bearings is in automobile wheel bearings because both radial forces (the weight of the vehicle) and thrust forces (the cornering force generated when the vehicle makes a turn) are present.

Tapered roller bearings

While straight roller bearings support forces that are directed mostly radially, and tapered roller bearings can support a combination of radial and thrust forces, the *thrust roller bearing* carries loads that are directed mostly along a shaft. One type of thrust bearing is shown in Figure 4.27 (see page 148). The rolling elements in this case are cylinders having a slight barrel shape to them. In contrast to the straight roller bearing of Figure 4.26(a), these rollers are oriented radially and perpendicular to the shaft. Thrust bearings are appropriate for such applications as a rotating table that must support the dead weight of cargo but that also needs to turn freely.

Thrust roller bearings

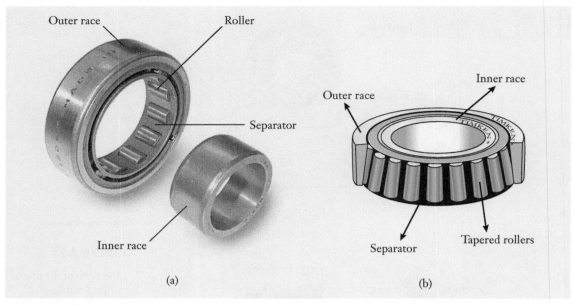

Figure 4.26

(a) A straight roller bearing. The inner race has been removed to show the rollers and separator. (b) A tapered roller bearing that is widely used in the front wheels of automobiles.

(a) Image courtesy of the authors. (b) Reprinted with permission by The Timken Company.

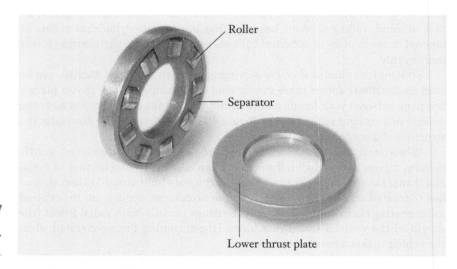

Figure 4.27

A thrust roller bearing.

Image courtesy of the authors.

Example 4.8 | *Treadmill's Belt Drive*

An electric motor is used to power an exercise treadmill. Forces are applied to the treadmill's shaft by the motor's drive belt and by the wide, flat belt that is the surface used for walking or running. The tight and loose spans

Example 4.8 | *continued*

of the drive belt together apply 110 lb to the shaft, and the treadmill's belt applies 70 lb. The shaft is supported by ball bearings on each side of the belt. Calculate the magnitudes and directions of the forces exerted by the shaft on the two bearings. (See Figure 4.28.)

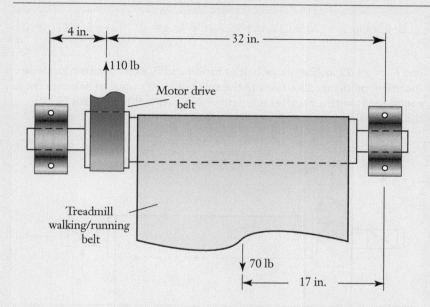

Figure 4.28

Approach

We are tasked with finding the forces on the two bearings from the two belts. We first assume that all the forces act parallel to the *y*-direction. The free body diagram of the shaft is drawn, along with the sign conventions for the coordinate directions and rotation. On the diagram, first label the 110 lb and 70 lb belt tensions, and then denote the forces exerted by the bearings on the shaft as F_A and F_B. At this point, we don't know whether those unknown forces act in the positive or negative *y*-directions. By drawing them on the free body diagram using our sign convention, we will rely on the calculation to determine the actual direction of the forces. If a numerical value turns out to be negative, the result will mean that the force acts in the negative *y*-direction. (See Figure 4.29 on page 150.)

Solution

Because there are two unknowns (F_A and F_B), two equilibrium equations are needed to solve the problem. By summing forces in the *y*-direction

$$(110 \text{ lb}) - (70 \text{ lb}) + F_A + F_B = 0 \quad \leftarrow \left[\sum_{i=1}^{N} F_{y,i} = 0 \right]$$

Example 4.8 | continued

or $F_A + F_B = -40$ lb. Apply a moment balance for the second equation. By choosing the pivot point to coincide with the center of bearing A, force F_A will be eliminated from the calculation. Summing moments about point A, we have

$$(110 \text{ lb})(4 \text{ in.}) - (70 \text{ lb})(19 \text{ in.}) + (F_B)(36 \text{ in.}) = 0 \qquad \leftarrow \left[\sum_{i=1}^{N} M_{A,i} = 0\right]$$

and $F_B = 24.72$ lb. The motor belt's tension and F_B exert counterclockwise (positive) moments about A, and the treadmill belt's tension balances those components with a negative moment. Substituting this value for F_B into the force balance returns $F_A = -64.72$ lb.

Figure 4.29

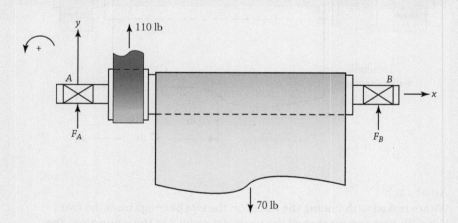

Discussion
These forces are the same order of magnitude as the applied forces, and that makes sense. Also, when in use, the treadmill will exert forces on the bearings in the x- and z-directions, but those are not accounted for in our analysis. Since the calculated value for F_A is negative, the force that is exerted by bearing A on the shaft acts in the negative y-direction with magnitude of 64.72 lb. Following the principle of action–reaction from Newton's third law, the directions of the forces exerted by the shaft on the bearings are opposite those exerted by the bearings on the shaft.

Bearing A: 64.72 lb in the negative y-direction

Bearing B: 24.72 lb in the positive y-direction

Example 4.9 | *Automobile Wheel Bearings*

A 13.5 kN automobile is being driven at 50 km/h through a turn radius of 60 m. Assuming that the forces are equally balanced among the four wheels, calculate the magnitude of the resultant force acting on the tapered roller bearings that support each wheel. To calculate the cornering force, apply Newton's second law ($F = ma$) with the centripetal acceleration ($a = v^2/r$) where m is the vehicle's mass, v denotes its speed, and r is the turn radius.

Approach
We are tasked with finding the resultant force from the applied radial and thrust forces on the vehicle wheel bearings. We assume that each wheel carries one-quarter of the vehicle's weight and that force component is oriented radially on the wheel's bearings. The cornering force acting on the entire vehicle is mv^2/r, directed toward the center of the turn, and the fraction carried by each wheel is a thrust force parallel to the wheel's axle. In terms of those variables, we will determine a general symbolic equation for the magnitude of a wheel's resultant force, and then substitute specific values to obtain a numerical result. (See Figure 4.30.)

Figure 4.30

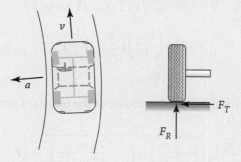

Solution
With w denoting the automobile's weight, each wheel carries the radial force

$$F_R = \frac{w}{4}$$

The thrust force carried by one wheel is given by the expression

$$F_R = \frac{1}{4}\frac{mv^2}{r} \qquad \leftarrow [F = ma]$$

where the vehicle's mass is

$$m = \frac{w}{g}$$

Example 4.9 | *continued*

The resultant of those two perpendicular force components is

$$F = \sqrt{F_R^2 + F_T^2} \quad \leftarrow \boxed{F = \sqrt{F_x^2 + F_y^2}}$$

$$= \frac{m}{4}\sqrt{g^2 + \left(\frac{v^2}{r}\right)^2}$$

Next, substitute the numerical values given in the problem's statement into this general expression. The vehicle's mass is

$$m = \frac{13.5 \times 10^3 \text{ N}}{9.81 \text{ m/s}^2} \quad \leftarrow \boxed{m = \frac{w}{g}}$$

$$= 1.376 \times 10^3 \left(\frac{\text{kg} \cdot \text{m}}{\text{s}^2}\right)\left(\frac{\text{s}^2}{\text{m}}\right)$$

$$= 1.376 \times 10^3 \text{ kg}$$

or 1.376 Mg. In consistent dimensions, the velocity is

$$v = \left(50 \frac{\text{km}}{\text{h}}\right)\left(10^3 \frac{\text{m}}{\text{km}}\right)\left(\frac{1}{3600} \frac{\text{h}}{\text{s}}\right)$$

$$= 13.89 \left(\frac{\text{km}}{\text{h}}\right)\left(\frac{\text{m}}{\text{km}}\right)\left(\frac{\text{h}}{\text{s}}\right)$$

$$= 13.89 \frac{\text{m}}{\text{s}}$$

With a 60 m turn radius, the magnitude of the resultant force acting on a wheel's bearings is

$$F = \left(\frac{1.376 \times 10^3 \text{ kg}}{4}\right)\sqrt{\left(9.81\frac{\text{m}}{\text{s}^2}\right)^2 + \left(\frac{(13.89 \text{ m/s})^2}{60 \text{ m}}\right)^2} \quad \leftarrow \boxed{F = \frac{m}{4}\sqrt{g^2 + \left(\frac{v^2}{r}\right)^2}}$$

$$= 3551\frac{\text{kg} \cdot \text{m}}{\text{s}^2}$$

$$= 3551 \text{ N}$$

$$= 3.551 \text{ kN}$$

Discussion

Because of the cornering force, the wheel bearings carry somewhat more than one-quarter of the vehicle's weight (3.375 kN). As a double check, we can verify the dimensional consistency of the calculations by noting that the term v^2/r, which is combined with the gravitational acceleration g, has the dimensions of acceleration.

$$F = 3.551 \text{ kN}$$

Summary

The objective of this chapter has been to introduce the engineering concepts of force systems, moments, and equilibrium in the context of engineering structures and machines. The primary variables, symbols, and units are summarized in Table 4.3, and the key equations are listed in Table 4.4 (see on page 154). After developing these concepts, we applied them to determine the magnitudes and directions of forces acting on and within simple structures and machines. Engineers often perform a force analysis to see whether a design will be feasible and safe. One of the skills that mechanical engineers develop is the ability to apply equations to physical problems clearly and consistently. Selecting the object to be included in a free body diagram, choosing the directions for coordinate axes, and picking the best point for balancing moments are some of the choices that you need to make in solving problems of this nature.

We also applied the concepts of force systems to several different types of rolling element bearings that are used in machine design. Bearings and the other machine components, which we will examine in later chapters, have special properties and terminology, and mechanical engineers need to be fluent with those building blocks to select the component that is best suited for particular hardware.

In the next chapter, we will take another step toward designing structures and machines so that they will be strong enough to support the forces acting on them. We will build on the properties of force systems and take into account the strength characteristics of the materials from which the mechanical components are made.

Quantity	Conventional Symbols	Conventional Units	
		USCS	SI
Force vector	$\mathbf{F}$	lb	N
Force components	$F_x, F_y, F_{x,i}, F_{y,i}$	lb	N
Force magnitude	F	lb	N
Force direction	θ	deg, rad	deg, rad
Resultant	$\mathbf{R}, R, R_x, R_y$	lb	N
Moment about O	$M_o, M_{o,i}$	in · lb, ft · lb	N · m
Perpendicular lever arm	d	in., ft	m
Moment component offset	$\Delta x, \Delta y$	in., ft	m

Table 4.3

Quantities, Symbols, and Units that Arise when Analyzing Forces in Structures and Machines

Force vector	
Rectangular-polar conversion	$F = \sqrt{F_x^2 + F_y^2}, \quad \theta = \tan^{-1}\left(\dfrac{F_y}{F_x}\right)$
Polar-rectangular conversion	$F_x = F \cos\theta, \quad F_y = F \sin\theta$
Resultant of N forces	$R_x = \sum_{i=1}^{N} F_{x,i}, \quad R_y = \sum_{i=1}^{N} F_{y,i}$
Moment about point O	
Perpendicular lever arm	$M_o = F_d$
Moment components	$M_o = \pm F_x \, \Delta y \pm F_y \, \Delta x$
Equilibrium	
Displacement	$\sum_{i=1}^{N} F_{x,i} = 0, \quad \sum_{i=1}^{N} F_{y,i} = 0$
Rotation	$\sum_{i=1}^{N} M_{o,i} = 0$

Table 4.4

Key Equations that
Arise when Analyzing
Forces in Structures
and Machines

Self-Study and Review

4.1. What are Newton's three laws of motion?

4.2. What are the conventional dimensions for forces and moments in the USCS and SI?

4.3. About how many newtons are equivalent to 1 lb?

4.4. How do you calculate the resultant of a force system by using the vector algebra and vector polygon methods? When do you think it is more expedient to use one method over the other?

4.5. How do you calculate a moment by using the perpendicular lever arm and moment components methods? When do you think it is more expedient to use one method over the other?

4.6. Why is a sign convention used when calculating moments using the component method?

4.7. What are the equilibrium requirements for particles and rigid bodies?

4.8. What steps are involved in drawing a free body diagram?

4.9. Describe some of the differences between ball, straight roller, tapered roller, and thrust bearings. Give examples of real situations in which you would select one to be used instead of another.

4.10. What is the purpose of a bearing's separator?

4.11. Make a cross-sectional drawing of a tapered roller bearing.

4.12. Give examples of situations where bearings are subjected to radial forces, thrust forces, or some combination of the two.

Problems

P4.1

In 1995, the Sampoong Department Store collapsed in South Korea, killing 501 people and injuring 937. First believed to be an act of terrorism, investigators later determined that it was the result of poor engineering and construction management. Research this failure and describe how a proper structural force analysis could have prevented the disaster.

P4.2

The cylindrical coordinate robot on a factory's assembly line is shown in a top view (Figure P4.2). The 50 N force acts on a workpiece being held at the end of the robot's arm. Express the 50 N force as a vector in terms of unit vectors **i** and **j** that are aligned with the x- and y-axes.

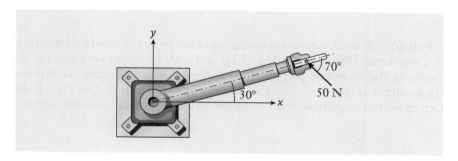

Figure P4.2

P4.3

During the power stroke of an internal-combustion engine, the 400 lb pressure force pushes the piston down its cylinder (Figure P4.3). Determine the components of that force in the directions along and perpendicular to the connecting rod AB.

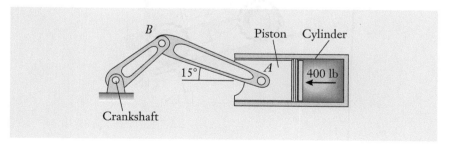

Figure P4.3

P4.4

A vector polygon for summing 2 kN and 7 kN forces is shown (Figure P4.4, see on page 156). Determine (a) the magnitude R of the resultant by using the law of cosines and (b) its angle of action θ by using the law of sines.

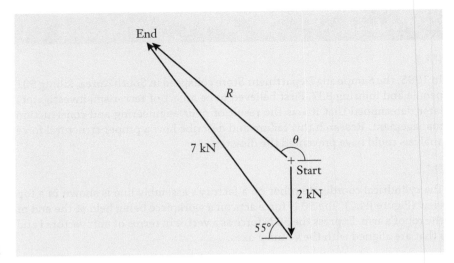

Figure P4.4

P4.5

A hydraulic-lift truck carries a shipping container on the inclined loading ramp in a warehouse (Figure P4.5). The 12 kN and 2 kN forces act on a rear tire as shown in the directions perpendicular and parallel to the ramp. (a) Express the resultant of those two forces as a vector using the unit vectors **i** and **j**. (b) Determine the magnitude of the resultant and its angle relative to the incline.

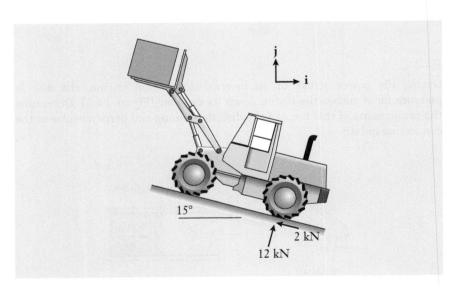

Figure P4.5

P4.6

Three tension rods are bolted to a gusset plate (Figure P4.6). Determine the magnitude and direction of their resultant. Use the (a) vector algebra and (b) vector polygon methods. Compare the answers from the two methods to verify the accuracy of your work.

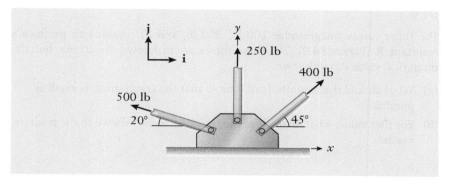

Figure P4.6

P4.7

The bucket of an excavator at a construction site is subjected to 1200 lb and 700 lb digging forces at its tip (Figure P4.7). Determine the magnitude and direction of their resultant. Use the (a) vector algebra and (b) vector polygon methods. Compare the answers from the two methods to verify the accuracy of your work.

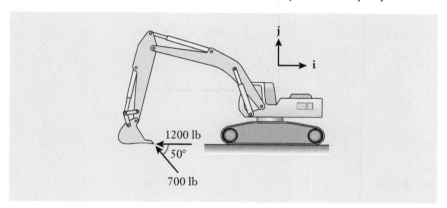

Figure P4.7

P4.8

Forces of 225 N and 60 N act on the tooth of a spur gear (Figure P4.8). The forces are perpendicular to one another, but they are inclined by 20° relative to the x-y coordinates. Determine the magnitude and direction of their resultant. Use the (a) vector algebra and (b) vector polygon methods. Compare the answers from the two methods to verify the accuracy of your work.

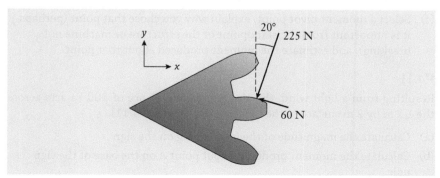

Figure P4.8

P4.9

The three forces (magnitudes 100 lb, 200 lb, and P) combine to produce a resultant **R** (Figure P4.9). The three forces act in known directions, but the numerical value P is unknown.

(a) What should the magnitude of **P** be so that the resultant is as small as possible?

(b) For that value, what angle does the resultant make relative to the positive x-axis?

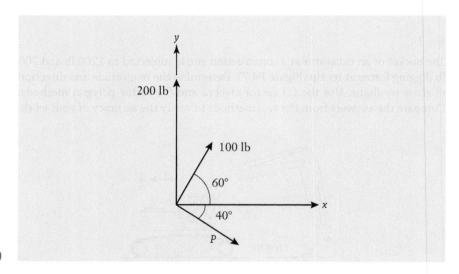

Figure P4.9

P4.10

Find a real physical example of a mechanical structure or machine that has a moment acting on it.

(a) Make a clear labeled drawing of the situation.

(b) Estimate the dimensions, and the magnitudes and directions of the forces that act on it. Show these on your drawing. Briefly explain why you estimate the dimensions and forces to have the numerical values that you assigned.

(c) Select a moment pivot point, explain why you chose that point (perhaps it is important from the standpoint of the structure or machine not breaking), and estimate the moment produced about that point.

P4.11

Resulting from a light wind, the air pressure imbalance of 100 Pa acts across the 1.2 m by 2 m surface of the highway sign (Figure P4.11).

(a) Calculate the magnitude of the force acting on the sign.

(b) Calculate the moment produced about point A on the base of the sign pole.

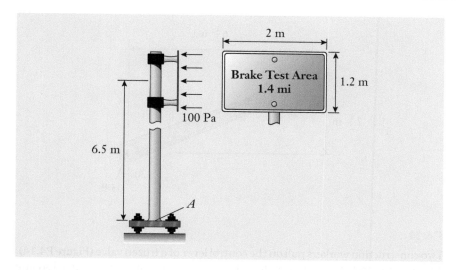

Figure P4.11

P4.12

The spur gear has a pitch radius of 2.5 in. (Figure P4.12). During a geartrain's operation, a 200 lb meshing force acts at 25° relative to horizontal. Determine the moment of that force about the center of the shaft. Use the (a) perpendicular lever arm and (b) moment components methods. Compare the answers of the two methods to verify the accuracy of your work.

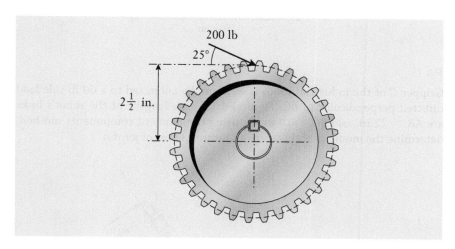

Figure P4.12

P4.13

Determine the moment of the 35 lb force about the center *A* of the hex nut (Figure P4.13, see on page 160).

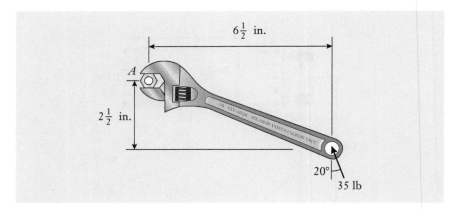

Figure P4.13

P4.14

Two construction workers pull on the control lever of a frozen valve (Figure P4.14). The lever connects to the valve's stem through the key that fits into partial square grooves on the shaft and handle. Determine the net moment about the center of the shaft.

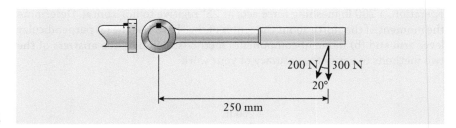

Figure P4.14

P4.15

Gripper C of the industrial robot is accidentally subjected to a 60 lb side load directed perpendicular to BC (Figure P4.15). The lengths of the robot's links are $AB = 22$ in. and $BC = 18$ in. By using the moment components method, determine the moment of this force about the center of joint A.

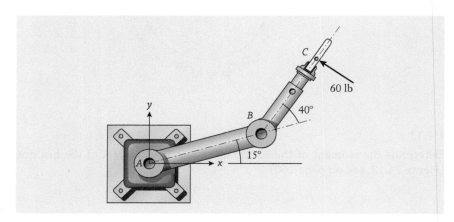

Figure P4.15

P4.16

The mobile boom lift is used in construction and maintenance applications (Figure P4.16). The hydraulic cylinder AB exerts a 10 kN force on joint B that is directed along the cylinder. By using the moment components method, calculate the moment of this force about the lower support point C of the boom.

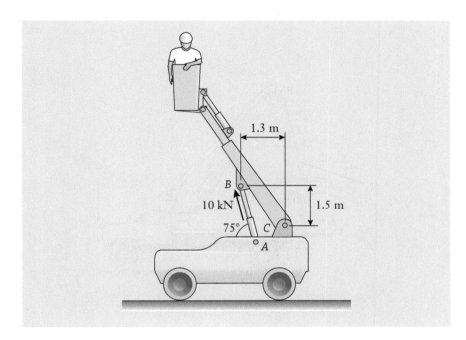

Figure P4.16

P4.17

A trough of concrete weighs 800 lb (Figure P4.17).

(a) Draw a free body diagram of the cable's ring A.

(b) Treating the ring as a particle, determine the tension in cables AB and AC.

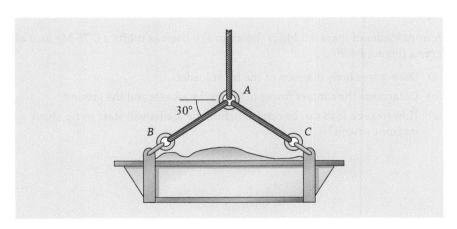

Figure P4.17

P4.18

Cable AB on the boom truck is hoisting the 2500 lb section of precast concrete (Figure P4.18). A second cable is under tension P, and workers use it to pull and adjust the position of the concrete section as it is being raised.

(a) Draw a free body diagram of hook A, treating it as a particle.

(b) Determine P and the tension in cable AB.

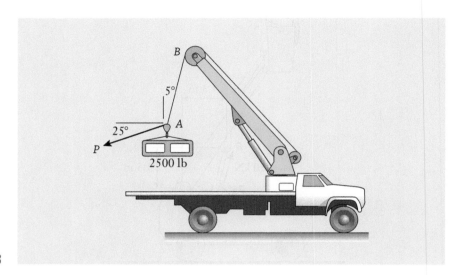

Figure P4.18

P4.19

Solve the problem of Example 4.5 by using the force components method. Replace the polar representation of the anchor strap's tension by the horizontal and vertical components T_x and T_y, and solve for them. Use your solution for T_x and T_y to determine the magnitude T and direction θ of the anchor strap's tension.

P4.20

A front loader of mass 4.5 Mg is shown in side view as it lifts a 0.75 Mg load of gravel (Figure P4.20).

(a) Draw a free body diagram of the front loader.

(b) Determine the contact forces between the wheels and the ground.

(c) How heavy a load can be carried before the loader will start to tip about its front wheels?

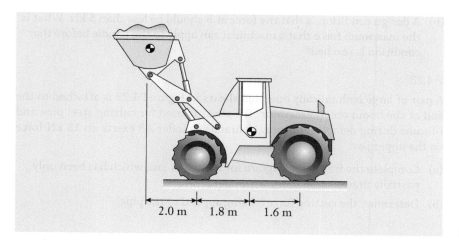

P4.21

Adjustable pliers hold a round metal bar as a machinist grips the handles with $P = 50$ N (Figure P4.21). Using the free body diagram shown for the combined lower jaw and upper handle, calculate the force A that is being applied to the bar.

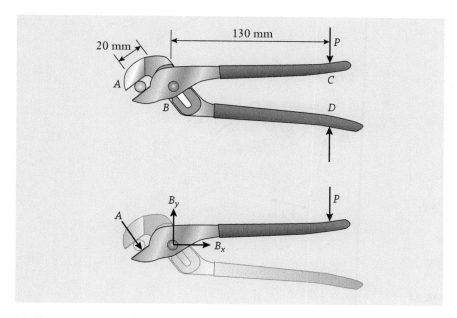

Figure P4.21

P4.22

Refer to P4.21.

(a) Measure the angle of force A directly from the diagram and use it to find the magnitude of the force at hinge B.

(b) A design condition is that the force at B should be less than 5 kN. What is the maximum force that a machinist can apply to the handle before that condition is reached?

P4.23

A pair of large hydraulically operated shears in Figure P4.23 is attached to the end of the boom on an excavator. The shear is used for cutting steel pipe and I-beams during demolition work. Hydraulic cylinder AB exerts an 18 kN force on the upper jaw.

(a) Complete the free body diagram for the upper jaw, which has been only partially drawn.

(b) Determine the cutting force F being applied to the pipe.

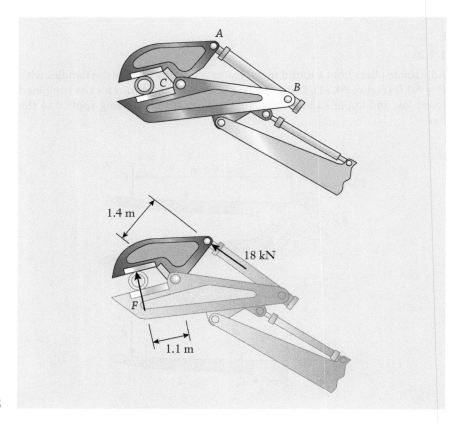

Figure P4.23

P4.24

A cross section of the original design for the double-decker skyways in the Kansas City Hyatt Regency hotel is shown, along with the forces acting on the nuts and washers that support the upper and lower walkways (Figure P4.24).

(a) Draw free body diagrams of the upper and lower walkways, including the weight w that acts on each.

(b) Determine the forces P_1 and P_2 between the washers and the walkways, and the tensions T_1 and T_2 in the hanger rod.

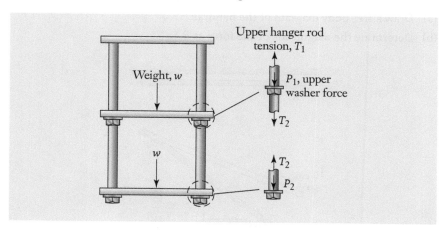

Figure P4.24

P4.25

A cross-sectional view is shown of the skyways in the Kansas City Hyatt Regency hotel in their as-constructed form, along with the forces acting on the nuts and washers that contact the two walkways (Figure P4.25).

(a) Draw free body diagrams of the upper and lower walkways, including the weight w that acts on each.

(b) Determine the forces P_1, P_2, and P_3 between the washers and the walkways, and the tensions T_1, T_2, and T_3 in the hanger rods.

(c) The skyways' collapse was associated with an excessive force P_1. How does the value you calculated here compare with the value of P_1 obtained in P4.24?

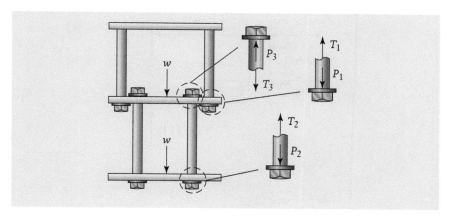

Figure P4.25

P4.26

A handrail, which weighs 120 N and is 1.8 m long, was mounted to a wall adjacent to a small set of steps (Figure P4.26, see page 166). The support at

A has broken, and the rail has fallen about the loose bolt at *B* so that one end now rests on the smooth lower step.

(a) Draw a free body diagram of the handrail.

(b) Determine the magnitude of the force at *B*.

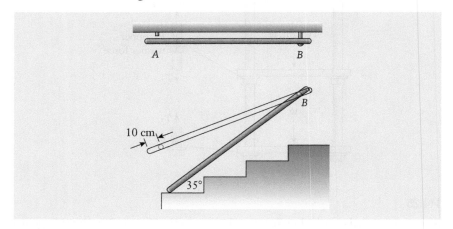

Figure P4.26

P4.27

A multipurpose utility tool grips a cotter pin at *A* while 15 lb forces are applied to the handles (Figure P4.27).

(a) Complete the free body diagram of the combined upper jaw and lower handle assembly, which has been only partially drawn.

(b) Calculate the force acting at *A*.

(c) Alternatively, how much greater would the force be if the pin was being cut at *B*?

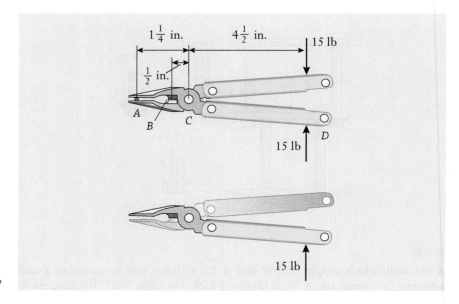

Figure P4.27

P4.28

The rolling-element bearings in a pillow-block bearing are contained within the housing block, which in turn can be bolted to another surface. Two radial forces act on the pillow-block bearing as shown (Figure P4.28).

(a) Can the value of F be adjusted so that the resultant of the two forces is zero?

(b) If not, for what value will the resultant be minimized?

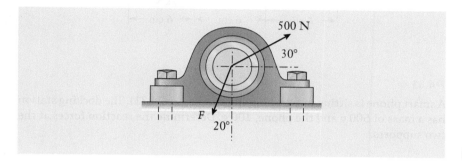

Figure P4.28

P4.29

Horizontal and vertical forces act on the pillow-block bearing as it supports a rotating shaft (Figure P4.29). Determine the magnitude of the resultant force and its angle relative to horizontal. Is the resultant force on the bearing a thrust or radial force?

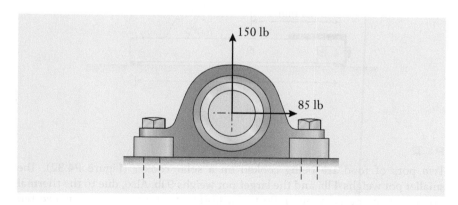

Figure P4.29

P4.30

Ball bearings support a shaft at points A and B (Figure P4.30, see on page 168). The shaft is used to transmit power between two V-belts that apply forces of 1 kN and 1.4 kN to the shaft. Determine the magnitudes and directions of the forces acting on the bearings.

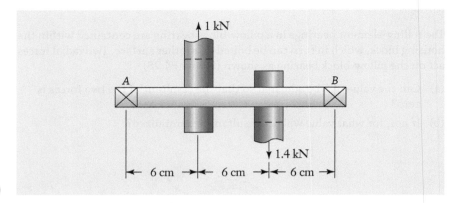

Figure P4.30

P4.31

A smart phone is sitting in a docking station (Figure P4.31). The docking station has a mass of 500 g and the phone, 100 g. Determine the reaction forces at the two supports.

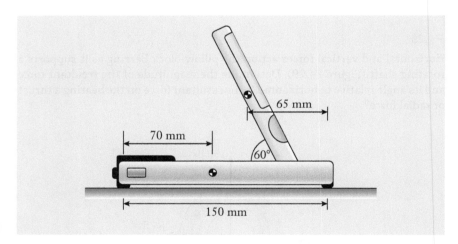

Figure P4.31

P4.32

Two pots of food are being cooked on a solar cooker (Figure P4.32). The smaller pot weighs 4 lb, and the larger pot weighs 9 lb. Also, due to the thermal expansion of the parabolic reflector, a horizontal force of 0.5 lb is exerted outward on the two supports. Determine the magnitude of the resultant force at the two supports, A and B, and the angle of each resultant force relative to horizontal.

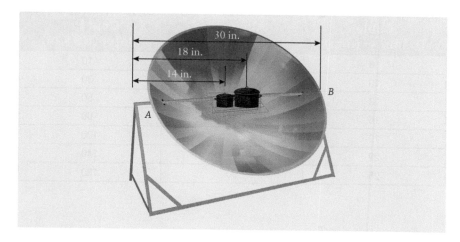

P4.33

Find an example of a structure or machine that has several forces acting on it.

(a) Make a clear labeled drawing of it.

(b) Estimate the dimensions and the magnitudes and directions of the forces that act on it. Show these on your drawing. Briefly explain why you estimate the dimensions and forces to have the numerical values that you assigned.

(c) Using a method of your choice, calculate the resultant of the system of forces.

P4.34*

Many worldwide cities are in locations where extreme weather conditions can damage critical urban infrastructures. Select a representative at-risk city and design a structural system to shield the city. As a group, develop a set of design requirements and at least ten different design concepts. Using your requirements as the criteria, select the top two concepts. For these two concepts, estimate the worst-case loading conditions and draw a free body diagram for them. Which concept do you think is better in the worst-case conditions and why?

P4.35*

Considering both safety and cost, determine the best cable option from Table P4.35 (see page 170) to support a cellular tower of a given height, H = 30 m, and maximum horizontal force, F = 20 kN (see Figure P4.35 on page 170). Also specify the radius R (to the nearest meter) of the supports at point B, C, and D measured from the base of the structure; the diameter of the steel cable chosen; and the total cost of the cable used. Present your approach, solution, and discussion in a formal report. Note the following starting assumptions your group should make (you will most likely have more assumptions):

• The force remains horizontal and acts at the top of the tower.
• The center of the base of the tower acts as the origin, O.

Cable Diameter (mm)	Allowable Load (kN)	Cost Per Meter (US$/m)
6	6	10
10	13	30
13	20	50
16	35	90
19	50	100
22	70	140
25	90	180

Table P4.35

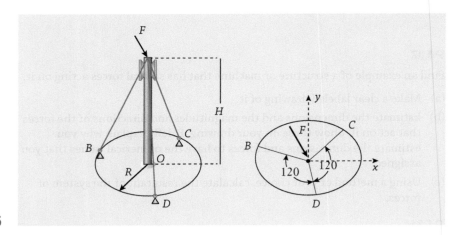

Figure P4.35

- The direction of the force in Figure P4.35 is arbitrary. Your cable structure should be designed to withstand this force acting at any angle on the tower.
- The cables support the tower only in tension. Otherwise, they are slack.
- The tower and supports are all on a flat horizontal plane.
- Supports B, C, and D should all be located at a distance, R, from the base and equally spaced from each other.
- Neglect the weight of the cables.

References

Meriam, J. L., and Kraige, L. G., *Engineering Mechanics: Statics*, 5th ed. Hoboken, NJ: Wiley, 2002.

Pytel, A., and Kiusalaas, J., *Engineering Mechanics: Statics*, 3rd ed., Mason, Ohio: Cengage, 2010.

Roddis, W. M. K., "Structural Failures and Engineering Ethics," *ASCE Journal of Structural Engineering*, 119(5), 1993, pp. 1539–1555.

CHAPTER 5

Materials and Stresses

- Identify circumstances in which a mechanical component is loaded in tension, compression, or shear, and calculate the stress present.

- Sketch a stress–strain curve, and use it to describe how a material responds to the loads applied to it.

- Explain the meaning of the material properties known as the elastic modulus and yield strength.

- Understand the differences between elastic and plastic responses of materials and between their ductile and brittle behaviors.

- Discuss some of the properties and uses for metals and their alloys, ceramics, polymers, and composite materials.

- Apply the concept of a factor of safety to the design of mechanical components that are subjected to tension or shear stress.

▶ 5.1 Overview

As one of their responsibilities, mechanical engineers design hardware so that it won't break when used and so that it can carry the forces acting on it reliably and safely. As an example, consider Boeing's 787 Dreamliner, which weighs up to 550,000 lb when fully loaded. When the airplane is parked on the ground, its weight is supported by the landing gear and wheels. During flight, the aircraft's wings create an upward lift force that exactly balances the weight. Each wing, therefore, carries a force that is equal to half of the airplane's weight, which in this case is equivalent to some 90 family-sized automobile sedans. Subjected to the lift force, the wings bend upward, and, if the flight happens to encounter rough weather, the wings will bend up and down by an additional considerable amount as the plane is buffeted by turbulence. When engineers selected the aircraft's materials, they took into account the facts that an aircraft's wings are subjected to large forces, that they sag under their self-weight, and that they bend upward in response to the lift forces. The wings are designed to be strong, safe, and reliable while being no heavier than necessary to meet the design requirements.

By applying the properties of force systems as described in Chapter 4, you have seen how to calculate the magnitudes and directions of forces that act on certain structures and machines. Knowing those forces alone, however, is not enough information to decide whether a certain piece of hardware will be strong enough and not fail in its task. By "fail" or "failure," we not only mean that the hardware will not break, but also that it will not stretch or bend so much as to become significantly distorted. A 5 kN force, for instance, might be large enough to break a small bolt or to bend a shaft so much that it would wobble and not spin smoothly. A larger-diameter shaft or one that is made from a higher grade material, on the other hand, might very well be able to support that force without incurring any damage.

With those thoughts in mind, you can see that the circumstances for a mechanical component to break, stretch, or bend depend not only on the forces applied to it, but also on its dimensions and the properties of the material from which it is made. Those considerations give rise to the concept of *stress* as a measure of the intensity of a force applied over a certain area. Conversely, the *strength* of a material describes its ability to support and withstand the stress applied to it. Engineers compare the stress present in a component to the strength of its material in order to determine whether the design is satisfactory. As an example, the broken crankshaft shown in Figure 5.1 was removed from a single-cylinder internal-combustion engine. This failure was accelerated by the presence of sharp corners in the shaft's rectangular keyway, which is used to transfer torque between the shaft and a gear or pulley. The spiral shape of the fracture surface indicates that the shaft had been overloaded by a high torque prior to breaking. Engineers are able to combine their knowledge of forces, materials, and dimensions in order to learn from past failures and to improve and evolve the design of new hardware.

In this chapter, we will discuss some of the properties of engineering *materials* and examine the *stresses* that can develop within them. Within these topics lies the discipline known as solid mechanics, and they fit into the hierarchy of mechanical engineering topics shown in Figure 5.2. Tension,

Stress

Strength

Element 4: Materials and stresses

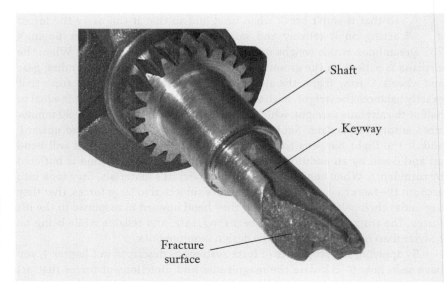

Figure 5.1

A broken crankshaft that was removed from a single-cylinder internal-combustion engine.

Image courtesy of the authors.

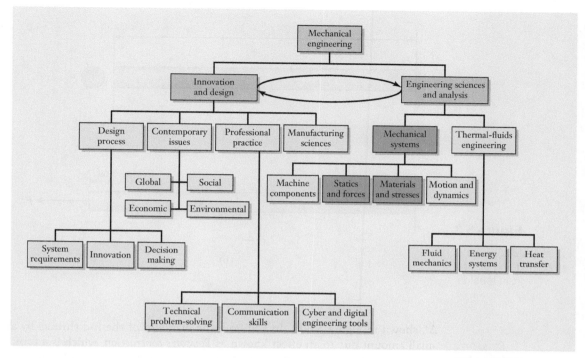

Figure 5.2

Relationship of the topics emphasized in this chapter (shaded blue boxes) relative to an overall program of study in mechanical engineering.

compression, and shear stress are quantities that engineers calculate when they relate the dimensions of a mechanical component to the forces acting on it. Those stresses are then compared to the material's physical properties to determine whether failure is expected to occur. When the strength exceeds the stress, we expect that the structure or machine component will be able to carry the forces without incurring damage. Engineers conduct these types of force, stress, materials, and failure analyses while they design products.

▶ 5.2 Tension and Compression

The type of stress that is most readily visualized and useful for you to develop intuition about materials and stresses is called tension and compression. Figure 5.3 (see page 174) shows a built-in round rod that is held fixed at its left end and then later placed in tension by the force F that pulls on the right end. Before the force is applied, the rod has original length L, diameter d, and cross-sectional area

$$A = \pi \frac{d^2}{4} \tag{5.1}$$

Engineers usually calculate the cross-sectional area of round rods, bolts, and shafts in terms of their diameter rather than their radius r ($A = \pi r^2$) since it's more practical to measure the diameter of a shaft using a caliper gauge. As the force F is gradually applied, the rod stretches along its length by the amount

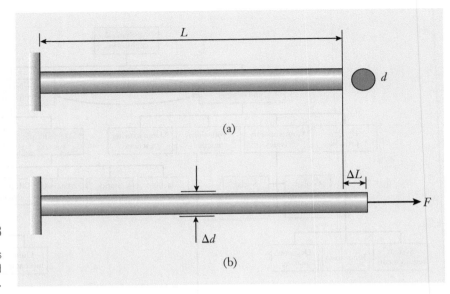

Figure 5.3

A straight rod is
stretched and placed
in tension.

ΔL shown in Figure 5.3(b). In addition, the diameter of the rod shrinks by a
small amount due to an effect known as *Poisson's contraction*, which is a topic
described in the next section. In any event, the change in diameter Δd is smaller
and usually less noticeable than the lengthwise stretch ΔL. To gauge the relative
amounts of ΔL and Δd, try stretching a rubber band to notice how its length,
width, and thickness change.

    If the force is not too great, the rod will return to its original diameter and
length (just like a spring) when F is removed. If the rod is not permanently
deformed after F has been applied, the stretching is said to occur *elastically*.
Alternatively, the force could have been great enough that the rod would
be *plastically* deformed, meaning that, when the force was applied and then
removed, the rod would be longer than it was initially. You can experiment with
a desktop paper clip to see firsthand the difference between the elastic and
plastic behavior of materials. Bend one end of the paper clip a small amount—
perhaps just a millimeter or two—and notice how it springs back to the original
shape when you release it. On the other hand, you could unwrap the paper clip
into a nearly straight piece of wire. In that case, the paper clip does not spring
back. The force was large enough to permanently change the material's shape
through plastic deformation.

    Although the force is applied at only one end of the rod in our example,
its influence is felt at each cross section along the rod's length. As shown in
Figure 5.4, imagine slicing through the rod at some interior point. The segment
that is isolated in the free body diagram of Figure 5.4(b) shows F applied to
the rod's right-hand end and an equivalent *internal force* on the segment's
left-hand end that balances F. Such must be the case; otherwise, the segment
shown in the free body diagram would not be in equilibrium. The location of our
hypothetical slice through the rod is arbitrary, and we conclude that a force of
magnitude F must be carried by the rod at each of its cross sections.

Poisson's contraction

Elastic behavior

Plastic behavior

Internal force

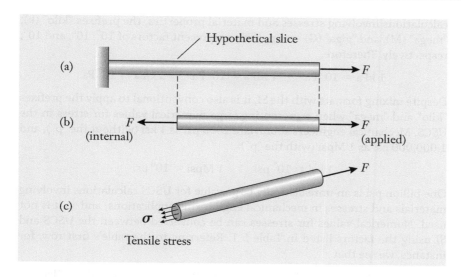

Figure 5.4

(a) A rod that has been stretched. (b) A section that is sliced from the rod to expose the internal force. (c) The tensile stress that is distributed over the rod's cross section.

Since the rod is formed of a continuous solid material, we do not realistically expect that the internal force will be concentrated at a point as depicted by the force vector arrow in Figure 5.4(b). Instead, the influence of the force will be spread out and smeared over the rod's cross section; this process is the basic idea behind stresses in mechanical components. *Stress* is essentially an internal force that has been distributed over the area of the rod's cross section [Figure 5.4(c)], and it is defined by the equation

Stress

$$\sigma = \frac{F}{A} \tag{5.2}$$

Like the force F, the stress σ (the lowercase Greek character sigma) is oriented perpendicularly to a hypothetical slice made through the cross section. When the stress tends to lengthen the rod, it is called *tension*, and $\sigma > 0$. On the other hand, when the rod is shortened, the stress is called *compression*. In that case, the direction of σ in Figure 5.4 reverses to point inward, and $\sigma < 0$. The orientations of forces in tension and compression are shown in Figure 5.5.

Tension
Compression

Similar to the pressure within a liquid or gas, stress is also interpreted as a force that has been distributed over an area. Therefore, stress and pressure have the same dimensions. In the SI, the derived unit of stress is the pascal (1 Pa = 1 N/m²), and the dimension pound per square inch (1 psi = 1 lb/in²) is used in the USCS. Because large numerical quantities frequently arise in

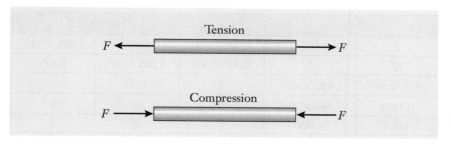

Figure 5.5

Directions of tensile and compressive forces.

calculations involving stresses and material properties, the prefixes "kilo" (k), "mega" (M), and "giga" (G) are applied to represent factors of 10^3, 10^6, and 10^9, respectively. Therefore,

$$1 \text{ kPa} = 10^3 \text{ Pa} \qquad 1 \text{ MPa} = 10^6 \text{ Pa} \qquad 1 \text{ GPa} = 10^9 \text{ Pa}$$

Despite mixing formats with the SI, it is also conventional to apply the prefixes "kilo" and "mega" when representing large numerical values for stress in the USCS. Mechanical engineers abbreviate 1000 psi as 1 ksi (without the "p"), and 1,000,000 psi as 1 Mpsi (with the "p"):

$$1 \text{ ksi} = 10^3 \text{ psi} \qquad 1 \text{ Mpsi} = 10^6 \text{ psi}$$

One billion psi is an unrealistically large value for USCS calculations involving materials and stresses in mechanical engineering applications, and so it is not used. Numerical values for stresses can be converted between the USCS and SI using the factors listed in Table 5.1. Referring to the table's first row, for instance, we see that

$$1 \text{ psi} = 10^{-3} \text{ ksi} = 6.895 \times 10^3 \text{ Pa} = 6.895 \text{ kPa} = 6.895 \times 10^{-3} \text{ MPa}$$

While stress is related to the intensity of a force's application, the engineering quantity called strain measures the amount that the rod stretches. *Elongation* The *elongation* ΔL in Figure 5.3 is one way to describe how the rod lengthens when F is applied, but it is not the only way, nor is it necessarily the best. If a second rod has the same cross-sectional area but is only half as long, then according to Equation (5.2), the stress within it would be the same as in the first rod. However, we expect intuitively that the shorter rod would stretch by a smaller amount. To convince yourself of this principle, hang a weight from two different lengths of rubber band, and notice how the longer band stretches more. Just as stress is a measure of force per unit area, the quantity called *Strain* *strain* is defined as the amount of elongation that occurs per unit of the rod's original length. Strain (the lowercase Greek character epsilon) is calculated from the expression

$$\varepsilon = \frac{\Delta L}{L} \tag{5.3}$$

Because the length dimensions cancel out in the numerator and denominator, strain is a dimensionless quantity. Strain is generally very small, and it can be expressed either as a decimal quantity (for instance, $\varepsilon = 0.005$) or as a percentage ($\varepsilon = 0.5\%$).

Table 5.1

Conversion Factors between USCS and SI Units for Stress

psi	ksi	Pa	kPa	MPa
1	10^{-3}	6.895×10^3	6.895	6.895×10^{-3}
10^3	1	6.895×10^6	6.895×10^3	6.895
1.450×10^{-4}	1.450×10^{-7}	1	10^{-3}	10^{-6}
0.1450	1.450×10^{-4}	10^3	1	10^{-3}
145.0	0.1450	10^6	10^3	1

Example 5.1 | *Hanger Rod in the Kansas City Hyatt Regency Hotel's Skyway*

In 1981, the Hyatt Regency Hotel in Kansas City, Missouri, was a one-year-old facility that included a 40-story tower and a spacious, four-story, open-air atrium. Suspended from the ceiling and hanging above the main lobby area, three floating walkways (called skyways) enabled guests on the hotel's first few floors to view and enjoy the expansive lobby from above. One Friday evening, during the party being held in the hotel's atrium, the connections that supported the fourth-floor's walkway suddenly broke. The entire skyway structure, comprising some 100,000 lb of debris, fell to the crowded atrium below, killing 114 people. During the investigation of the collapse, the connection between the hanger rod and the upper walkway was found to have broken at a load of approximately 20,500 lb. In the units of ksi, determine the stress in the 1.25 in. diameter hanger rod when the tension was 20,500 lb.

Approach

The hanger rod carries a tension force in the manner shown in Figure 5.3 with $F = 2.05 \times 10^4$ lb and $d = 1.25$ in. We are to calculate the stress by applying Equation (5.2).

Solution

By using Equation (5.1), the cross-sectional area of the hanger rod is

$$A = \frac{\pi(1.25 \text{ in.})^2}{4} \qquad \leftarrow \left[A = \frac{\pi d^2}{4}\right]$$

$$= 1.227 \text{ in}^2$$

The tensile stress is

$$\sigma = \frac{2.05 \times 10^4 \text{ lb}}{1.227 \text{ in}^2} \qquad \leftarrow \left[\sigma = \frac{F}{A}\right]$$

$$= 1.670 \times 10^4 \frac{\text{lb}}{\text{in}^2}$$

$$= 1.670 \times 10^4 \text{ psi}$$

where we have substituted the definition of the derived unit psi (pound per square inch) for stress from Table 3.5. It is conventional to apply the abbreviation "ksi" to represent the factor of 1000 psi in a more compact form:

$$\sigma = (1.670 \times 10^4 \text{ psi})\left(10^{-3} \frac{\text{ksi}}{\text{psi}}\right)$$

$$= 16.70 \, \cancel{(\text{psi})}\left(\frac{\text{ksi}}{\cancel{\text{psi}}}\right)$$

$$= 16.70 \text{ ksi}$$

Example 5.1 | *continued*

Discussion
As we'll see in the following section, this level of stress is not particularly high when compared to the strength of steel materials. Although the tension may not have been sufficient to break the hanger rod, it was great enough to destroy the connection between the rod and the pedestrian walkway at the Hyatt Hotel. As was the case in the skyway's collapse, it is not an uncommon situation in mechanical engineering for the connections between components to be weaker than the components themselves.

$$\sigma = 16.70 \text{ ksi}$$

Example 5.2 | *Bolt Clamp*

The U-bolt is used to attach the body (formed with I-beam construction) of a commercial moving van to its chassis (formed from hollow box channel). (See Figure 5.6.) The U-bolt is made from a 10 mm diameter rod, and the nuts on it are tightened until the tension in each straight section of the U-bolt is 4 kN. (a) Show how forces are transferred through this assembly by drawing free body diagrams of the U-bolt and its nuts, the body and chassis stack, and the clamping plate. (b) In the units of MPa, calculate the tensile stress in a straight section of the U-bolt.

Figure 5.6

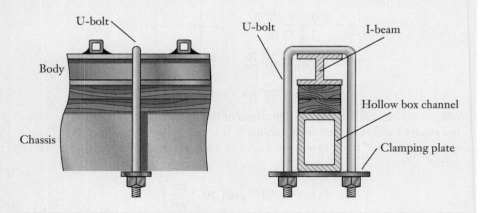

Approach
In part (a), we will isolate three free bodies, and each must be in equilibrium: the U-bolt and its nuts, the body and chassis stack, and the

Example 5.2 | *continued*

clamping plate. The forces acting within the assembly will be equal in magnitude but oppositely directed on adjacent components. The straight sections of the U-bolt are placed in tension in the manner shown in Figure 5.3 with $F = 4$ kN and $d = 10$ mm. We will calculate the stress by using Equation (5.2).

Solution

(a) Because each of the two straight sections of the U-bolt carries 4 kN of tension, their 8 kN resultant is transferred as compression to the body and chassis stack. (See Figure 5.7.) The 8 kN load is likewise applied by the box channel to the clamping plate. The equal and oppositely directed 4 kN forces act between the clamping plate and the nuts threaded onto the U-bolt.

Figure 5.7

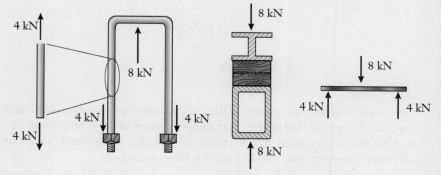

Straight section   U-bolt and nuts   Body and chassis stack   Clamping plate

(b) The cross-sectional area of the bolt is

$$A = \frac{\pi(10 \text{ mm})^2}{4} \qquad \leftarrow \left[A = \frac{\pi d^2}{4} \right]$$

$$= 78.54 \text{ mm}^2$$

which must be converted to dimensionally consistent units for stress in the SI:

$$A = (78.54 \text{ mm}^2) \left(10^{-3} \frac{\text{m}}{\text{mm}} \right)^2$$

$$= 7.854 \times 10^{-5} (\text{mm}^2) \left(\frac{\text{m}^2}{\text{mm}^2} \right)$$

$$= 7.854 \times 10^{-5} \text{m}^2$$

Example 5.2 | *continued*

The tensile stress becomes

$$\sigma = \frac{4000\,\text{N}}{7.854 \times 10^{-5}\,\text{m}^2} \quad \leftarrow \left[\sigma = \frac{F}{A} \right]$$

$$= 5.093 \times 10^7 \frac{\text{N}}{\text{m}^2}$$

$$= 5.093 \times 10^7\,\text{Pa}$$

where we have applied the definition of the derived unit pascal from Table 3.2. The SI prefix "mega" (Tables 3.3 and 5.1) condenses the large power-of-ten exponent in order to write the result in a more conventional form:

$$\sigma = (5.093 \times 10^7\,\text{Pa})\left(10^{-6} \frac{\text{MPa}}{\text{Pa}} \right)$$

$$= 50.93\,(\text{Pa})\left(\frac{\text{MPa}}{\text{Pa}} \right)$$

$$= 50.93\,\text{MPa}$$

Discussion

This is a significant tensile stress in the rod's straight sections; this makes sense because the applied load is large. In the upper portion of the U-bolt where the rod has 90° corners and is in contact with the I-beam, a more complicated state of stress is present, which would require a different analysis.

$$\sigma = 50.93\,\text{MPa}$$

Focus On

EXTREME ENVIRONMENTS

Many innovative products are successful because of their structural integrity in extreme environments. When designing products for use in extreme environments, engineers must consider many sources of applied loads to simulate accurate use conditions. For example, while the devastation of the 2011 tsunami disaster in Japan still impacts the entire country, innovative products are being created to help victims in future disasters. In Figure 5.8(a), a "Jinriki" is shown, which turns a wheelchair into a modern rickshaw and allows for much easier transport up and down stairs and over snow, sand, or mud. The additional forces on the handles when pulling someone through an extreme environment must be considered when designing the shape, size,

(a)

Figure 5.8

(a) Custom-made
handles that allow
wheelchairs to be
pulled over adverse
environments. (b) A
telemark ski binding
from Bishop Bindings
with enhanced
structural and
material performance.

Courtesy of Wilderness Inquiry; Courtesy
of Bishop Bindings LLC

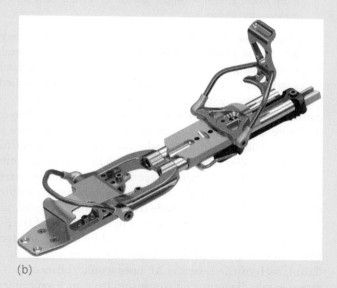

(b)

and materials for the handles. In Figure 5.8(b), an innovative telemark ski binding, designed by two mechanical engineers in Colorado includes a solid stainless steel main pivot to eliminate the structural failure of the previous design which used two threaded components. Also, a decision to increase the wall thickness in a particular area of a critical component reduced the stress in that area by 30%, according to model simulations (finite element analysis). This was critical because it significantly increased the life of the high strength aluminum being used for this component. Whether it is commercial space travel, natural disaster mitigation, biomedical device development, or high-performance sporting, these extreme environments require mechanical engineers to consider many forms of loading, stress, and failure of the materials they choose for the product components.

▶ 5.3 Material Response

T he definitions of stress and strain, in contrast to force and elongation, are useful because they are scaled with respect to the rod's size. Imagine conducting a sequence of experiments with a collection of rods made of identical material but having various diameters and lengths. As each rod is pulled in tension, the force and elongation would be measured. In general, for a given level of force, each rod would stretch by a different amount because of the variations in diameter and length.

For each individual rod, however, the applied force and elongation would be proportional to one another, following

$$F = k\Delta L \tag{5.4}$$

where the parameter k is called the *stiffness*. This observation is the basis of the concept known as *Hooke's law*. In fact, British scientist Robert Hooke wrote in 1678 that

the power of any spring is in the same proportion with the tension thereof; that is, if one power stretch or bend it one space, two will bend it two, and three will bend it three, and so forward.

Note that Hooke used the term "power" for what we today call "force." In this respect, any structural component that stretches or bends can be viewed as a spring having stiffness k, even though the component itself might not necessarily look like a "spring" in the sense of being a coiled wire helix.

Continuing with our hypothetical experiment, we next imagine constructing a graph of F versus ΔL for each of the different rods. As indicated in Figure 5.9(a), the lines on these graphs would have different slopes (or stiffnesses) depending on the values of d and L. For a given force, longer rods and ones with smaller cross sections stretch more than the other rods. Conversely, shorter rods and ones with larger cross-sectional areas stretch less. Our graph would show a number of straight lines, each having a different slope. Despite the fact that the rods are made from the same material, they would appear to be quite different from one another in the context of the F-versus-ΔL graph.

On the other hand, the rods would behave in an identical manner when their stretching is instead described by stress and strain. As shown in Figure 5.9(b), each of the F-versus-ΔL lines would collapse onto a single line in the stress–strain diagram. Our conclusion from such experiments is that while the stiffness depends on the rod's dimensions, the stress–strain relationship is a property of the material alone and is independent of the test specimen's size.

Figure 5.10 shows an idealized *stress–strain curve* for a typical structural-quality steel. The stress–strain diagram is broken down into two regions: the low-strain *elastic region* (where no permanent set remains after the force has been applied and removed) and the high-strain *plastic region* (where the force is large enough that, upon removal, the material has permanently elongated). For strains below the *proportional limit* (point A), you can see from the diagram that stress and strain are proportional to one another and that they therefore satisfy the relationship

$$\sigma = E\varepsilon \tag{5.5}$$

Stiffness
Hooke's law

Stress–strain curve

Elastic and plastic regions

Proportional limit

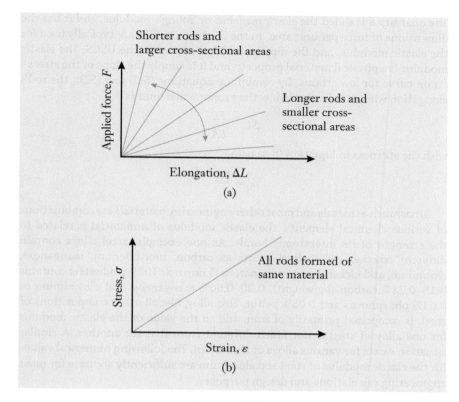

Figure 5.9

(a) Force–elongation behaviors of rods having various cross-sectional areas and lengths. (b) Each rod has similar stress–strain behavior.

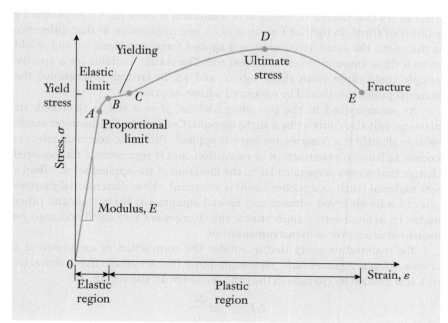

Figure 5.10

Idealized stress–strain curve for structural steel.

Elastic modulus   The quantity E is called the *elastic modulus*, or Young's modulus, and it has the dimensions of force per unit area. In the SI, the units GPa are typically used for the elastic modulus, and the dimension Mpsi is used in the USCS. The elastic modulus is a physical material property, and it is simply the slope of the stress–strain curve for low strains. By combining Equations (5.2) and (5.3), the rod's elongation when it is loaded below the proportional limit is

$$\Delta L = \frac{FL}{EA} \tag{5.6}$$

with the stiffness in Equation (5.4) being

$$k = \frac{EA}{L} \tag{5.7}$$

Inasmuch as metals and most other engineering materials are combinations of various chemical elements, the elastic modulus of a material is related to the strength of its interatomic bonds. As one example, steel alloys contain different fractions of such elements as carbon, molybdenum, manganese, chromium, and nickel. A common material known as 1020 grade steel contains 0.18–0.23% carbon (by weight), 0.30–0.60% manganese, and a maximum of 0.04% phosphorus and 0.05% sulfur. This alloy, like all other compositions of steel, is comprised primarily of iron, and so the value of the elastic modulus for one alloy of steel is not much different from that for another. A similar situation exists for various alloys of aluminum. The following numerical values for the elastic modulus of steel and aluminum are sufficiently accurate for most engineering calculations and design purposes:

$$E_{steel} \approx 210 \text{ GPa} \approx 30 \text{ Mpsi}$$

$$E_{aluminum} \approx 70 \text{ GPa} \approx 10 \text{ Mpsi}$$

You can see that the elastic modulus of aluminum is lower than that of steel by a factor of three. In light of Equation (5.6), one implication of that difference is that, with the same dimensions and applied force, an aluminum rod would stretch three times as much as a steel rod. The elastic modulus for a specific sample could differ from these values, and so, in critical applications, the material properties should be measured whenever practical.

As we described in the preceding section, after a rod is stretched, its diameter will also contract by a slight amount. Conversely, the diameter would enlarge slightly if a compressive force is applied. This cross-sectional effect is known as Poisson's contraction or expansion, and it represents a dimensional change that occurs perpendicular to the direction of the applied force. When a soft material (such as a rubber band) is stretched, those dimensional changes can often be observed without any special equipment. For metals and other materials in engineering applications, the changes are very small and must be measured using precision instrumentation.

Poisson's ratio   The material property that quantifies the contraction or expansion of a cross section is *Poisson's ratio*, represented by ν (the lowercase Greek character nu). It is defined by changes in the rod's diameter Δd and length ΔL:

$$\Delta d = -\nu d \frac{\Delta L}{L} \tag{5.8}$$

The negative sign in this equation sets the sign convention that tension (with $\Delta L > 0$) causes the diameter to contract ($\Delta d < 0$), and compression causes the diameter to expand. For many metals, $\nu \approx 0.3$, with numerical values generally falling in the range 0.25–0.35.

Returning to our discussion of the stress–strain diagram of Figure 5.10, point B is called the *elastic limit*. For loading between points A and B, the material continues to behave elastically, and it will spring back upon removal of the force, but the stress and strain are no longer proportional. As the load increases beyond B, the material begins to show a permanent set. *Yielding* starts to occur in the region between B and C; that is, even for small changes in stress, the rod experiences a large change in strain. In the yielding region, the rod stretches substantially even as the force grows only slightly because of the shallow slope in the stress–strain curve. For that reason, the onset of yielding is often taken by engineers as an indication of failure. The value of stress in region B–C defines the material property called the *yield strength*, S_y. As the load is even further increased beyond point C, the stress grows to the *ultimate strength*, S_u at point D. That value represents the largest stress that the material is capable of sustaining. As the test continues, the stress in the figure actually decreases, owing to a reduction in the rod's cross-sectional area, until the sample eventually fractures at point E.

Stress–strain curves are measured on a device called a *materials testing machine*. Figure 5.11 shows an example of equipment in which a computer both controls the test stand and records the experimental data. During tension

Elastic limit

Yielding

Yield strength
Ultimate strength

Materials testing machine

Figure 5.11

The engineer is using a materials testing machine to bend a metal bar between two supports. The computer controls the experiment and records the force and deflection data.

Photo Courtesy of MTS Systems Corporation.

testing, a specimen such as a rod of steel is clamped between two jaws that are gradually pulled apart, placing the specimen in tension. A load cell that measures force is attached to one of the jaws, and a second sensor (called an extensometer) measures how much the sample is stretched. A computer records the force and elongation data during the experiment, and those values are then converted to stress and strain by using Equations (5.2) and (5.3). When the stress–strain data is drawn on a graph, the slope in the low-strain region is measured to determine E, and the value of the stress S_y at the yield point is read off the curve.

A stress–strain diagram for a sample of structural steel is depicted in Figure 5.12. The stress is shown in the USCS dimensions of ksi, and the dimensionless strain is shown as a percentage. We can use this diagram to

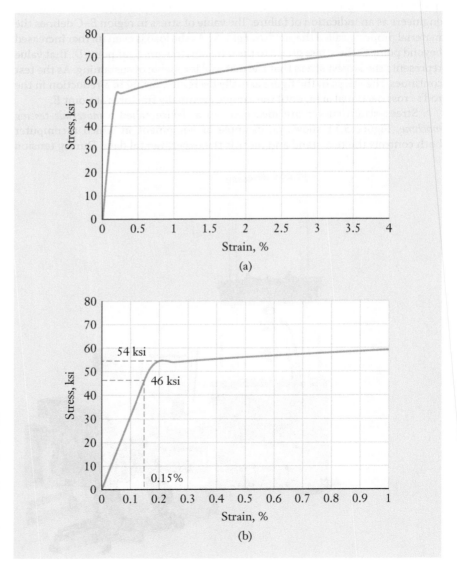

Figure 5.12

A measured stress–strain curve for a sample of structural steel (a) shown over a wide range of strain, and (b) magnified to highlight the low-strain proportional and yielding regions.

determine the elastic modulus and yield strength of this sample. The relation between σ and ε is nearly straight for small strains (up to about 0.2%), and the modulus E is determined from the slope of the line in that region. The strain was zero when no stress was applied, and, as read off the axes of the graph, the specimen was subjected to 46 ksi at a strain of 0.15%. Following Equation (5.5), the elastic modulus is

$$E = \frac{46 \times 10^3 \text{ psi}}{0.0015} = 3.06 \times 10^7 \text{ psi} = 30.6 \text{ Mpsi}$$

which is quite close to the rule-of-thumb value for steel (30 Mpsi). The yield point is also evident in Figure 5.12(b), and we measure $S_y = 54$ ksi directly from the graph's ordinate.

For aluminum and other nonferrous metals, the sharp corner appearing at the yield point in the stress–strain diagram in Figure 5.10 and the narrow yielding region B–C are generally not seen. Instead, such materials tend to exhibit a smoother and more gradual transition between the elastic and plastic regions. For such cases, a technique called the *0.2% offset method* is instead used to define the yield strength. As an illustration, Figure 5.13 shows a stress–strain curve that was measured for an aluminum alloy on a materials testing machine. The elastic modulus is again determined by drawing a straight line through the origin to match the stress–strain curve in the proportional region. When the stress is 15 ksi, the strain is 0.14%, and the elastic modulus therefore becomes

0.2% offset method

$$E = \frac{15 \times 10^3 \text{ psi}}{0.0014} = 1.1 \times 10^7 \text{ psi} = 11 \text{ Mpsi}$$

which is close to aluminum's nominal value (10 Mpsi). However, unlike the steel alloy of Figure 5.12, the onset of yielding is not sharp and evident. In the 0.2% offset method, the yield stress is determined from the curve's intersection with a line that is drawn at slope E but offset from the origin by 0.2%. In Figure 5.13,

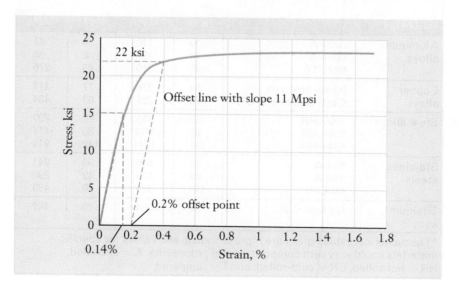

Figure 5.13

A measured stress–strain curve of an aluminum sample. The elastic modulus is determined from the slope of the curve at low strain, and the yield stress is found by using the 0.2% offset yield method.

the straight construction line drawn with slope 11 Mpsi from the offset point intersects the stress–strain diagram at the yield point, and $S_y = 22$ ksi. That value is taken as the level of stress where the material begins to yield appreciably and becomes unacceptable for further use.

Tables 5.2 and 5.3 list material property data for several metals, including the elastic modulus, Poisson's ratio, weight density, and ultimate and yield strengths. The properties of any given metallic sample, however, could differ from those listed in the table. Whenever possible, particularly for applications where failure could result in an injury, the material properties should be measured directly, or the material's supplier should be contacted. Common uses for the metals and alloys listed in Table 5.3 are discussed in Section 5.5.

Table 5.2

Elastic Modulus and Weight Density of Selected Metals*

Material	Elastic Modulus, E		Poisson's Ratio, ν	Weight Density, ρ_w	
	Mpsi	GPa		lb/ft³	kN/m³
Aluminum alloys	10	72	0.32	172	27
Copper alloys	16	110	0.33	536	84
Steel alloys	30	207	0.30	483	76
Stainless steels	28	190	0.30	483	76
Titanium alloys	16	114	0.33	276	43

*The numerical values given are representative, and values for specific materials could vary with composition and processing.

Table 5.3

Ultimate and Yield Strengths of Selected Metals*

Material		Ultimate Strength, S_u		Yield Strength, S_y	
		ksi	MPa	ksi	MPa
Aluminum alloys	3003-A	16	110	6	41
	6061-A	18	124	8	55
	6061-T6	45	310	40	276
Copper alloys	Naval brass-A	54	376	17	117
	Cartridge brass-CR	76	524	63	434
Steel alloys	1020-HR	66	455	42	290
	1045-HR	92	638	60	414
	4340-HR	151	1041	132	910
Stainless steels	303-A	87	600	35	241
	316-A	84	579	42	290
	440C-A	110	759	70	483
Titanium alloy	Commercial	80	551	70	482

*The numerical values given are representative, and values for specific materials could vary with composition and processing. A = annealed, HR = hot-rolled, CR = cold-rolled, and T = tempered.

Example 5.3 | *U-Bolt's Dimensional Changes*

For the 10 mm diameter steel U-bolt in Example 5.2, determine the
(a) strain, (b) change in length, and (c) change in diameter of the
bolt's 325 mm long straight section. Use the rule-of-thumb value
$E = 210$ GPa for the elastic modulus, and take the Poisson's ratio as
$v = 0.3$. (See Figure 5.14.)

Figure 5.14

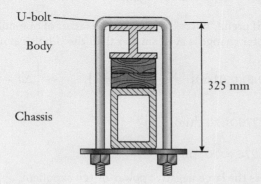

U-bolt

Body

325 mm

Chassis

Approach
The stress in the U-bolt's straight section was previously found in Example
5.2 to be $\sigma = 5.093 \times 10^7$ Pa. Calculate the strain, change in length, and
change in diameter by applying Equations (5.5), (5.3), and (5.8) to parts
(a), (b), and (c), respectively.

Solution
(a) The strain in the straight section is

$$\varepsilon = \frac{5.093 \times 10^7 \, \text{Pa}}{210 \times 10^9 \, \text{Pa}} \qquad \leftarrow \left[\varepsilon = \frac{\sigma}{E} \right]$$

$$= 2.425 \times 10^{-4} \frac{\text{Pa}}{\text{Pa}}$$

$$= 2.425 \times 10^{-4}$$

Because this value is a small dimensionless number, we will write it as
the percentage $\varepsilon = 0.02425\%$.

(b) The change in the U-bolt's length (the elongation) is

$$\Delta L = (2.425 \times 10^{-4})(0.325 \, \text{m}) \qquad \leftarrow [\Delta L = \varepsilon L]$$

$$= 7.882 \times 10^{-5} \, \text{m}$$

Example 5.3 | *continued*

Convert this numerical value to the derived SI unit of microns following the definition in Table 3.2:

$$\Delta L = (7.882 \times 10^{-5}\,\text{m})\left(10^6\,\frac{\mu\text{m}}{\text{m}}\right)$$

$$= 78.82\,(\text{m})\left(10^6\,\frac{\mu\text{m}}{\text{m}}\right)$$

$$= 78.82 \times \mu\text{m}$$

Here the SI prefix "micro" represents the factor of one-millionth.

(c) The diameter change is even smaller than the U-bolt's elongation:

$$\Delta d = -(0.3)(0.01\,\text{m})\left(\frac{7.882 \times 10^{-5}\,\text{m}}{0.325\,\text{m}}\right) \qquad \leftarrow \left[\Delta d = -vd\,\frac{\Delta L}{L}\right]$$

$$= -7.276 \times 10^{-7}\,(\text{m})\left(\frac{\text{m}}{\text{m}}\right)$$

$$= -7.276 \times 10^{-7}\,\text{m}$$

To suppress the large negative power-of-ten exponent, apply the "nano" prefix from Table 3.3:

$$\Delta d = (-7.276 \times 10^{-7}\,\text{m})\left(10^9\,\frac{\text{nm}}{\text{m}}\right)$$

$$= -727.6\,(\text{m})\left(\frac{\text{nm}}{\text{m}}\right)$$

$$= -727.6\,\text{nm}$$

Discussion

The U-bolt's elongation is small indeed; this is expected because the U-bolt is made of steel. The elongation is approximately the same as the diameter of a human hair or slightly larger than the 632.8 nm wavelength of light in a helium–neon laser. The bolt's diameter therefore contracts by an amount only somewhat more than one wavelength of light. Even though the bolt carries a load of 4 kN (approximately 900 lb), the changes in its dimensions are imperceptible to the eye and would require specialized equipment to measure.

$$\varepsilon = 0.02425\%$$

$$\Delta L = 78.82\,\mu\text{m}$$

$$\Delta d = -727.6\,\text{nm}$$

Example 5.4 | *Rod Stretching*

A round rod is made from the steel alloy having the stress–strain characteristics shown in Figure 5.12. (See Figure 5.15.) When it is subjected to 3500 lb of tension (approximately equal to the weight of an automobile sedan), calculate (a) the stress and strain in the rod, (b) the amount that it stretches, (c) its change in diameter, and (d) its stiffness. (e) If the force was only 1000 lb, by what amount would the rod have stretched? Use the value $\nu = 0.3$ for the Poisson's ratio.

Figure 5.15

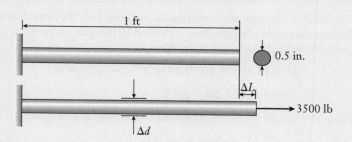

Approach

First calculate the stress by applying Equation (5.2). Then, calculate the strain, change in length, change in diameter, and stiffness by applying Equations (5.5), (5.3), (5.8), and (5.7) respectively.

Solution

(a)   The rod's cross-sectional area is

$$A = \frac{\pi(0.5 \text{ in.})^2}{4} \qquad \leftarrow \left[A = \frac{\pi d^2}{4} \right]$$

$$= 0.1963 \text{ in}^2$$

and the tensile stress is

$$\sigma = \frac{3500 \text{ lb}}{0.1963 \text{ in}^2} \qquad \leftarrow \left[\sigma = \frac{F}{A} \right]$$

$$= 1.783 \times 10^4 \frac{\text{lb}}{\text{in}^2}$$

$$= 1.783 \times 10^4 \text{ psi}$$

The USCS dimension ksi is next used to represent the factor of 1000 psi:

$$\sigma = (1.783 \times 10^4 \text{ psi}) \left(10^{-3} \frac{\text{ksi}}{\text{psi}} \right)$$

$$= 17.83 \, (\text{psi}) \left(\frac{\text{ksi}}{\text{psi}} \right)$$

$$= 17.83 \text{ ksi}$$

Example 5.4 | *continued*

The strain within the rod is given by

$$\varepsilon = \frac{1.783 \times 10^4\,\text{psi}}{30.6 \times 10^6\,\text{psi}} \qquad \leftarrow \left[\varepsilon = \frac{\sigma}{E}\right]$$

$$= 5.825 \times 10^{-4}\left(\frac{\text{psi}}{\text{psi}}\right)$$

$$= 5.825 \times 10^{-4}$$

or 0.05825%. Because stress and elastic modulus have the same units, their dimensions cancel when strain is calculated by using Equation (5.5).

(b) The 3500 lb load elongates the rod by

$$\Delta L = (5.825 \times 10^{-4})(12\ \text{in.}) \qquad \leftarrow [\Delta L = \varepsilon L]$$

$$= 6.990 \times 10^{-3}\ \text{in.}$$

(c) Using $\nu = 0.3$ for steel, the diameter changes by the amount

$$\Delta d = -(0.3)(0.5\,\text{in.})\left(\frac{6.990 \times 10^{-3}\,\text{in.}}{12\ \text{in.}}\right) \qquad \leftarrow \left[\Delta d = -\nu d\frac{\Delta d}{L}\right]$$

$$= -8.738 \times 10^{-5}\,(\text{in.})\left(\frac{\text{in.}}{\text{in.}}\right)$$

$$= -8.738 \times 10^{-5}\,\text{in.}$$

and the negative sign convention indicates that the diameter contracts.

(d) The rod's stiffness is determined from the material's elastic modulus, cross-sectional area, and length:

$$k = \frac{(30.6 \times 10^6\,\text{psi})(0.196\,\text{in}^2)}{12\,\text{in.}} \qquad \leftarrow \left[k = \frac{EA}{L}\right]$$

$$= 5.007 \times 10^5\left(\frac{\text{lb}}{\text{in}^2}\right)(\text{in}^2)\left(\frac{1}{\text{in.}}\right)$$

$$= 5.007 \times 10^5\,\frac{\text{lb}}{\text{in.}}$$

In reconciling the units, we have expanded the stress unit psi as lb/in².

(e) With a force of only 1000 lb, the rod would stretch by

$$\Delta L = \frac{1000\ \text{lb}}{5.007 \times 10^5\,\text{lb/in.}} \qquad \leftarrow \left[\Delta L = \frac{F}{k}\right]$$

$$= 1.997 \times 10^{-3}\,(\text{lb})\left(\frac{\text{in.}}{\text{lb}}\right)$$

$$= 1.997 \times 10^{-3}\ \text{in.}$$

Example 5.4 | *continued*

Discussion
In the USCS, the derived unit mil (Table 3.5) is equivalent to one-thousandth of an inch, and it is convenient for representing small length and diameter changes. A sheet of writing paper is only 3–4 mils thick, and so the rod stretches by about the thickness of two sheets of standard paper, which is expected with the given load and choice of steel for the rod. The diameter contracts by a far smaller amount, less than 0.10 mil. To place that small change in perspective, if the rod's initial diameter had been measured to five significant digits as 0.50000 in., the diameter after extension would be 0.49991 in. Indeed, measuring a change at the fifth decimal place requires sensitive and well-calibrated instrumentation.

$$\sigma = 17.83 \text{ ksi}$$

$$\varepsilon = 0.05825\%$$

$$\Delta L \text{ (at 3500 lb)} = 6.990 \text{ mils}$$

$$\Delta d = -0.08738 \text{ mil}$$

$$k = 5.007 \times 10^5 \text{ lb/in.}$$

$$\Delta L \text{ (at 1000 lb)} = 1.997 \text{ mils}$$

▶ 5.4 Shear

I n Figure 5.4, the tensile stress σ acts along the length of the rod, and it is also oriented perpendicular to the rod's cross section. Roughly speaking, a tensile stress will stretch a mechanical component and tend to pull it apart. However, excessive forces can damage hardware in other ways as well. One example is a shear stress, which develops when a force tends to slice or cut through a structure or machine component.

Shear differs from tension and compression in that the stress is oriented in the same plane as the rod's cross section. That is, a shear stress is associated with a force that acts parallel to the surface of the cross section. Consider the block of elastic material in Figure 5.16 (see page 194) that is being pressed downward and pushed between the two rigid supports. As the force F is applied, the material tends to be sliced, sheared, or cut along the two edges that are marked as *shear planes* in the figure. A free body diagram of the block is shown in Figure 5.16(b), and equilibrium in the vertical direction requires that $V = F/2$. The two forces V are called *shear forces*, and you can see that they lie in the shear planes and are parallel to them.

Shear plane

Shear force

Like tension and compression, shear forces are also spread continuously within material over any cross section that we can imagine making. The shear force V results from the combination of the shear stresses in Figure 5.16(c)

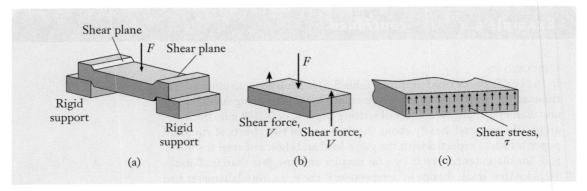

Figure 5.16

Shear forces and stresses act on material that is being pressed between two rigid supports.

acting over the entire exposed area. In this case, the shear stress τ (the lowercase Greek character tau) is defined

$$\tau = \frac{V}{A} \tag{5.9}$$

where A is the exposed area of the cross section.

Shear stress is often associated with connections that are made between components in a structure or machine, including bolts, pins, rivets, welds, and adhesives. Two types of attachments that are seen in practice are known as single and double shear. The terminology refers to the manner in which the shear forces are transmitted between the two objects joined together. Figure 5.17

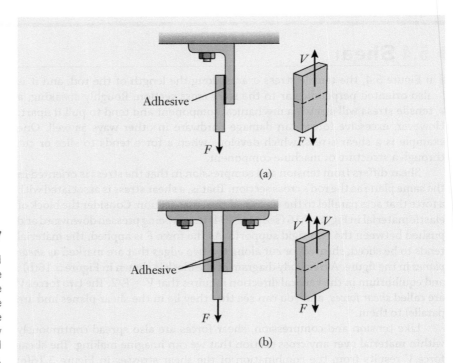

Figure 5.17

Adhesively bonded connections that are placed in (a) single shear and (b) double shear. The adhesive layers are indicated by the heavily weighted lines.

shows these configurations for the illustrative case of adhesively bonded lap joints. By using free body diagrams, we imagine disassembling the pieces to expose the shear forces within the adhesive layers. For the case of *single shear* in Figure 5.17(a), the full load is carried by just one layer of adhesive, and $V = F$. *Double shear* is illustrated in Figure 5.17(b). Because two surfaces share the load, the shear stress is halved and $V = F/2$. Double shear connections transfer shear forces along two planes simultaneously.

Single shear
Double shear

Example 5.5 | *Wire Cutter's Hinge Connection*

In Example 4.6, we found that the wire cutter's hinge pin B must support a 385 N force when the handles are pressed together. If the diameter of the hinge pin is 8 mm, determine the pin's shear stress in the SI dimensions of MPa. (See Figure 5.18.)

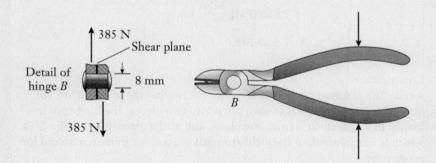

Figure 5.18

Approach
The force transmitted by pin B between one combined jaw/handle, and the other was found in Example 4.6 by applying the requirements of static equilibrium. We now extend that analysis to examine how intensely the hinge pin's material is being loaded. Since the 385 N force is transmitted from one jaw/handle to the other across only one shear plane, the hinge pin is loaded in single shear and $V = 385$ N. Calculate the shear stress by applying Equation (5.9).

Solution
The pin's cross-sectional area is found from Equation (5.1):

$$\Delta L = \frac{\pi (0.008\,\text{m})^2}{4} \quad \leftarrow \left[A = \frac{\pi d^2}{4} \right]$$

$$= 5.027 \times 10^{-5}\,\text{m}^2$$

Example 5.5 | *continued*

The shear stress is

$$\tau = \frac{385\,\text{N}}{5.027 \times 10^{-5}\,\text{m}^2} \qquad \leftarrow \left[\tau = \frac{V}{A}\right]$$

$$= 7.659 \times 10^6 \frac{\text{N}}{\text{m}^2}$$

$$= 7.659 \times 10^6\,\text{Pa}$$

where in the last step we used the definition of the derived unit pascal from Table 3.2. Convert the numerical value to the conventional stress dimensions of MPa as follows:

$$\tau = (7.659 \times 10^6\,\text{Pa})\left(10^{-6}\frac{\text{MPa}}{\text{Pa}}\right)$$

$$= 7.659\,(\text{Pa})\left(\frac{\text{MPa}}{\text{Pa}}\right)$$

$$= 7.659\,\text{MPa}$$

Discussion

As the 385 N force is transmitted from one jaw/handle to the other, the action of that force tends to slice pin B into two pieces. The hinge pin is loaded in single shear, across one plane, and at the intensity of 7.659 MPa, which is much less than the yield strength of steel, a common material for hinge pins.

$$\tau = 7.659\,\text{MPa}$$

Example 5.6 | *Clevis Joint*

The threaded rod is subjected to 350 lb of tension, and that force is transferred through the clevis joint to the fixed base. In the USCS dimensions of ksi, determine the shear stress acting within the $\frac{3}{8}$ in. diameter hinge pin. (See Figure 5.19.)

Example 5.6 | *continued*

Figure 5.19

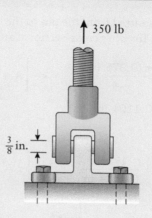

350 lb

$\frac{3}{8}$ in.

Approach
The threaded rod and the bolts attached to the fixed base are loaded
along their length in tension. The clevis joint's hinge pin carries forces
perpendicular to its length, and it is loaded in shear. To determine the
magnitude of the shear force, draw free body diagrams of the clevis joint
to show how forces are carried within it. (See Figure 5.20.)

Figure 5.20

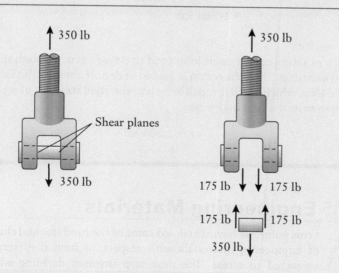

350 lb

Shear planes

350 lb

350 lb

175 lb   175 lb

175 lb   175 lb

350 lb

Example 5.6 | *continued*

Solution

The 350 lb force is transmitted from the pin to the base through two shear planes, and $V = 175$ lb. The pin's cross-sectional area is

$$A = \frac{\pi(0.375 \text{ in.})^2}{4} \quad \leftarrow \left[A = \frac{\pi d^2}{4} \right]$$

$$= 0.1104 \text{ in}^2$$

The shear stress is

$$\tau = \frac{175 \text{ lb}}{0.1104 \text{ in}^2} \quad \leftarrow \left[\tau = \frac{V}{A} \right]$$

$$= 1.584 \times 10^3 \frac{\text{lb}}{\text{in}^2}$$

$$= 1.584 \times 10^3 \text{ psi}$$

where we have used the abbreviation psi to denote lb/in^2. To place the numerical value in conventional form, we convert the stress to the dimension ksi by using Table 5.1:

$$\tau = (1.584 \times 10^3 \text{ psi})\left(10^{-3} \frac{\text{ksi}}{\text{psi}} \right)$$

$$= 1.584 \, (\cancel{\text{psi}})\left(\frac{\text{ksi}}{\cancel{\text{psi}}} \right)$$

$$= 1.584 \text{ ksi}$$

Discussion

The forces acting on the clevis joint tend to cleave or cut through the pin at two locations. The connection is loaded in double shear at the intensity of 1.584 ksi, which again is much less than the yield strength of steel, a common material for hinge pins.

$$\boxed{\tau = 1.584 \text{ ksi}}$$

▶ 5.5 Engineering Materials

At this point, we have discussed some of the fundamental characteristics of engineering materials with respect to how they respond when subjected to stress. The next step involves deciding what type of material should be used in a particular design application. A wide variety of materials is available for engineering products, and choosing the correct ones is an important aspect of the design process. Mechanical engineers select materials in the context of both the product's purpose and the processes that will be used during its manufacture. The main classes of materials encountered in mechanical engineering are:

- Metals and their alloys
- Ceramics
- Polymers
- Composite materials

Electronic materials comprise another class that includes the semiconductors that are used widely in electronic, computer, and telecommunication systems. Devices such as microprocessors and memory chips use metal materials, such as electrical conductors and ceramic materials as insulators.

Engineers select materials based on their performance, cost, availability, and past track record in similar applications. Because the production of engineering materials involves the consumption of natural resources and energy, environmental concerns are also factors involved in the selection process. The more manufacturing steps that are required to produce the material and form it into a final product, the greater the material's cost both economically and environmentally. The full life cycle of a material involves the use of natural resources, for instance, in the form of ores, processing raw materials, fabricating and manufacturing products, using the product, and then either disposing of the product or recycling its materials.

In selecting the materials to be used in a product, an engineer first needs to decide on the class of materials to use. Once the class has been chosen (for instance, metals and their alloys), an engineer next determines which material within the class is most appropriate (for instance, steel or aluminum). Many products are designed to use combinations of different classes of materials, each one best suited for a specific task. Automobiles, for instance, contain about 50–60% steel in the frame, engine, and drivetrain components; 5–10% aluminum for engine and body components; and 10–20% plastics used for trim and interior components. The remaining fraction includes glass for the windows, lead for the battery, rubber for the tires, and other materials.

Metals and Their Alloys

Metals are relatively stiff and heavy materials; in other words, from a technical standpoint, they generally have large values for their elastic modulus and density. The strength of metals can be increased by mechanical and heat treatments and by *alloying*, which is the process of adding small amounts of other carefully chosen elements to a base metal. From a design standpoint, metals are a good choice to use in structures and machines that must carry large forces. On the negative side, metals are susceptible to corrosion, and, as a result, they can deteriorate and weaken over time. Another attractive feature of metals is that many methods exist to make them, shape them, and attach them. Metals are versatile materials because they can be manufactured by casting, extrusion, forging, rolling, cutting, drilling, and grinding.

Some metals, by virtue of their processing and alloys, have high degrees of *ductility*, that is, the ability of a material to withstand a significant amount of

Alloying

Ductility

stretching before it fractures. In the stress–strain curve of Figure 5.10, a ductile material has a wide region over which plastic deformation occurs; the steel that is used in paper clips is a good example of a ductile metal. A brittle material like glass, on the other hand, exhibits essentially no plastic deformation. For obvious reasons, ductile metals are well suited for use in structures and machines because, when they become overloaded, the materials give prior warning by noticeably stretching or bending before they break.

Metals include a number of important alloys such as aluminum, copper, steel, and titanium (Table 5.3):

- The 3003 grade aluminum alloy is often produced in the form of wide flat sheets, and it can be easily bent and shaped to form boxes and covers for electronic equipment, among other products. Used for machined mechanical components that are subjected to moderate forces, the 6061 alloy is available either annealed (A) or tempered (T6). Annealing and tempering are processing steps that involve heat treatment to improve the material's strength.

- Copper alloys include brasses (which are yellowish alloys of copper and zinc) and bronzes (which are brownish alloys of copper and tin). These materials do not have particularly high strengths, but they are resistant to corrosion and can be easily joined by soldering. Copper alloys are used in gears, bearings, and the tubing in condensers and heat exchangers.

- Grade 1020 steel is a common, easily worked, and relatively inexpensive medium-grade alloy of steel. The 4340 grade is a higher strength and more expensive material than grade 1020. Although all steels are comprised primarily of iron and a small fraction of carbon, they are differentiated based on mechanical and heat treatments and on the presence of additional alloying metals, including carbon, manganese, nickel, chromium, and molybdenum.

- For stainless steels, the 316 alloy is used for corrosion-resistant nuts, bolts, and pipe fittings, and the higher-strength material 440C is used to form the raceways of rolling element bearings (Section 4.6).

- Titanium alloys are strong, lightweight, and corrosion resistant. On the other hand, they are also far more expensive and difficult to machine than other metals. Titanium is used in chemical industrial pipes, gas turbine blades, high-performance aircraft structures, submarines, and other demanding material applications.

Ceramics

When you think of ceramics, images of coffee mugs, dinner plates, and artwork probably come to mind. Engineering ceramics, on the other hand, are used in the automotive, aerospace, electronics, telecommunications, computer, and medical industries for applications encompassing high temperatures, corrosion, electrical insulation, and wear resistance. Ceramics are produced by heating naturally occurring minerals and chemically treated powders in a furnace to form a rigid mechanical component.

Ceramics are hard, brittle, crystalline materials that can comprise metals and nonmetals. Ceramics have large elastic modulus values, but, because they are brittle and tend to break suddenly when overloaded, ceramics are not appropriate for supporting large tensile forces. Mechanical components made from ceramics become significantly weakened by the presence of small defects, cracks, holes, bolted connections, and so forth.

An important characteristic of ceramics is that they can withstand extreme temperatures and insulate other mechanical components from heat. Ceramics are used as thermal barrier coatings to protect turbine blades from the high temperatures developed in jet engines. They are also used in thermal protection systems in rocket exhaust cones and in the windshields of many aircraft, and they were used in the space shuttle to insulate the spacecraft's structural frame during reentry.

Some examples of ceramics are the compositions silicon nitride (Si_3N_4), alumina (Al_2O_3), and titanium carbide (TiC). Alumina is sometimes formed into a honeycomb-like support structure that is used in an automobile's exhaust system and catalytic converter. Because of its mechanical, electrical, and thermal characteristics, the advanced ceramic AlTiC (64% Al_2O_3 and 36% TiC) is used in computer hard disk drives to support the recording heads above the surface of the rotating disks.

Engineers and physicians are finding an increasing range of applications for ceramic materials in medical applications, including the repair or replacement of human hips, knees, fingers, teeth, and diseased heart valves. Ceramics are one of the few materials that can withstand the corrosive environment of the human body's interior for an extended length of time. Ceramic implants and coatings of ceramics on metallic joint replacements have been found to stimulate bone growth and to protect the metallic portions of an implant from the immune system.

Polymers

Plastics and elastomers are two types of polymers. The root of "polymer" is a Greek word meaning "of many parts," and it emphasizes the fact that polymers are giant molecules formed as long chains of smaller, building-block molecules. These polymer macromolecules have enormous molecular weights, and they can contain hundreds of thousands of atoms. Each macromolecule is made up of a large number of simpler units that are joined together in a regular repeating pattern. Polymers are organic compounds; that is, their chemical formulation is based on the properties of the element carbon. Carbon atoms are able to attach themselves to one another more than other elements can, and other atoms (such as oxygen, hydrogen, nitrogen, and chlorine) are attached to those carbon chains. From a chemical standpoint, therefore, engineering polymers are formed from large-chained molecules having a regular pattern and based on carbon.

Rubber and silk are two naturally occurring macromolecules, but chemists and chemical engineers have developed hundreds of useful macromolecular materials. Synthetic polymers are classified into two groups: *plastics* (which can be extruded into sheets and pipes or molded to form a wide range of products) and *elastomers* (which are compliant in a manner characteristic of rubber). Unlike the first two classes of materials—metals and their alloys and ceramics—plastics and elastomers are relatively soft materials. They typically have an elastic modulus that is many times smaller than metals. In addition, their properties also change significantly with temperature. At room temperature, polymers may stretch and behave elastically, but, as the temperature is lowered, they become brittle. These materials are not well suited for applications where strength is required or for operation at elevated temperatures. Nevertheless, plastics and elastomers are widely used and remarkable engineering materials. They are relatively inexpensive, lightweight, good insulators against heat and electricity, and easy to shape and mold into complex parts.

Plastics are one of the most utilized engineering materials in any industry, and some of the most common forms are polyethylene, polystyrene, epoxy, polycarbonate, polyester, and nylon. Elastomers, the second category of polymers, are the synthetic rubber-like macromolecules that are elastic and stretchable in a manner that is characteristic of rubber. Elastomers can be greatly deformed and still return to their original shape after being released.

In one of their largest applications, elastomers are used to make tires for vehicles ranging from mountain bikes to aircraft. Other elastomers include the polyurethane foam that is used to insulate buildings, silicone sealant and adhesive, and neoprene, which is resistant to chemicals and oils. Elastomers are also used to make supports and mounting blocks that can reduce the vibration produced by a machine. Vibration isolation mounts that incorporate elastomers are used to attach an automobile's engine to its chassis and to isolate the hard disk drive in a laptop or tablet computer from shock and vibration in the event the computer is accidentally dropped.

Composite Materials

As their name implies, composites are mixtures of several materials, and their formulation can be customized and tailored for specific applications. Composite materials are generally comprised of two components: the matrix and the reinforcement. The matrix is a relatively ductile material that holds and binds together the strong reinforcing particles or fibers embedded in it. Some composite materials comprise a polymer matrix (usually epoxy or polyester) that is reinforced by many small-diameter fibers of glass, carbon, or Kevlar®.

Composites are not well suited for high temperatures because, like plastics and elastomers, the polymer matrix softens as the temperature increases. The main idea behind fiber-reinforced composites is that the strong fibers carry most of the applied force. Other examples of composite materials are concrete that has been reinforced with steel rods, automobile tires that include steel reinforcing belts in an elastomer matrix, and power transmission belts that use fiber or wire cords to carry the belt's tension (Section 8.6).

Composite materials are an example of the adage that "the whole is greater than the sum of its parts" in that their mechanical properties are superior to those of the constituent materials alone. The primary advantages of composites are that they can be made very stiff, strong, and lightweight. However, the additional processing steps necessary to produce these materials raise their cost.

The widespread usage of fiber-reinforced composite materials began in the aerospace industry (Figure 5.21) where weight is at a premium. A substantial amount of an aircraft's weight can be reduced by incorporating composite materials into the airframe, horizontal and vertical stabilizers, flaps, and wing skins. Approximately 30% of the external surface area of the Boeing 767

Figure 5.21

This aircraft is fabricated from aluminum, titanium, composites, and other advanced materials in order to meet the competing performance requirements of low weight and high strength.

jiawangkun/Shutterstock.com

commercial airliner is formed from composite materials. As the technology of composites has matured and costs have decreased, these materials have been adopted in automobiles, spacecraft, boats, architectural structures, bicycles, skis, tennis rackets, and other consumer products.

Focus On

THE DESIGN OF NEW MATERIALS

Engineers and scientists are continually developing new materials that can be used in the design of innovative products. In conventional materials such as metals, ceramics, and plastics, the material weight, strength, and density are correlated. For example many high-strength materials are strong but also heavy, limiting their application in environments where keeping weight to a minimum is important. But engineers are developing new materials that are challenging the traditional assumptions about materials.

As one example, nanostructured ceramics combine tiny struts and joints allowing the structural and mechanical properties of ceramics to be changed and as a result less

dependent on weight. These nanostructures are allowing for the development of one of the strongest and lightest substances ever made with applications in the automotive, aerospace, and energy fields. They are also allowing for materials that can recover their shape after being crushed, much like a sponge.

Other materials are being developed, taking inspiration from the design of biological systems. At the University of Southern California, bi-thermal materials consisting of two materials that respond differently to temperature changes can change their configuration to help reduce energy costs. For example, in Figure 5.22,

Heating

Figure 5.22

Thermobimetals such as this nickel-magnesium lattice consist of two thin sheet metals with different expansion coefficients. When heated the material curls, acting as a type of skin for a building allowing the building to ventilate automatically.

Doris Kim Sung

this structure can regulate the temperature in its local environment by dynamically ventilating a space and shading it from sun. This responsive material behavior simulates human skin and other natural systems that open pores for cooling or close them to retain energy. Researchers at the Royal Institute of Technology in Sweden have developed so-called nanopaper by tightly weaving together nanosized cellulose threads, found in the cell walls of plants and algae. The resulting structure is stronger and tougher than cast iron and can be manufactured from renewable materials at relatively low temperature and pressure. Another form of nanopaper is made from metal nanowires. With its ability to soak up to 20 times its weight in oil, this nanopaper is being used to help clean oil spills and other environmental toxins. Researchers at the Massachusetts Institute of Technology have developed a new type of self-assembling photovoltaic cell. These cells mimic the repair mechanisms found in plants and are expected to be 40% more efficient at energy conversion than current photovoltaic cells. These new

cells could potentially prolong the currently limited life span of solar panels indefinitely.

Other newly developed materials include a bio-inspired plastic from Harvard University that is light, thin, and strong, mimicking an insect wing and providing a biocompatible option for trash bags, packaging, and diapers. At Northwestern University and Michigan State University, a new thermoelectric material converts waste heat to electricity at record-setting efficiencies allowing engineers to reuse previously wasted heat in factories, power plants, vehicles, and ships.

These discoveries will continue to impact the types of products and systems that mechanical engineers can design and develop. For example, imagine being able to work with materials that can repair themselves over time. You might design a product with glass capable of repairing cracks or with metals that can renew their protective coating to prevent corrosion. These are some of the material technologies that could soon become realistic resources for engineers in the design of products, systems, and machines.

Example 5.7 | *Selecting Materials to Minimize Weight*

In designing a truss structure, an engineer finds that the metallic material for a rod should be chosen subject to three constraints. (See Figure 5.23.) The rod must:

- Support a force of magnitude F
- Have length L
- Elongate by less than ΔL

The rod's cross-sectional area and the material from which it will be made remain to be selected by the designer. Among steel, aluminum, and titanium, which material would you recommend to minimize the rod's weight?

Example 5.7 | *continued*

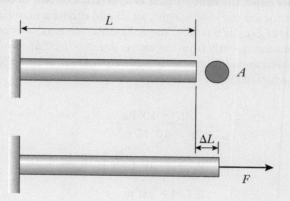

Figure 5.23

Approach
To determine the best material, we must consider two material properties—density and elastic modulus—in the calculations involving the rod's weight and elongation. The rod's weight is $w = \rho_w AL$, where ρ_w is the metal's weight per unit volume listed in Table 5.2. The length and cross-sectional area, the elastic modulus of the rod's material, and the applied force are related by Equation (5.6).

Solution
Although the rod's area is not known, it can be expressed in terms of the given quantities (F, L, and ΔL) and the elastic modulus E:

$$A = \frac{FL}{E\Delta L} \qquad \leftarrow \left[\Delta L = \frac{FL}{EA} \right]$$

By substituting into the expression for the rod's weight and making some algebraic manipulations, we have

$$w = (\rho_w L)\left(\frac{FL}{E\Delta L}\right) \qquad \leftarrow [w = \rho_w AL]$$

$$= \left(\frac{FL^2}{\Delta L}\right)\left(\frac{\rho_w}{E}\right)$$

$$= \frac{FL^2/\Delta L}{E/\rho_w}$$

The numerator in this expression is a grouping of variables that are given in the problem's statement: the applied force, the rod's length, and the allowable elongation. Thus, the numerical value of the numerator is fixed.

Example 5.7 | *continued*

On the other hand, the denominator E/ρ_w is characteristic of the material alone. To minimize the rod's weight, we should choose a material having as large a value of E/ρ_w as possible. The rod's cross-sectional area can then be determined subsequently from the expression $A = FL/E\Delta L$. By using the elastic modulus and weight-density values that are listed in Table 5.2, the ratio's value for steel is

$$\frac{E}{\rho_w} = \frac{207 \times 10^9 \,\text{Pa}}{76 \times 10^3 \,\text{N/m}^3}$$

$$= 2.724 \times 10^6 \left(\frac{\text{N}}{\text{m}^2}\right)\left(\frac{\text{m}\cdot\text{m}^2}{\text{N}}\right)$$

$$= 2.724 \times 10^6 \,\text{m}$$

and in a similar fashion, we find the values for aluminum (2.667×10^6 m) and titanium (2.651×10^6 m). By tabulating these numbers, we can see that steel offers a slight advantage over aluminum and titanium, but the difference amounts to only a few percent.

Table 5.4

Material	E/ρ_w (m)
Steel	2.724×10^6
Aluminum	2.667×10^6
Titanium	2.651×10^6

Discussion

The parameter E/ρ_w is a measure of a material's stiffness per unit weight. For an application involving tension, we should choose a material that maximizes E/ρ_w for the structure to have high stiffness and low weight. As it turns out, steel has a slight advantage, but the E/ρ_w ratio is nearly the same for aluminum and titanium. Once the material's cost is factored in, however, steel would clearly be the preferable choice since aluminum is somewhat more expensive than steel, and titanium is significantly more expensive. Of course, our analysis has not taken into account corrosion, strength, ductility, and machinability. Those factors could be important for the task at hand and influence the final recommendation. You can see how selecting a material involves making trade-offs between cost and performance.

Steel has a slight advantage.

▶ 5.6 **Factor of Safety**

Mechanical engineers determine the shape and dimensions and select materials for a wide range of products and types of hardware. The analysis that supports their design decisions takes into account the tensile, compressive, and shear stresses that are present, the material properties, and other factors that you will encounter later in your study of mechanical engineering. Designers are aware that a mechanical component can break or otherwise be rendered useless by a variety of causes. For instance, it could yield and be permanently deformed, fracture suddenly into many pieces because the material is brittle, or become damaged through corrosion. In this section, we introduce a simple model that engineers use to determine whether a mechanical component is expected to yield because of either tension or shear stress. This analysis predicts the onset of yielding in ductile materials, and it is a useful tool that enables mechanical engineers to prevent the material in a piece of hardware from being loaded to, or above, its yield stress. Yielding is only one of many possible failure mechanisms, however, and so our analysis in this section obviously will not offer predictions regarding other kinds of failure.

From a practical standpoint, engineers recognize that, despite their analysis, experiments, experience, and design codes, nothing can be built to perfection. In addition, no matter how carefully they might try to estimate the forces that are going to act on a structure or machine, the mechanical component could be accidentally overloaded or misused. For those reasons, a catchall parameter called the factor of safety is generally introduced to account for unexpected effects, imprecision, uncertainty, potential assembly flaws, and material degradation. The *factor of safety* is defined as the ratio of the stress at failure to the stress during ordinary use. The factor of safety to guard against ductile yielding is often chosen to fall in the range of 1.5–4.0; that is, the design is intended to be between 150% and 400% as strong as it needs to be for ordinary use. For engineering materials with better-than-average reliability or for well-controlled operating conditions, the lower values for the factor of safety are appropriate. When new or untried materials will be used or other uncertainties are present, the larger factors will result in a safer design.

Factor of safety

When a straight rod is subjected to tension as in Figure 5.3, the possibility of it yielding is assessed by comparing the stress σ to the material's yield strength S_y. Failure due to ductile yielding is predicted if $\sigma > S_y$. Engineers define the tensile-stress factor of safety as

$$n_{\text{tension}} = \frac{S_y}{\sigma} \qquad (5.10)$$

If the factor of safety is greater than one, this viewpoint predicts that the component will not yield, and if it is less than one, failure is expected to occur. For components that are instead loaded in pure shear, the shear stress τ is compared to the yield strength in shear, which is denoted by S_{sy}. As developed in more advanced treatments of stress analysis, one perspective on the failure of materials relates the yield strength in shear to the value in tension according to the expression

$$S_{sy} = \frac{S_y}{2} \qquad (5.11)$$

The shear yield strength can therefore be determined from the strength values obtained from standard tensile testing, as listed in Table 5.3. To evaluate the possibility of a ductile material yielding in shear, we compare stress and strength in terms of the shear factor of safety

$$n_{\text{shear}} = \frac{S_{\text{sy}}}{\tau}$$

(5.12)

The numerical value for the factor of safety that an engineer chooses for a particular design will depend on many parameters, including the designer's own background, experience with components similar to the one being analyzed, the amount of testing that will be done, the material's reliability, the consequences of failure, maintenance and inspection procedures, and cost. Certain spacecraft components might be designed with the factor of safety being only slightly greater than one in order to reduce weight, which is at a premium in aerospace applications. To counterbalance that seemingly small margin for error, those components will be extensively analyzed and tested, and they will be developed and reviewed by a team of engineers having a great deal of collective experience. When forces and load conditions are not known with certainty or when the consequences of a component's failure would be significant or endanger life, large values for the factor of safety are appropriate. Engineering handbooks and design codes generally recommend ranges for safety factors, and those references should be used whenever possible. In part, design codes set safety standards for many mechanical products, as we saw when discussing the top ten achievements of the mechanical engineering profession in Section 1.3.

Example 5.8 | *Designing a Gear-to-Shaft Connection*

The spur gear is used in a transmission, and it is attached to the 1 in. diameter shaft by a key having a 0.25 × 0.25 in. cross section and length of 1.75 in. (See Figure 5.24.) The key is made from grade 1045 steel, and it fits into matching slots that are machined in the shaft and gear. When the gear is driven on its tooth by the 1500 lb force, determine (a) the shear stress in the key and (b) the factor of safety against yielding.

Figure 5.24

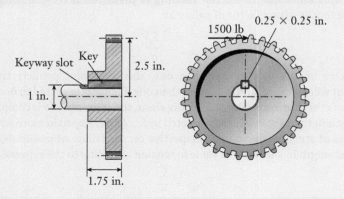

Example 5.8 | *continued*

Approach

Torque is transmitted between the gear and the shaft through the key, which tends to be sliced along a single shear plane by the force transmitted between the gear and the shaft. (See Figure 5.25.) We determine the magnitude of the shear force by applying the condition of rotational equilibrium to the gear. Although the 1500 lb force tends to rotate the gear in the clockwise direction, the shear force between the gear and key balances that torque in the counterclockwise direction.

Figure 5.25

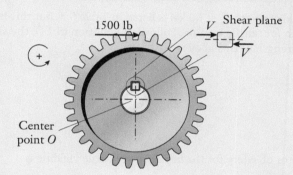

Solution

(a) The shear force on the key is determined from the balance between its torque and that of the 1500 lb tooth force about the shaft's center:

$$-(1500\ \text{lb})(2.5\ \text{in.}) + V(0.5\ \text{in.}) = 0 \qquad \leftarrow \left[\sum_{i=1}^{N} M_{o,i} = 0\right]$$

and $V = 7500$ lb. This force is distributed over the shear plane of the key having cross-sectional area

$$A = (1.75\ \text{in.})(0.25\ \text{in.}) = 0.4375\ \text{in}^2$$

The shear stress is

$$\tau = \frac{7500\ \text{lb}}{0.4375\ \text{in}^2} \qquad \leftarrow \left[\tau = \frac{V}{A}\right]$$

$$= 1.714 \times 10^4\ \frac{\text{lb}}{\text{in}^2}$$

$$= 1.714 \times 10^4\ \text{psi}$$

Example 5.8 | *continued*

To place the numerical value in more conventional form, we convert to the dimension ksi:

$$\tau = (1.714 \times 10^4 \, \text{psi}) \left(10^{-3} \, \frac{\text{ksi}}{\text{psi}} \right)$$

$$= 17.14 \, (\cancel{\text{psi}}) \left(\frac{\text{ksi}}{\cancel{\text{psi}}} \right)$$

$$= 17.14 \, \text{ksi}$$

(b) The yield strength $S_y = 60$ ksi is listed in Table 5.3, but this is the strength for tensile loading. Following Equation (5.11), the shear yield strength of this alloy is

$$S_{sy} = 60 \, \frac{\text{ksi}}{2} \quad \leftarrow \left[S_{sy} = \frac{S_y}{2} \right]$$

$$= 30 \, \text{ksi}$$

The factor of safety for the key against ductile yielding is

$$n_{shear} = \frac{30 \, \text{ksi}}{17.14 \, \text{ksi}} \quad \leftarrow \left[n_{shear} = \frac{S_{sy}}{\tau} \right]$$

$$= 1.750 \, \frac{\text{ksi}}{\text{ksi}}$$

$$= 1.750$$

which is a dimensionless number.

Discussion

Since the factor of safety is greater than one, our calculation indicates that the connection between the gear and shaft is strong enough to prevent yielding by an additional margin of 75%. If the force on the gear's tooth would grow by more than that amount, then the key would be expected to yield, and the design would be unsatisfactory. In that event, the key's cross-sectional area would need to be increased, or a higher strength material would need to be selected.

$$\tau = 17.14 \, \text{ksi}$$

$$n_{shear} = 1.750$$

Summary

Ⓞne of the primary functions of mechanical engineers is to design structures and machine components so that they are reliable and do not break. Engineers analyze stress, strain, and strength to determine whether a component will be safe or risks being overloaded to the point that it might deform excessively or fracture. The important quantities introduced in this chapter, common symbols for them, and their units are summarized in Table 5.5, and the key equations are reviewed in Table 5.6 (see page 212).

Generally, engineers conduct a stress analysis during the design process, and the results of those calculations are used to guide the choice of materials and dimensions. As we introduced in Chapter 2, when the shape of a component or the loading circumstances are particularly complicated, engineers will use computer-aided design tools to calculate the stresses and deformations. Figure 5.26 illustrates the results of one such computer-aided stress analyses for the blades of a wind turbine.

Figure 5.26

Computer-aided stress analysis of the blades on a wind turbine during operation. The color levels indicate the magnitude of stress.

Image courtesy Autodesk, Inc.

Quantity	Conventional Symbols	Conventional Units	
		USCS	SI
Tensile stress	σ	psi, ksi, Mpsi	Pa, kPa, MPa
Shear stress	τ	psi, ksi, Mpsi	Pa, kPa, MPa
Elastic modulus	E	Mpsi	GPa
Yield strength Tension Shear	S_y S_{sy}	ksi ksi	MPa MPa
Ultimate strength	S_u	ksi	MPa
Strain	ε	—	—
Poisson's ratio	ν	—	—
Factor of safety	$n_{tension}, n_{shear}$	—	—
Stiffness	k	lb/in.	N/m

Table 5.5

Quantities, Symbols, and Units that Arise when Analyzing Stresses and Material Properties

Tension and compression	
Stress	$\sigma = \dfrac{F}{A}$
Strain	$\varepsilon = \dfrac{\Delta L}{L}$
Material response	$\sigma = E\varepsilon$
Rod deformation	
Elongation	$\Delta L = \dfrac{FL}{EA}$
Diameter change	$\Delta d = -\nu d \dfrac{\Delta L}{L}$
Hooke's law	$F = k\Delta L$
Stiffness	$k = \dfrac{EA}{L}$
Shear	
Stress	$\tau = \dfrac{V}{A}$
Yield strength	$S_{sy} = \dfrac{S_y}{2}$
Factor of safety	
Tension	$n_{tension} = \dfrac{S_y}{\sigma}$
Shear	$n_{shear} = \dfrac{S_{sy}}{\tau}$

Table 5.6

Key Equations that Arise when Analyzing Materials and Stresses

In this chapter, we discussed the loading conditions known as tension, compression, and shear, as well as a failure mechanism called ductile yielding. The following concepts are central to selecting materials and setting dimensions in the design of structures and machines:

- *Stress* is the intensity of a force distributed over an exposed area of material. Depending on the direction in which the stress acts, it can be tension or shear.

- *Strain* is the change in length per unit of original length. Because of its definition as a ratio of two lengths, strain is a dimensionless quantity, and it is often expressed as a decimal percentage. At a strain of 0.1%, a rod that was 1 m long will have stretched by a factor of 0.001, or 1 mm.

- *Strength* captures the ability of a material to withstand the stresses acting on it. Mechanical engineers compare stresses to the strength of materials to assess whether yielding will occur.

An important part of mechanical engineering involves choosing the materials that will be used in the design of a structure or machine component. This choice is an important aspect of the design process, and mechanical engineers must consider performance, economic, environmental, and manufacturing issues when making this decision. In this chapter, we introduced some of the characteristics of the main classes of materials encountered in mechanical engineering: metals and their alloys, ceramics, polymers, and composite materials. Each class of materials has its own advantages, special characteristics, and preferred applications.

Once the forces acting on a mechanical component have been determined, the dimensions of the component have been assigned, and the materials to be used in production have been identified, the reliability of the design is assessed. The factor of safety is the ratio of the stress at failure to the stress during ordinary use. It has been said that there are *knowns* (things that we know and that we can account for in a design), *known unknowns* (things that we don't know but that we at least know we don't know them), and *unknown unknowns* (an unknown that might surprise us because we are unaware of it and that might unexpectedly compromise a design). The factor of safety is intended to improve a design's reliability and to account for known and unknown unknowns in the forms of uncertainty in usage, materials, and assembly.

Self-Study and Review

5.1. How are stress and strain defined for a rod that is loaded in tension?

5.2. In the SI and USCS, what are the conventional dimensions for stress and strain?

5.3. Sketch a stress–strain diagram, and label some of its important features.

5.4. What is the difference between the elastic and plastic behavior of materials?

5.5. Define the following terms: elastic modulus, proportional limit, elastic limit, yield point, and ultimate point.

5.6. What are the approximate numerical values for the elastic modulus of steel and aluminum?

5.7. How is the yield strength found using the 0.2% offset method?

5.8. What is Poisson's ratio?

5.9. In what ways do tensile and shear stresses differ?

5.10. Discuss some of the characteristics and applications for metals and their alloys, ceramics, polymers, and composite materials.

5.11. How is the shear yield strength S_{sy} related to the yield strength S_y that is obtained from a tensile test?

5.12. What is the factor of safety? When is the factor of safety too small? Can it be too large?

5.13. Discuss some of the trade-offs that an engineer would consider when deciding whether a design's factor of safety is too large or too small.

Problems

P5.1

Find a real physical example of a mechanical structure or machine that has tensile stress present.

(a) Make a clear labeled drawing of the situation.

(b) Estimate the dimensions of the structure or machine and the magnitudes and directions of the forces that act on it. Show these on the drawing. Briefly explain why you estimate the dimensions and forces to have the numerical values that you assigned.

(c) Calculate the magnitude of the stress.

P5.2

A 1 Mg container hangs from a 15 mm diameter steel cable. What is the stress in the cable?

P5.3

Grade 1020 steel has a yield strength of 42 ksi and an elastic modulus of 30 Mpsi. Another grade of steel has a yield strength of 132 ksi. What is its elastic modulus?

P5.4

A steel cable of diameter $\frac{3}{16}$ in. is attached to an eyebolt and tensioned to 500 lb (Figure P5.4). Calculate the stress in the cable and express it in the dimensions psi, ksi, Pa, kPa, and MPa.

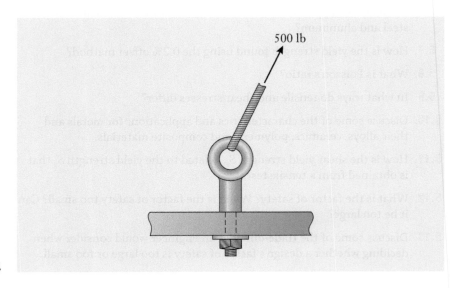

500 lb

Figure P5.4

P5.5

When a 120 lb woman stands on a snow-covered trail, she sinks slightly into the snow because the compressive stress between her ski boots and the snow is larger than the snow can support without crumbling. Her cross-country skis are 65 in. long and $1\frac{7}{8}$ in. wide. After estimating the dimensions of the boot's sole, calculate the percentage reduction in stress applied to the snow when she is wearing skis instead of boots.

P5.6

As a machinist presses the handles of the compound-action bolt cutters, link *AB* carries a 7.5 kN force (Figure P5.6). If the link has a 14 × 4-mm rectangular cross section, calculate the tensile stress within it.

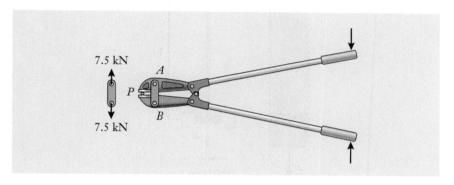

Figure P5.6

P5.7

See Figure P5.7.

(a) By using either the vector algebra or polygon methods for finding a resultant, determine the magnitude of *F* that will cause the net effect of the three forces to act vertically.

(b) For that value of *F*, determine the stress in the bolt's $\frac{3}{8}$ in. diameter straight shank.

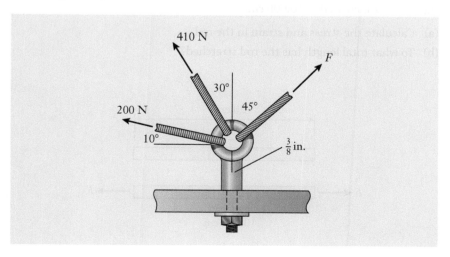

Figure P5.7

P5.8

In a machine shop, the band saw blade is cutting through a work piece that is fed between the two guide blocks B (Figure P5.8). To what tension P should the blade be set if its stress is to be 5 ksi during use? Neglect the small size of the teeth relative to the blade's width.

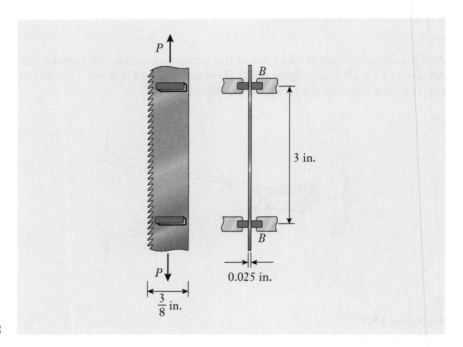

Figure P5.8

P5.9

An 8 mm diameter aluminum rod has lines scribed on it that are exactly 10 cm apart (Figure P5.9). After a 2.11 kN force has been applied, the separation between the lines has increased to 10.006 cm.

(a) Calculate the stress and strain in the rod.

(b) To what total length has the rod stretched?

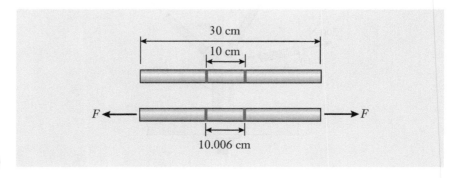

Figure P5.9

P5.10

The tires of the 4555 lb sedan are 6.5 in. wide (Figure P5.10). Each tire contacts the ground over a distance of 4.25 in. as measured along the vehicle's length. Calculate the compressive stress between each tire and the road. The locations of the vehicle's mass center and the wheelbase dimensions are shown.

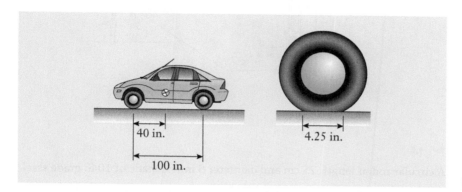

40 in.

4.25 in.

100 in.

Figure P5.10

P5.11

Determine the elastic modulus and the yield strength for the material having the stress–strain curve shown in Figure P5.11. Use the 0.2% offset method.

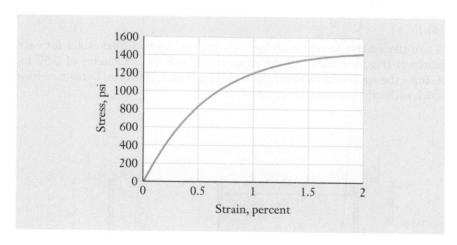

Figure P5.11

P5.12

A 1 ft long rod is made from the material of P5.11. By what amount must the rod be stretched from its original length for it to begin yielding?

P5.13

The steel bolt and anchor assembly is used to reinforce the roof of a passageway in an underground coal mine (Figure P5.13, see page 218). In installation, the bolt is tensioned to 5000 lb. Calculate the stress, strain, and extension of the bolt if it is formed from 1045 steel alloy.

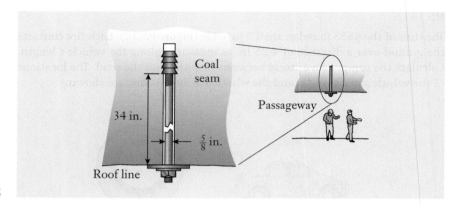

Figure P5.13

P5.14

A circular rod of length 25 cm and diameter 8 mm is made of 1045 grade steel.

(a) Calculate the stress and strain in the rod, and its extension, when it is subjected to 5 kN of tension.

(b) At what force would the rod begin to yield?

(c) By what amount would the rod have to be stretched beyond its original length in order to yield?

P5.15

A two-tier system to repair bridges is supported by two steel cables for each platform (Figure P5.15). All the support cables have a diameter of 0.50 in. Assume the applied loads are at the midpoint of each platform. Determine how much each cable stretches at points A–D.

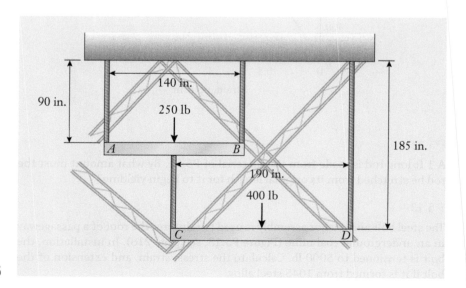

Figure P5.15

P5.16

For the system in P5.15, assume that two 180 lb workers are going to be standing on the system to repair a bridge, one on each platform. The following design requirement must be met: the stretch of each cable must be less than 0.01 in. at the worst-case loading condition for each cable. Determine the minimum diameter required for each cable.

P5.17

An engineer determines that a 40 cm long rod of 1020 grade steel will be subjected to a tension of 20 kN. The following two design requirements must be met: the stress must remain below 145 MPa, and the rod must stretch less than 0.125 mm. Determine an appropriate value for the rod's diameter to meet these two requirements. Round up to the nearest millimeter when reporting your answer.

P5.18

Find a real physical example of a mechanical structure or machine that has shear stress present.

(a) Make a clear labeled drawing of the situation.

(b) Estimate the dimensions of the structure or machine and the magnitudes and directions of the forces that act on it. Show these on the drawing. Briefly explain why you estimated the dimensions and forces to have the numerical values that you assigned.

(c) Calculate the magnitude of the stress.

P5.19

The small steel plate is connected to the right angle bracket by a 10 mm diameter bolt (Figure P5.19). Determine the tensile stress at point A in the plate and the shear stress in the bolt.

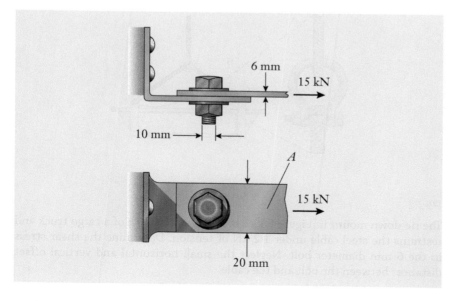

Figure P5.19

P5.20

A 600 lb force acts on the vertical plate, which in turn is connected to the horizontal truss by five $\frac{3}{16}$ in. diameter rivets (Figure P5.20).

(a) If the rivets share the load equally, determine the shear stress in them.

(b) In a worst-case scenario, four of the rivets have corroded, and the load is carried by only one rivet. What is the shear stress in this case?

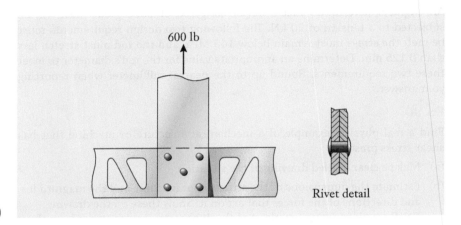

600 lb

Rivet detail

Figure P5.20

P5.21

A detailed view is shown for the connection B in the concrete trough from P4.17 in Chapter 4 (Figure P5.21). Determine the shear stress acting in the shackle's $\frac{3}{8}$ in. bolt.

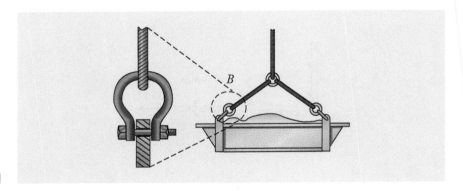

B

Figure P5.21

P5.22

The tie-down mount in Figure P5.22 is bolted to the deck of a cargo truck and restrains the steel cable under 1.2 kN of tension. Determine the shear stress in the 6 mm diameter bolt. Neglect the small horizontal and vertical offset distances between the bolt and the cable.

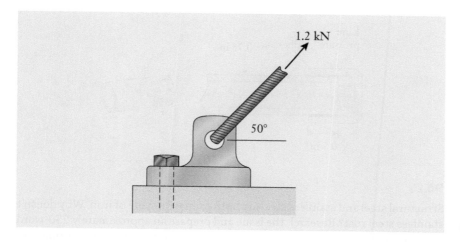

Figure P5.22

P5.23

A spur gear transmits 35 N·m of torque to the 20 mm diameter driveshaft. The 5 mm diameter setscrew threads onto the gear's hub, and it is received in a small hole that is machined in the shaft. Determine the shear stress in the screw along the shear plane B–B (Figure P5.23).

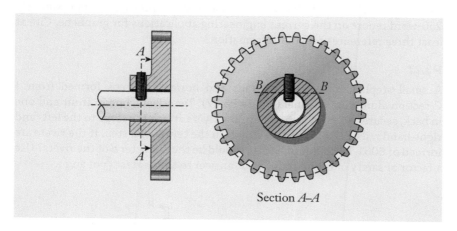

Section *A–A*

Figure P5.23

P5.24

A plastic pipe carries deionized water in a microelectronics clean room, and one end of it is capped (Figure P5.24). The water pressure is $p_0 = 50$ psi, and the cap is attached to the end of the pipe by an adhesive. Calculate the shear stress τ present in the adhesive.

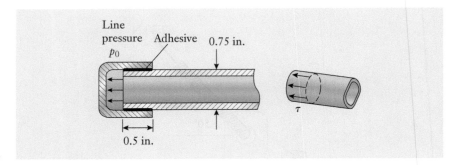

Figure P5.24

P5.25

Structural steel and stainless steel are both primarily made of iron. Why doesn't stainless steel rust? Research the issue and prepare an approximately 250-word report describing the reason. Cite at least three references for your information.

P5.26

Recent technological advances have allowed for the production of graphene, which is a sheet of carbon atoms one atom thick, densely packed into a honeycomb structure. This material has excellent electrical, thermal, and optical properties, making it an ideal candidate for the design of electronic components. These characteristics are further complemented by graphene's ability to kill bacteria and by its breaking strength, which initial tests demonstrate is 200 times greater than steel. Research this material and prepare an approximately 250-word report on the current engineering applications for graphene. Cite at least three references for your information.

P5.27

A small stepladder has vertical rails and horizontal steps formed from a C-section aluminum channel (Figure P5.27). Two rivets, one in front and one in back, secure the ends of each step. The rivets attach the steps to the left- and right-hand rails. A 200 lb person stands in the center of a step. If the rivets are formed of 6061-T6 aluminum, what should be the diameter d of the rivets? Use a factor of safety of 6, and round your answer to the nearest $\frac{1}{16}$ of an in.

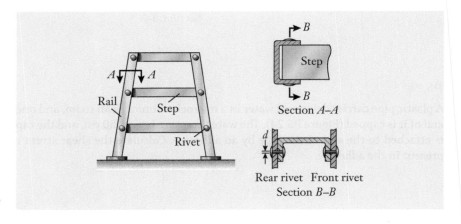

Figure P5.27

P5.28

For the exercise of P5.27, and for the most conservative design, at what location on the step should you specify in your calculation that the 200 lb person stand? Determine the rivet diameter for that loading condition.

P5.29

A $\frac{3}{8}$ in. diameter bolt connects the marine propeller to a boat's $1\frac{1}{4}$ in. driveshaft (Figure P5.29). To protect the engine and transmission if the propeller accidentally strikes an underwater obstacle, the bolt is designed to be cut when the shear stress acting on it reaches 25 ksi. Determine the contact force between the blade and obstacle that will cause the bolt to be sheared, assuming a 4 in. effective radius between the point of contact on the blade and the center of the driveshaft.

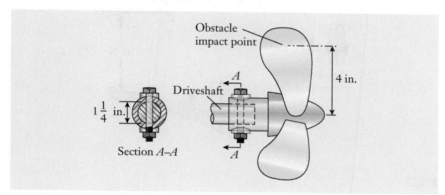

Figure P5.29

P5.30

A machinist squeezes the handle of locking pliers while loosening a frozen bolt (Figure P5.30). The connection at point A, which is shown in a magnified cross-sectional view, supports a 4.1 kN force.

(a) Determine the shear stress in the 6 mm diameter rivet at A.

(b) Determine the factor of safety against yielding if the rivet is formed from 4340 steel alloy.

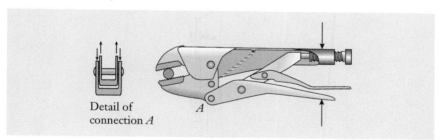

Figure P5.30

P5.31

Compound lever shears cut through a piece of wire at A (Figure P5.31, see on page 224).

(a) By using the free body diagram of handle CD, determine the magnitude of the force at rivet D.

(b) Referring to the magnified cross-sectional drawing of the connection at D, determine the shear stress in the rivet.

(c) If the rivet is formed from 4340 steel alloy, what is the factor of safety?

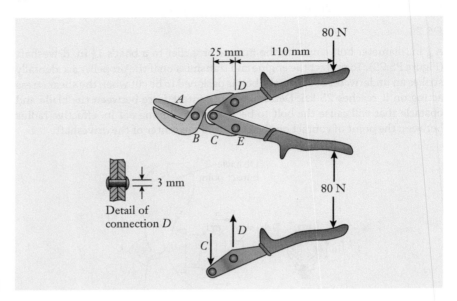

Figure P5.31

P5.32

Plates and rods are frequently used to help rehabilitate broken bones (Figure P5.32). Calculate the shear stress in the lower 5 mm diameter biomedical bolt if it is supporting a 1300 N force from the bone.

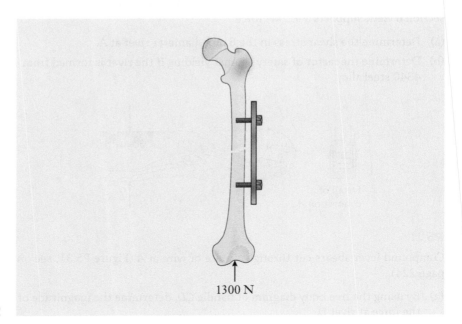

Figure P5.32

P5.33*

Adhesive tape is capable of supporting relatively large shear stress, but it is not able to support significant tensile stress. In this problem, your group will measure the shear strength of a piece of tape. Refer to Figure P5.33.

(a) Cut about a dozen segments of tape having identical length L and width b. The exact length isn't important, but the segments should be easily handled.

(b) Develop a means to apply and measure the pull force F on the tape. Use, for instance, dead weights (cans of soda or exercise weights) or a small fishing scale.

(c) Attach a segment of tape to the edge of a table, with only a portion of tape adhering to the surface. In your tests, consider attachment lengths ranging between a fraction of an inch and several inches.

(d) Being careful to apply the pull force straight along the tape, measure the value F necessary to cause the adhesive layer to slide or shear off the table. Tabulate pull-force data for a half dozen different lengths a.

(e) Make graphs of pull force and shear stress versus a. From the data, estimate the value of the shear stress above which the tape will slide and come loose from the table.

(f) At what length a did the tape break before it sheared off the table?

(g) Repeat the tests for the orientation in which F is applied perpendicular to the surface, tending to peel the tape instead of shearing it. Compare the tape's strengths for shear and peeling.

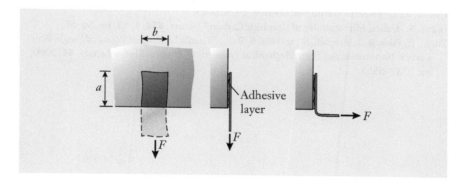

Figure P5.33

P5.34*

Using the bridge repair system in P5.15, as a group develop three additional two-dimensional configurations that utilize different suspension arrangements. The following design requirements must be met: Two-level access to the bridge must be provided with the top platform at 90 in. below the road and the lower level 185 in. below the road; each level must accommodate no more than two 200 lb workers; standard structural steel cables ($S_y = 250$ MPa) of 0.50 in. diameter with varying lengths are to be used. Evaluate the four designs (the three from this problem and the one from P5.15) with respect to their factor of safety in worst-case loading conditions and identify which configuration has the highest factor of safety.

P5.35*

A suspended walkway is being designed for a new multi-story shopping mall. Having a width of 3.5 m it spans 36 m and has an evenly distributed weight of 300 kN. The walkway will be supported only by 6 m long suspension rods in order to leave the area below clear of cumbersome supports. Determine the number of rods, diameter of the rods, and location of the rods noting the following guidelines:

- You need to consider price in your determination.
- You need to consider at least two different materials for your analysis of the rod size and location.
- You need to estimate worst case loading conditions including using a factor of safety of 2.

Present your approach, solution, and discussion in a formal report.

References

Ashby, M. F., *Materials Selection in Mechanical Design*. Butterworth-Heinemann, 1999.

Askeland, D. R., and Phulé, P. P., *The Science and Engineering of Materials*, 4th ed. Thomson-Brooks/Cole, 2003.

Dujardin, E., Ebbesen, T. W., Krishan, A., Yianilos, P. N. and Treacy, M. M. J., "Young's Modulus of Single-Walled Nanotubes," *Physical Review B*, **58**(20), 1998, pp. 14013–14019.

Gere, J. M., and Timoshenko, S. P., *Mechanics of Materials*, 4th ed. PWS Publishing, 1997.

Hooke, R., *De Potentia Restitutiva*. London, 1678, p. 23.

Iijima, S. "Helical Microtubules of Graphitic Carbon," *Nature*, **354**, 1991, pp. 56–58.

Yu, M. F., Files, B. S., Arepalli, S., and Ruoff, R. S., "Tensile Loading of Ropes of Single Wall Carbon Nanotubes and Their Mechanical Properties," *Physical Review Letters*, **84**, 2000, pp. 5552–5555.

CHAPTER 6

Fluids Engineering

CHAPTER OBJECTIVES

- Recognize the application of fluids engineering to such diverse fields as microfluidics, aerodynamics, sports technology, and medicine.
- Explain in technical terms the differences between a solid and a fluid, and the physical meanings of a fluid's density and viscosity properties.
- Understand the characteristics of laminar and turbulent fluid flows.

- Calculate the dimensionless Reynolds number, which is the most significant numerical value in fluids engineering.
- Determine the magnitudes of the fluid forces known as buoyancy, drag, and lift in certain applications.
- Analyze the volumetric flow rate and pressure drop of fluids flowing through pipes.

▶ 6.1 Overview

In this chapter, we introduce the subject of fluids engineering and its role in applications as diverse as aerodynamics, biomedical and biological engineering, piping systems, microfluidics, and sports engineering. The study of fluids, which are classified as either liquids or gases, is further broken down into the areas of fluid statics and dynamics. Mechanical engineers apply the principles of fluid statics to calculate the pressure and buoyancy force of fluids acting on stationary objects, including ships, tanks, and dams. Fluid dynamics refers to the behavior of liquids or gases when they are moving or when an object is moving through an otherwise stationary fluid.

Hydrodynamics and aerodynamics are the specializations focusing on the motions of water and air, which are the most common fluids encountered in engineering. Those fields encompass not only the design of high-speed vehicles but also the motions of oceans and the atmosphere. Some engineers and scientists apply sophisticated computational models to simulate and understand interactions among the atmosphere, oceans, and global climates (Figure 6.1, see page 228). The motion of fine pollutant particles in the air, improved weather forecasting, and the precipitation of raindrops

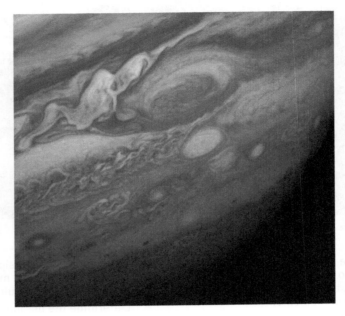

Figure 6.1

The field of fluids engineering can involve the motion of fluids on very large— even planetary— scales. Storms on the Earth, as well as the Great Red Spot on Jupiter shown here, form and move according to the principles of fluid mechanics.

NASA/JPL

Element 5:
Fluids engineering

and hailstones are some of the key issues that are addressed. The field of fluid mechanics is an exacting one, and many advances in it have occurred in conjunction with developments in applied mathematics and computer science. *Fluids engineering* fits within the broader context of the mechanical engineering topics shown in Figure 6.2.

Reflecting on the top ten list of the mechanical engineering profession's achievements (Table 1.1), some 88% of the electricity in the United States is produced by a process that involves continuously cycling water between liquid and steam, and back again. Coal, oil, natural gas, and nuclear fuels are used to heat water into steam, which in turn drives turbines and electrical generators. Another 7% of America's electricity is produced by hydroelectric power plants, and wind power provides even a smaller fraction. As we will explore later in Section 7.7, collectively speaking, over 98% of the electricity in the United States is produced through processes that involve fluids engineering in one form or another. The properties of fluids, the forces they generate, and the manner in which they flow from one location to another are key aspects of mechanical engineering.

Fluid mechanics also plays a central role in biomedical engineering, a field that was ranked as one of the mechanical engineering profession's top ten achievements. Biomedical applications include the design of devices that deliver medicine by inhaling an aerosol spray and the flow of blood through arteries and veins. These devices are capable of performing chemical and medical diagnostics by exploiting the properties of fluids on microscopic scales.

Microfluidics

This emerging field, which is known as *microfluidics*, offers the potential for advances in genomic research and pharmaceutical discovery. Just as the field of electronics has undergone a revolution in miniaturization, chemical and medical laboratory equipment that presently fill an entire room are being miniaturized and made more economical.

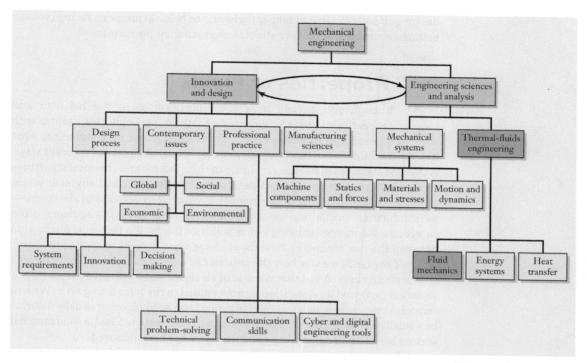

Figure 6.2

Relationship of the topics emphasized in this chapter (shaded blue boxes) relative to an overall program of study in mechanical engineering.

The forces generated by stationary or moving fluids are important to the hardware designed by mechanical engineers. Up to this point, we have considered mechanical systems where the forces arise from gravity or the interactions between connected components. Liquids and gases also generate forces, and in this chapter we will examine the fluid forces known as buoyancy, drag, and lift. As shown in Figure 6.3, mechanical engineers apply sophisticated computer-aided engineering tools to understand complex airflows around aircraft and automobiles. In fact, those same methods have been applied to

Figure 6.3

Mechanical engineers apply computer simulations of three-dimensional airflow including these vortex structures generated by aircraft landing gear.

Courtesy of ANSYS, Inc.

design golf balls capable of longer flight and to help ski jumpers, racing cyclists, marathon runners, and other athletes improve their performances.

▶ 6.2 Properties of Fluids

Although you already have some intuition as to the behavior and properties of fluids in everyday situations, we begin this chapter with a seemingly simple question: From an engineering standpoint, what exactly is a fluid? Scientists categorize compositions of matter in different ways. A chemist classifies materials according to their atomic and chemical structures in the context of the periodic table of elements. An electrical engineer might group materials according to the manner in which they respond to electricity— as conductors, insulators, or semiconductors. Mechanical engineers often categorize substances as being either solids or fluids. The technical distinction between the two centers on how they behave when forces are applied to them.

In Chapter 5, we saw how the behavior of a solid material is described by a stress–strain curve. A rod that is made of an elastic solid will satisfy Hooke's law, Equation (5.4), and its elongation is proportional to the force acting on it. When a tension, compression, or shear force is applied to a solid object, it usually deforms by a small amount. As long as the yield stress has not been reached, a solid material springs back to its original shape once the force has been removed.

A fluid, on the other hand, is a substance that is unable to resist a shear force without continuously moving (shear forces and stresses are discussed in Section 5.4). No matter how small, any amount of shear stress applied to a fluid will cause it to move, and it will continue to flow until the force is removed. Fluid substances are further categorized as being either liquids or gases, and the distinction here depends on whether the fluid easily can be *compressed* (Figure 6.4). When forces are applied to a liquid, the volume does not change appreciably, even though the liquid may move and change its shape. For the purposes of most engineering applications, a liquid is an incompressible fluid. The hydraulic systems that control flight surfaces in aircraft, power off-road construction equipment, and control automotive brakes develop their large forces by transmitting pressure from the liquid hydraulic fluid to pistons and other actuators. Gases, the second category of fluids, have molecules that separate from one another widely in order to expand and fill an enclosure. A gas can be easily compressed, and, when it is compressed, its density and pressure increase accordingly.

Figure 6.4

(a) For most practical purposes in engineering, liquids are incompressible and retain their original volume when forces act on them. (b) The gas within the cylinder is compressed by the piston and force *F*.

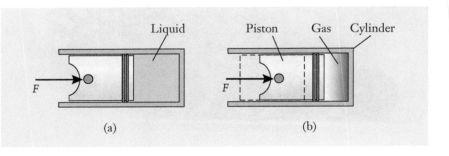

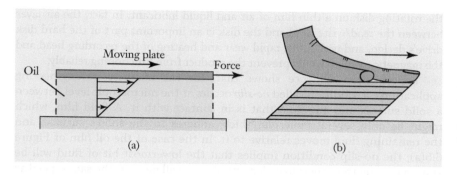

Figure 6.5

(a) A layer of oil is sheared between a moving plate and a stationary surface. (b) The shearing motion of the fluid is conceptually similar to a deck of cards that is pressed and slid between hand and a tabletop.

The primary difference between a solid and a fluid is the manner in which each behaves when subjected to a shear force. Figure 6.5(a) depicts a thin layer of fluid that is being sheared between a stationary surface and a flat plate that is moving horizontally. The plate is separated from the surface by a small distance, and the fluid between them might be a thin layer of machine oil. When a force is applied to the upper plate, it will begin to slide over and shear the oil layer. A fluid responds to shear stress by a continuous motion that is called a *flow*. As an analogy, place a deck of playing cards on a tabletop, and as you press your hand against the top of the deck, also slide your hand horizontally [Figure 6.5(b)]. The uppermost card moves with your hand, and the lowermost card sticks to the table. The playing cards in between are sheared, with each one slipping slightly relative to its neighbors. The oil layer in Figure 6.5(a) behaves in a similar manner.

Flow

A fluid layer is also sheared between two surfaces when a puck slides over an air hockey table, an automobile tire hydroplanes over water on a road's surface, and a person takes a plunge down a water slide. In the field of computer data storage, the read/write head in a hard disk drive (Figure 6.6) floats above the surface of

Figure 6.6

The read/write heads in this computer hard disk drive slide above the surface of the rotating disk on an exceptionally thin film of air and lubricant.

Bragin Alexey/Shutterstock.com

the rotating disk on a thin film of air and liquid lubricant. In fact, the air layer between the read/write head and the disk is an important part of the hard disk drive's design, and without it, rapid wear and heating of the recording head and the magnetic medium would prevent the product from functioning reliably.

Experimental evidence shows that, for the majority of engineering applications, a condition called *no-slip* occurs at the microscopic level between a solid surface and any fluid that is in contact with it. A fluid film, which might be only several molecules thick, adheres to the solid's surface, and the remaining fluid moves relative to it. In the case of the oil film of Figure 6.5(a), the no-slip condition implies that the lowermost bit of fluid will be stationary, and the uppermost element of fluid will move at the same speed as the adjacent plate. As we look across the thickness of the oil film, each layer of fluid moves at a different speed, with the velocity of the oil changing gradually across its thickness.

No-slip condition

When the upper plate in Figure 6.7 slides over the fluid layer at constant speed, it is in equilibrium in the context of Newton's second law of motion. The applied force F is balanced by the cumulative effect of the shear stress

$$\tau = \frac{F}{A} \qquad (6.1)$$

Viscosity

exerted by the fluid on the plate. The property of a fluid that enables it to resist a shear force by developing steady motion is called *viscosity*. This parameter is a physical property of all gases and liquids, and it measures the stickiness, friction, or resistance of a fluid. When compared to water, honey and maple

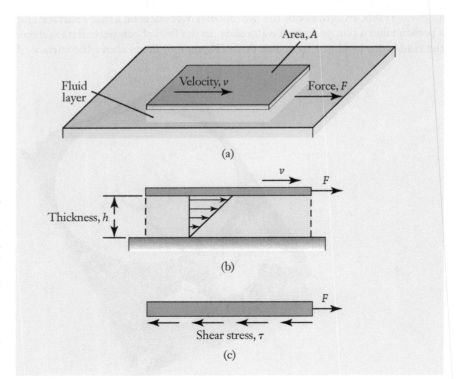

Figure 6.7

(a) A fluid layer is sheared between a stationary surface and a moving plate.
(b) The velocity of the fluid changes across its thickness.
(c) The applied force is balanced by the shear stress exerted on the plate by the fluid.

Fluid	Density, ρ		Viscosity, μ	
	kg/m³	slug/ft³	kg/(m · s)	slug/(ft · s)
Air	1.20	2.33×10^{-3}	1.8×10^{-5}	3.8×10^{-7}
Helium	0.182	3.53×10^{-4}	1.9×10^{-5}	4.1×10^{-7}
Freshwater	1000	1.94	1.0×10^{-3}	2.1×10^{-5}
Seawater	1026	1.99	1.2×10^{-3}	2.5×10^{-5}
Gasoline	680	1.32	2.9×10^{-4}	6.1×10^{-6}
SAE 30 oil	917	1.78	0.26	5.4×10^{-3}

Table 6.1

Density and Viscosity Values for Several Gases and Liquids at Room Temperature and Pressure

syrup, for instance, have relatively high viscosity values. All fluids have some internal friction, and experiments show that, in many cases, the magnitude of the shear stress is directly proportional to the plate's sliding velocity. Those substances are called *Newtonian fluids*, and they satisfy the relation

Newtonian fluid

$$\tau = \mu \frac{v}{h} \qquad (6.2)$$

The parameter μ (the lowercase Greek character mu, as listed in Appendix A) is called the fluid's viscosity, and it relates the fluid's shear stress to the plate's speed. For Equation (6.2) to be dimensionally consistent, we can see that viscosity has the units of mass/(length-time).

Viscosity values are listed in Table 6.1 for several common fluids. In both the SI and USCS, the numerical values for μ are generally small. Because the viscosity property arises frequently in fluids engineering, a special unit called the *poise* (P) was created and named in recognition of the French physician and scientist Jean Poiseuille (1797–1869), who studied the flow of blood through capillaries in the human body. The poise is defined as

Poise

$$1\,\text{P} = 0.1 \frac{\text{kg}}{\text{m} \cdot \text{s}}$$

Focus On FLUIDS IN THE DESIGN OF MICRO AND MACRO SYSTEMS

Fluid mechanics plays a critical role in the design of a wide range of mechanical systems across many orders of magnitude. Large laboratory equipment is now being replaced by microfluidic devices that combine pipes, valves, and pumps to deliver and process fluid samples on a single silicon chip, such as the one in Figure 6.8 (see page 234). These devices, sometimes referred to as a *lab-on-a-chip*, are based on the principle of processing a tiny quantity of fluid, thousands of times smaller than a raindrop. Handling such small volumes is desirable when the sample is either very expensive or when it could be hazardous in larger quantities.

In a microfluidic device, pipes and channels are designed and manufactured to dimensions that are smaller than the diameter of a human hair, and minute quantities of chemical and biological compounds are

Figure 6.8

(a) A microfluidic lab-on-a-chip to study fluid flow in cells, tissues and organ systems.
(b) Imaging of the flow of atmospheric rivers to better understand the phenomena that determine local precipitation rates.

Crump Institute for Molecular Imaging; NOAA ESRL

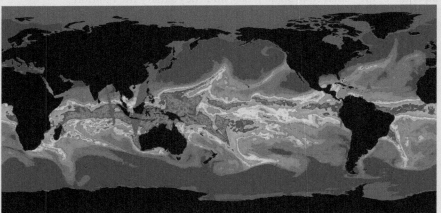

pumped through them. Volumes in a microfluidic device are so small that they are often measured in the units of nanoliters (10^{-9} L) or even picoliters (10^{-12} L). Manipulating fluids on such small dimensional scales offers technological opportunities for automating pharmaceutical experiments, detecting biological and chemical agents in the environment, analyzing and mapping DNA, controlling laminar fuel and oxidant flow in fuel cells, performing at-home diagnostic tests for infections, sorting biological cells, and even delivering precise dosages of a medication.

On the other end of the spectrum, new understandings of turbulent flows are helping engineers design more efficient airplanes, ships, and spacecraft; better ways to control

and eliminate urban pollution dispersal; and improved global weather prediction systems (Figure 6.8(b)). Physicists and engineers have long struggled to understand the complexities of the turbulent flows near a surface, including air moving over a fuselage or water moving across an underwater torpedo. However, researchers at Princeton University and the University of Melbourne in Australia have recently found a relationship between the unpredictable flows close to a surface and the smooth, predictable flow patterns away from the surface. Engineers, equipped with this improved understanding of large-scale flow environments, will be able to design large systems that operate more effectively and efficiently in these environments.

The units of kg/(m · s), slug/(ft · s), and P can each be used for viscosity. In addition to the poise, because the numerical values for μ often involve a power-of-ten exponent, the smaller dimension called the *centipoise* (cP) is sometimes used. Following the SI prefixes of Table 3.3, the centipoise is defined as 1 cP = 0.01 P. The centipoise is a relatively convenient dimension to remember, since the viscosity of freshwater at room temperature is about 1 cP.

Centipoise

Example 6.1 | *Machine Tool Guideways*

A milling machine is used in factories and metalworking shops to cut slots and grooves in metal workpieces (Figure 6.9). The material to be machined is held and moved beneath a rapidly rotating cutting tool. The workpiece and its holder slide over smooth guideways that are lubricated with oil having a viscosity of 240 cP. The two guideways are each of length 40 cm and width 8 cm (Figure 6.10). While setting up for a particular cut,

---

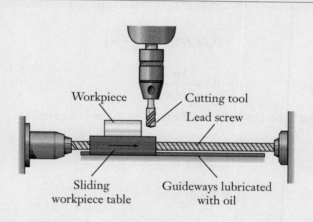

Figure 6.9

Workpiece   Cutting tool
             Lead screw

Sliding   Guideways lubricated
workpiece table   with oil

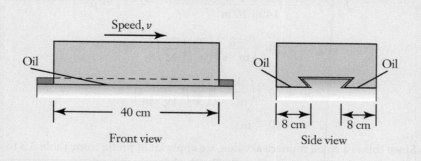

Figure 6.10

Speed, v

Oil

40 cm

Front view

Oil   Oil

8 cm   8 cm

Side view

Example 6.1 | *continued*

a machinist disengages the drive mechanism, applies 90 N of force to the table holding the workpiece, and is able to push it 15 cm in 1 s. Calculate the thickness of the oil film between the table and the guideways.

Approach
When the machinist pushes the table, the oil film is sheared in a manner similar to that shown in Figure 6.7. The table is pushed at the speed $v = (0.15 \text{ m})/(1 \text{ s}) = 0.15$ m/s. The area of contact between the table and the guideways is $A = 2(0.08 \text{ m})(0.4 \text{ m}) = 0.064 \text{ m}^2$. We can calculate the film thickness by applying Equations (6.1) and (6.2) to relate the force, speed, contact area, and film thickness.

Solution
We first convert the oil's viscosity to dimensionally consistent units using the definition of the unit centipoise:

$$\mu = (240 \text{ cP})\left(0.001 \frac{\text{kg}/(\text{m} \cdot \text{s})}{\text{cP}}\right)$$

$$= 0.24(\cancel{\text{cP}})\left(\frac{\text{kg}/(\text{m} \cdot \text{s})}{\cancel{\text{cP}}}\right)$$

$$= 0.24 \frac{\text{kg}}{\text{m} \cdot \text{s}}$$

By applying Equation (6.1), the shear stress in the layer of oil is

$$\tau = \frac{90 \text{ N}}{0.064 \text{ m}^2} \qquad \leftarrow \left[\tau = \frac{F}{A}\right]$$

$$= 1406 \frac{\text{N}}{\text{m}^2}$$

The thickness of the oil film then follows from Equation (6.2):

$$h = \frac{(0.24 \text{ kg}/(\text{m} \cdot \text{s}))(0.15 \text{ m/s})}{1406 \text{ N/m}^2} \qquad \leftarrow \left[\tau = \mu \frac{v}{h}\right]$$

$$= 2.56 \times 10^{-5}\left(\frac{\text{kg}}{\text{m} \cdot \text{s}}\right)\left(\frac{\text{m}}{\text{s}}\right)\left(\frac{\text{m}^2}{\text{N}}\right)$$

$$= 2.56 \times 10^{-5}\left(\frac{\text{kg}}{\text{m} \cdot \text{s}}\right)\left(\frac{\text{m}}{\text{s}}\right)\left(\frac{\text{m}^2 \cdot \text{s}^2}{\text{kg} \cdot \text{m}}\right)$$

$$= 2.56 \times 10^{-5} \text{ m}$$

Since this is a small numerical value, we apply an SI prefix from Table 3.3 to represent a factor of one-millionth. The thickness becomes

Example 6.1 | *continued*

$$h = (2.56 \times 10^{-5} \text{ m})\left(10^6 \frac{\mu\text{m}}{\text{m}}\right)$$

$$= 25.6 \ (\cancel{\text{m}})\left(\frac{\mu\text{m}}{\cancel{\text{m}}}\right)$$

$$= 25.6 \ \mu\text{m}$$

Discussion

When compared to the thickness of a human hair (approximately 70–100 μm in diameter), the oil film is thin indeed, but not atypical of the amount of lubrication present between moving parts in machinery. By inspecting Equation (6.2), we see that the shear stress is inversely proportional to the oil film's thickness. With only half as much oil, the table would be twice as hard to push.

$$h = 25.6 \ \mu\text{m}$$

▶ 6.3 Pressure and Buoyancy Force

The forces known as buoyancy, drag, and lift arise when fluids interact with a solid structure or vehicle. Discussed in Sections 6.6 and 6.7, drag and lift forces arise when there is relative motion between a fluid and a solid object. A vehicle can either move through the fluid (as an aircraft moves through air, for instance) or the fluid can flow around the structure (such as a gust of wind impinging on a skyscraper). However, forces between fluids and solid objects can arise even if there is no relative motion. The force that develops when an object is simply immersed in a fluid is called *buoyancy*, and it is related to the weight of the fluid displaced.

Buoyancy

The weight of a quantity of fluid is determined by its density ρ (the lowercase Greek character rho) and volume. Table 6.1 lists the density values of several gases and liquids in the SI and USCS. The weight of a volume V of fluid is given by the expression

$$w = \rho g V \tag{6.3}$$

where g is the gravitational acceleration constant of 9.81 m/s^2 or 32.2 ft/s^2. For this equation to be dimensionally consistent in the USCS, ρ must have the units of slugs (not pound-mass) per unit volume.

As you swim to the bottom of a pool or travel in the mountains, the pressure changes in the water or air that surrounds you, and your ears "pop" as they adjust to the rising or falling pressure. Our experience is that the pressure in a liquid or gas increases with depth. Referring to the beaker of liquid shown in Figure 6.11 (see page 238), the difference in pressure p between levels 0 and 1 arises because of the intervening liquid's weight. With the two levels separated

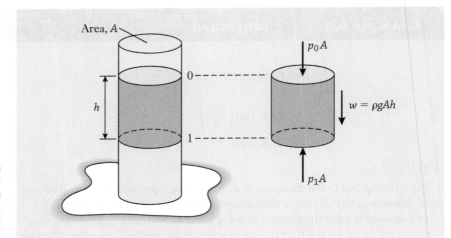

Figure 6.11

Equilibrium of a beaker filled with liquid. Pressure increases with depth because of the weight of the fluid above.

by depth h, the weight of the liquid column is $w = \rho g A h$, where Ah is the enclosed volume. By using the free body diagram of Figure 6.11, the equilibrium-force balance of the liquid column shows that the pressure at depth 1 is

$$p_1 = p_0 + \rho g h \qquad (6.4)$$

Pressure   The *pressure* grows in direct proportion to the depth and the density of the fluid. Similar to stress in Chapter 5, pressure has the dimensions of force-per-

Pascal   unit-area. In the SI, the unit of pressure is the *pascal* (1 Pa = 1 N/m²), named after the seventeenth-century scientist and philosopher Blaise Pascal, who conducted chemical experiments involving air and other gases. As listed in Table 3.5, the derived dimensions psi = lb/in² (pounds per square inch) and psf = lb/ft² (pounds per square foot) are generally used for pressure in the USCS, as is the unit of atmosphere (atm). Table 6.2 provides conversion factors between these conventional units. In the table's first row, for instance, we see that the pascal is related to the other three dimensions as follows:

$$1\ \text{Pa} = 1.450 \times 10^{-4}\ \text{psi} = 2.089 \times 10^{-2}\ \text{psf} = 9.869 \times 10^{-6}\ \text{atm}$$

Buoyancy force   When ships are docked in port and hot air balloons hover above the ground, they are subjected to the *buoyancy forces* created by the surrounding fluid. As shown in Figure 6.12, when a submarine is submerged and floating at a steady

Table 6.2

Conversion Factors Between USCS and SI Units for Pressure

Pa (N/m²)	psi (lb/in²)	psf (lb/ft²)	atm
1	1.450×10^{-4}	2.089×10^{-2}	9.869×10^{-6}
6895	1	144	6.805×10^{-2}
47.88	6.944×10^{-3}	1	4.725×10^{-4}
1.013×10^{5}	14.70	2116	1

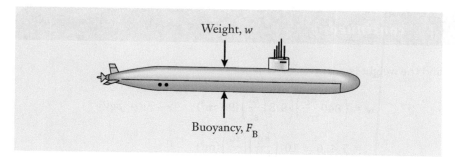

Weight, w

Buoyancy, F_B

Figure 6.12

Buoyancy force acting on a submerged submarine.

depth, the net force on it is zero since the upward buoyancy force balances the submarine's weight. The buoyancy force F_B equals the weight of fluid displaced by an object according to the equation

$$F_B = \rho_{\text{fluid}} g V_{\text{object}} \qquad (6.5)$$

where ρ represents the fluid's density, and V is the volume of fluid displaced by the object. Historically, this result is attributed to Archimedes, a Greek mathematician and inventor, who is said to have uncovered a fraud in the manufacture of a golden crown that had been commissioned by King Hieros II. The king suspected that an unscrupulous goldsmith had replaced some of the crown's gold with silver. Archimedes recognized that the principle embodied in Equation (6.5) could be used to determine whether the crown had been produced from pure gold or from a less dense (and less valuable) alloy of gold and silver (see Problem P6.9 at the end of the chapter).

Example 6.2 | *Aircraft's Fuel Capacity*

A commercial passenger aircraft is loaded to its maximum fuel rating of 90,000 L. By using the density of 840 kg/m³ for jet fuel, calculate the fuel's weight.

Approach
Since the volume and density of the jet fuel are given, we will apply Equation (6.3) in dimensionally consistent units to calculate the weight of the fuel. The definition of the derived unit liter is listed in Table 3.2.

Solution
The fuel's volume is first converted to the units of m³:

$$V = (90{,}000 \text{ L})\left(0.001 \frac{\text{m}^3}{\text{L}}\right)$$

$$= 90 \, (\cancel{\text{L}})\left(\frac{\text{m}^3}{\cancel{\text{L}}}\right)$$

$$= 90 \text{ m}^3$$

Example 6.2 | *continued*

and the weight is

$$w = \left(840\,\frac{\text{kg}}{\text{m}^3}\right)\left(9.81\,\frac{\text{m}}{\text{s}^2}\right)(90\ \text{m}^3) \qquad \leftarrow [w = \rho g V]$$

$$= 7.416 \times 10^5 \left(\frac{\text{kg}}{\text{m}^3}\right)\left(\frac{\text{m}}{\text{s}^2}\right)(\text{m}^3)$$

$$= 7.416 \times 10^5\,\frac{\text{kg} \cdot \text{m}}{\text{s}^2}$$

$$= 7.416 \times 10^5\ \text{N}$$

Since this numerical value has a large power-of-ten exponent, we will apply an SI prefix to represent it more compactly:

$$w = (7.416 \times 10^5\ \text{N})\left(10^{-3}\,\frac{\text{kN}}{\text{N}}\right)$$

$$= 741.6\ (\text{N})\left(\frac{\text{kN}}{\text{N}}\right)$$

$$= 741.6\ \text{kN}$$

Discussion
To place this quantity of fuel into perspective with the USCS, the fuel weighs some 167,000 lb, or over 80 tons. Passenger aircraft typically have a fuel fraction of 25–45%, putting the takeoff weight of this aircraft between 370,000 and 670,000 lb. For example, the Boeing 767-200 aircraft holds approximately 90,000 L of fuel and has a takeoff weight of almost 400,000 lb.

$$\boxed{w = 741.6\ \text{kN}}$$

Example 6.3 | *Deep Submergence Rescue Vehicle*

Intended for rescue missions in the event of a submarine accident, the Deep Submergence Rescue Vehicle can dive to a maximum depth of 5000 ft in the ocean. In dimensions of psi, by how much greater is the water's pressure at that depth than at the ocean's surface?

Approach
To find the difference in pressure, we will apply Equation (6.4), where the water's pressure increases in proportion to depth. We read the density of seawater from Table 6.1 as 1.99 slugs/ft^3 and assume that the density of the seawater is constant.

Example 6.3 | *continued*

Solution

We denote the difference in pressure between the ocean's surface (p_0) and the submarine (p_1) by $\Delta p = p_1 - p_0$. The pressure increase is given by

$$\Delta p = \left(1.99 \frac{\text{slugs}}{\text{ft}^3}\right)\left(32.2 \frac{\text{ft}}{\text{s}^2}\right)(5000 \text{ ft}) \quad \leftarrow [p_1 = p_0 + \rho g h]$$

$$= 3.204 \times 10^5 \left(\frac{\text{slug}}{\text{ft}^2 \cdot \text{ft}}\right)\left(\frac{\text{ft}}{\text{s}^2}\right)(\text{ft})$$

$$= 3.204 \times 10^5 \frac{\text{slug}}{\text{s}^2 \cdot \text{ft}}$$

We don't immediately recognize the dimensions of this quantity, and so we manipulate it slightly by multiplying both the numerator and denominator by the dimension of foot:

$$\Delta p = 3.204 \times 10^5 \left(\frac{\text{slug} \cdot \text{ft}}{\text{s}^2}\right)\left(\frac{1}{\text{ft}^2}\right)$$

$$= 3.204 \times 10^5 \frac{\text{lb}}{\text{ft}^2}$$

$$= 3.204 \times 10^5 \text{ psf}$$

We next convert this quantity to the desired unit of psi by using the conversion factor listed in the third row of Table 6.2:

$$\Delta p = (3.204 \times 10^5 \text{ psf})\left(6.944 \times 10^{-3} \frac{\text{psi}}{\text{psf}}\right)$$

$$= 2225 \text{ psf}\left(\frac{\text{psi}}{\text{psf}}\right)$$

$$= 2225 \text{ psi}$$

Discussion

At this depth, the water's pressure is over 150 times greater than the standard atmospheric pressure of 14.7 psi. Over 300,000 lb of force act on each square foot of the Deep Submergence Rescue Vehicle's hull, and the force over each square inch is equivalent to the weight of a small automobile. The density of the seawater will probably vary across this depth, but this gives a good estimation for the pressure difference.

$$\Delta p = 2225 \text{ psi}$$

Example 6.4 | *Great White Shark Attack*

In the classic thriller movie *Jaws*, Captain Quint manages to shoot harpoons into the great white shark that is attacking his boat. Each harpoon is attached to a cable, which in turn is tied to an empty watertight barrel. Quint's intention is to fatigue the shark by forcing it to drag the barrels through the water. For a sealed 55 gal barrel weighing 35 lb, what force must the shark overcome when it dives beneath the boat and fully submerges the barrel? (See Figure 6.13.)

Figure 6.13

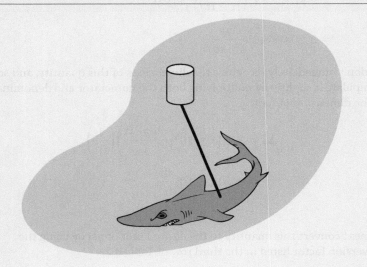

Approach

To find the force that the shark must overcome, we must consider the three forces acting on the barrel: its weight w, the tension T in the cable, and the buoyancy force F_B (Figure 6.14). The shark must overcome the cable's tension, which depends on the other two forces. We begin by drawing a free

Figure 6.14

Weight, w Buoyancy, F_B

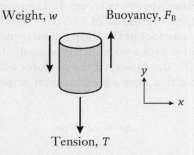

Tension, T

Example 6.4 | *continued*

body diagram of the barrel and indicate on it that we have chosen upward as the positive direction. The buoyancy force is proportional to the density of seawater, which is listed as 1.99 slugs/ft^3 in Table 6.1.

Solution

Since the barrel's weight is given, we will first calculate the magnitude of the buoyancy force by applying Equation (6.5). The volume of the barrel is converted to the dimensionally consistent units of ft^3 by using the conversion factor in Table 3.5:

$$V = (55 \text{ gal}) \left(0.1337 \frac{\text{ft}^3}{\text{gal}} \right)$$

$$= 7.354 \; (\text{gal}) \left(\frac{\text{ft}^3}{\text{gal}} \right)$$

$$= 7.354 \; \text{ft}^3$$

The buoyancy force becomes

$$F_B = \left(1.99 \frac{\text{slugs}}{\text{ft}^3} \right) \left(32.2 \frac{\text{ft}}{\text{s}^2} \right) (7.354 \text{ ft}^3) \quad \leftarrow [F_B = \rho_{\text{fluid}} g V_{\text{object}}]$$

$$= 471.2 \left(\frac{\text{slug}}{\text{ft}^3} \right) \left(\frac{\text{ft}}{\text{s}^2} \right) (\text{ft}^3)$$

$$= 471.2 \frac{\text{slug} \cdot \text{ft}}{\text{s}^2}$$

$$= 471.2 \text{ lb}$$

where we have used the definition of the pound from Equation (3.3). By referring to the free body diagram and the positive sign convention shown on it, the equilibrium-force balance for the barrel becomes

$$F_B - T - w = 0 \quad \leftarrow \left[\sum_{i=1}^{N} F_{y,i} = 0 \right]$$

which we solve for the cable's tension:

$$T = 471.2 \text{ lb} - 35 \text{ lb}$$

$$= 436.2 \text{ lb}$$

Discussion

As far as the shark is concerned, it feels the cable's tension, which is the difference between the buoyancy force and the barrel's weight. If the barrel instead weighed the same amount as F_B, then the cable's tension would be zero. In that event, the barrel would be neutrally buoyant and the shark would have to work much less to drag the barrel.

$$T = 436.2 \text{ lb}$$

▶ 6.4 Laminar and Turbulent Fluid Flows

I f you've ever traveled on an airplane, you might recall the pilot instructing you to fasten your seat belt because of the turbulence associated with severe weather patterns or airflow over mountain ranges. You may also have had other first-hand experiences with laminar and turbulent fluid flows. Try opening the valve on a garden hose (without a nozzle) by just a small amount, and watch how water streams out of it in an orderly fashion. The shape of the water stream doesn't change much from moment to moment, which is a classic example of laminar water flow. As you gradually open the valve, you'll eventually reach a point where the smooth stream of water starts to oscillate, break up, and become turbulent. What was once glassy-looking water is now disrupted and uneven. In general, slowly flowing fluids appear laminar and smooth, but at a high enough speed, the flow pattern becomes turbulent and random-looking.

When fluid flows smoothly around an object, as in the sketch of airflow around a sphere in Figure 6.15(a), the fluid is said to move in a laminar manner. *Laminar* flow occurs when fluid is moving relatively slowly (the exact definition of "relative" being given shortly). As fluid moves faster past the sphere, the flow's pattern begins to break up and become random, particularly on the sphere's trailing edge. The irregular flow pattern shown in Figure 6.15(b) is said to be *turbulent*. Small eddies and whirlpools develop behind the sphere, and the fluid downstream of the sphere has been severely disrupted by its presence.

The criterion to determine whether a fluid moves in a laminar or turbulent pattern depends on several factors: the size of the object moving through the fluid (or the size of the pipe or duct in which the fluid is flowing), the speed of the object (or of the fluid), and the density and viscosity properties of the fluid. The exact relationship among those variables was discovered in the latter half of the nineteenth century by British engineer Osborne Reynolds, who conducted experiments on the transition between laminar and turbulent flow through pipes. A dimensionless parameter, which is now recognized as being the most important variable in fluids engineering, was found to describe that transition. The *Reynolds number* (Re) is defined by the equation

Laminar

Turbulent

Reynolds number

Figure 6.15

(a) Laminar and (b) turbulent flow of a fluid around a sphere.

$$Re = \frac{\rho v l}{\mu} \tag{6.6}$$

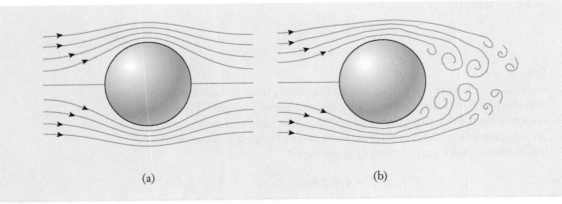

(a)   (b)

in terms of the fluid's density and viscosity, its speed v, and a *characteristic length l* that is representative of the problem at hand. For crude oil that is being pumped through a pipe, the characteristic length l is the pipe's diameter; for fluid flowing past the sphere in Figure 6.15, l is the sphere's diameter; for the ventilation system in a building, l is the diameter of the air duct; and so forth.

Characteristic length

The Reynolds number has the physical interpretation of being the ratio between the inertia and viscous forces acting within a fluid; the inertia force is proportional to density (Newton's second law), and the viscous force is proportional to viscosity [Equation (6.2)]. When the fluid moves quickly, is not very viscous, or is very dense, the Reynolds number will be large. The inertia force of a fluid tends to disrupt the fluid and cause it to flow irregularly. On the other hand, viscous effects are similar to friction, and by dissipating energy, they can stabilize the fluid so that the fluid flows smoothly.

From the standpoint of calculations, situations that arise in mechanical engineering involving laminar flow often can be described by relatively straightforward mathematical equations; that is generally not the case for turbulent flows. The usefulness of those equations, however, is limited to low speeds and ideal shapes such as spheres, flat plates, and cylinders. Experiments and detailed computer simulations are often necessary for engineers to understand the complexity of fluids flowing in real hardware and at actual operating speeds.

Example 6.5 | Reynolds Number

Calculate the Reynolds number for the following situations. (a) A Winchester .30–30 bullet leaves the muzzle of a rifle at 2400 ft/s. (b) Freshwater flows through a 1 cm diameter pipe at the average speed of 0.5 m/s. (c) SAE 30 oil flows under the same conditions as (b). (d) A fast attack submarine with hull diameter of 33 ft cruises at 15 knots. One knot is equivalent to 1.152 mph.

Approach
To calculate the Reynolds number for each situation, we apply the definition in Equation (6.6), making sure that the numerical quantities are dimensionally consistent. The density and viscosity values for air, freshwater, oil, and seawater are listed in Table 6.1.

Solution
(a) The bullet's diameter is 0.3 in., which we convert to the consistent dimension of ft (Table 3.5):

$$d = (0.3 \text{ in.})\left(0.0833 \frac{\text{ft}}{\text{in.}}\right)$$

$$= 0.025 \ (\text{in.})\left(\frac{\text{ft}}{\text{in.}}\right)$$

$$= 0.025 \text{ ft}$$

Example 6.5 | *continued*

The Reynolds number becomes

$$Re = \frac{(2.33 \times 10^{-3} \text{ slug/ft}^3)(2400 \text{ ft/s})(0.025 \text{ ft})}{3.8 \times 10^{-7} \text{ slug/(ft} \cdot \text{s})} \quad \leftarrow \left[Re = \frac{\rho v l}{\mu} \right]$$

$$= 3.679 \times 10^5 \left(\frac{\text{slug}}{\text{ft}^3} \right) \left(\frac{\text{ft}}{\text{s}} \right) (\text{ft}) \left(\frac{\text{ft} \cdot \text{s}}{\text{slug}} \right)$$

$$= 3.679 \times 10^5$$

(b) With the numerical values in the SI, the Reynolds number for water flowing in the pipe is

$$Re = \frac{(1000 \text{ kg/m}^3)(0.5 \text{ m/s})(0.01 \text{ m})}{1.0 \times 10^{-3} \text{ kg/(m} \cdot \text{s})} \quad \leftarrow \left[Re = \frac{\rho v l}{\mu} \right]$$

$$= 5000 \left(\frac{\text{kg}}{\text{m}^3} \right) \left(\frac{\text{m}}{\text{s}} \right) (\text{m}) \left(\frac{\text{m} \cdot \text{s}}{\text{kg}} \right)$$

$$= 5000$$

(c) When SAE 30 oil is pumped through the pipe instead of water, the Reynolds number is reduced to

$$Re = \frac{(917 \text{ kg/m}^3)(0.5 \text{ m/s})(0.01 \text{ m})}{0.26 \text{ kg/(m} \cdot \text{s})} \quad \leftarrow \left[Re = \frac{\rho v l}{\mu} \right]$$

$$= 17.63 \left(\frac{\text{kg}}{\text{m}^3} \right) \left(\frac{\text{m}}{\text{s}} \right) (\text{m}) \left(\frac{\text{m} \cdot \text{s}}{\text{kg}} \right)$$

$$= 17.63$$

(d) We need to convert the submarine's speed into consistent dimensions. The first step is to convert from knots to mph:

$$v = (15 \text{ knots}) \left(1.152 \frac{\text{mph}}{\text{knot}} \right)$$

$$= 17.28 \text{ (knot)} \left(\frac{\text{mph}}{\text{knot}} \right)$$

$$= 17.28 \text{ mph}$$

and then from mph to ft/s:

$$v = \left(17.28 \frac{\text{mi}}{\text{h}} \right) \left(5280 \frac{\text{ft}}{\text{mi}} \right) \left(\frac{1 \text{ h}}{3600 \text{ s}} \right)$$

$$= 25.34 \left(\frac{\text{mi}}{\text{h}} \right) \left(\frac{\text{ft}}{\text{mi}} \right) \left(\frac{\text{h}}{\text{s}} \right)$$

$$= 25.34 \frac{\text{ft}}{\text{s}}$$

Example 6.5 | *continued*

The submarine's Reynolds number becomes

$$Re = \frac{(1.99 \text{ slug/ft}^3)(25.34 \text{ ft/s})(33 \text{ ft})}{2.5 \times 10^{-5} \text{ slug/(ft} \cdot \text{s)}} \qquad \leftarrow \boxed{Re = \frac{\rho v l}{\mu}}$$

$$= 6.657 \times 10^7 \left(\frac{\text{slug}}{\text{ft}^3}\right)\left(\frac{\text{ft}}{\text{s}}\right)\text{ft}\left(\frac{\text{ft} \cdot \text{s}}{\text{slug}}\right)$$

$$= 6.657 \times 10^7$$

Discussion

As expected for a dimensionless quantity, the units of the numerator in
Re exactly cancel those of the denominator. Laboratory measurements
have shown that fluid flows through a pipe in a laminar pattern when Re is
smaller than approximately 2000. The flow is turbulent for larger values of
Re, as we see in the case of (a) the flow around the bullet and (d) the flow
around the submarine. In (b), we would expect the flow of water in the pipe
to be turbulent, while the flow in (c) would certainly be laminar because the
oil is so much more viscous than water.

$$Re_{\text{bullet}} = 3.679 \times 10^5$$
$$Re_{\text{water pipeline}} = 5000$$
$$Re_{\text{oil pipeline}} = 17.63$$
$$Re_{\text{submarine}} = 6.657 \times 10^7$$

Focus On   DIMENSIONLESS NUMBERS

Mechanical engineers often work with
dimensionless numbers. These are either pure
numbers that have no units or groupings of
variables in which the units exactly cancel
one another—again leaving a pure number.
A dimensionless number can be the ratio of
two other numbers, and in that instance, the
dimensions of the numerator and denominator
will cancel. Two dimensionless numbers that
we have already encountered are the Reynolds
number Re and Poisson's ratio v (from
Chapter 5).

Another example that might be familiar
to you is the Mach number Ma, which is used
to measure an aircraft's speed. It is named
after the nineteenth-century physicist Ernst
Mach. The Mach number is defined by the
equation $Ma = v/c$, and it is simply the ratio of
the aircraft's speed v to the speed of sound c
in air. At ground level, the speed of sound is
approximately 700 mph, but it decreases at high
altitude where the atmospheric pressure and
temperature are lower. The numerical values
for both v and c need to be expressed in the
same dimensions (for instance, mph) so that
the units will cancel in the equation for Ma.
A commercial airliner might cruise at a speed
of $Ma = 0.7$, while a supersonic fighter could
travel at $Ma = 1.4$.

▶ 6.5 **Fluid Flow in Pipes**

A practical application for the concepts of pressure, viscosity, and the Reynolds number is the flow of fluids through pipes, hoses, and ducts. In addition to distributing water, gasoline, natural gas, air, and other fluids, pipe flow is also an important topic for biomedical studies of the human circulatory system (Figure 6.16). Blood flows through the arteries and veins in your body in order to transport oxygen and nutrients to tissue and to remove carbon dioxide and other waste products. The vascular system comprises relatively large arteries and veins that branch out into many, much smaller capillaries extending throughout the body. In some respects, the flow of blood through those vessels is similar to that encountered in such engineering applications as hydraulics and pneumatics.

Fluids tend to flow from a location of high pressure to one of lower pressure. As the fluid moves in response, it develops viscous shear stresses that balance the pressure differential and produce steady flow. In the human circulatory system, with all other factors being equal, the greater the difference in pressure between the heart and femoral artery, the faster the blood will flow. The change in pressure along the length of a pipe, hose, or duct is called the *pressure drop*, denoted by Δp. The more viscous a fluid is, the greater the pressure differential that is necessary to produce motion. Figure 6.17 depicts a free body diagram of a volume of fluid that has been conceptually removed from a pipe. Since the pressure drop is related to the shear stress, we expect that Δp will increase with the fluid's viscosity and speed.

Pressure drop

In a section of the pipe that is away from disturbances (such as an inlet, pump, valve, or corner) and for low enough values of the Reynolds number,

Figure 6.16

The flow of blood in the human circulatory system is similar in many respects to the flow of fluids through pipes in other engineering applications. Images such as this of the human pulmonary system are obtained through magnetic resonance imaging and digital modeling, and they provide physicians and surgeons with the information they need to make accurate diagnoses and devise treatment plans.

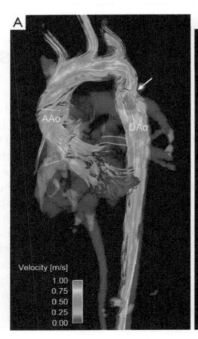

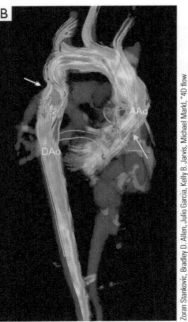

Velocity [m/s]
1.00
0.75
0.50
0.25
0.00

Zoran Stankovic, Bradley D. Allen, Julio Garcia, Kelly B. Jarvis, Michael Markl, "4D flow imaging with MRI," *Cardiovascular Diagnosis and Therapy*, Vol 4, No 2 (April 2014).

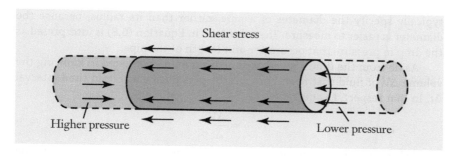

Figure 6.17

Free body diagram of a volume of fluid within a pipe. The pressure difference between two locations balances the viscous shear stresses between the fluid and the pipe's inner surface. The fluid is in equilibrium, and it moves with a constant speed.

the flow in the pipe is laminar. Experimental evidence shows that laminar flow occurs in pipes for $Re < 2000$. Recalling the no-slip condition (page 232), the velocity of the fluid is precisely zero on the inner surface of the pipe. By the principle of symmetry, the fluid will move fastest along the pipe's centerline, and decrease to zero velocity at the pipe's radius R (Figure 6.18). In fact, the velocity distribution in laminar flow is a parabolic function of radius, as given by the equation

$$v = v_{max}\left(1 - \left(\frac{r}{R}\right)^2\right) \qquad \text{(Special case: } Re < 2000\text{)} \qquad (6.7)$$

where r is measured outward from the pipe's centerline. The maximum velocity of the fluid

$$v_{max} = \frac{d^2 \Delta p}{16 \mu L} \qquad \text{(Special case: } Re < 2000\text{)} \qquad (6.8)$$

occurs at the pipe's centerline, and it depends on the pressure drop, the pipe's diameter $d = 2R$, the fluid's viscosity μ, and the pipe's length L. Engineers will

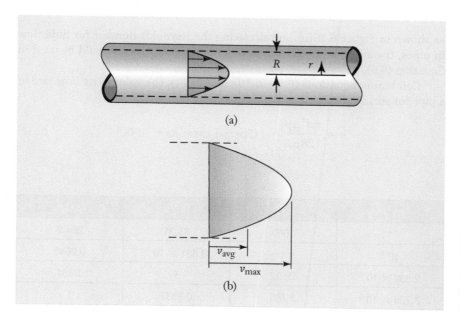

(a)

(b)

Figure 6.18

Steady laminar flow of a fluid in a pipe. The fluid's velocity is the greatest along the pipe's centerline, changes parabolically over the cross section, and falls to zero on the surface of the pipe.

typically specify the diameter of a pipe, rather than its radius, because the diameter is easier to measure. The term $\Delta p / L$ in Equation (6.8) is interpreted as the drop in pressure that occurs per unit length of the pipe.

Aside from the fluid's speed, we are often more interested in knowing the volume ΔV of fluid that flows through the pipe during a certain time interval Δt. In that respect, the quantity

$$q = \frac{\Delta V}{\Delta t} \tag{6.9}$$

Volumetric flow rate is called the *volumetric flow rate*, and it has the dimensions of m³/s or L/s in the SI and ft³/s or gal/s in the USCS. Conversion factors between those dimensions are given in Table 6.3. We can read off the conversion factors for the dimension m³/s from the first row of this table:

$$1\frac{m^3}{s} = 1000\frac{L}{s} = 35.31\frac{ft^3}{s} = 264.2\frac{gal}{s}$$

The volumetric flow rate is related to the pipe's diameter and to the velocity of the fluid flowing through it. Figure 6.19 depicts a cylindrical element of fluid having cross-sectional area A and length Δx flowing through a pipe. In the time interval Δt, the volume of fluid that flows past any cross-section of the pipe is given by $\Delta V = A\Delta x$. Since the average speed of the fluid in the pipe is $v_{avg} = \Delta x / \Delta t$, the volumetric flow rate is also given by

$$q = Av_{avg} \tag{6.10}$$

When the flow is laminar, the fluid's average velocity and the maximum velocity in Equation (6.8) are related by

$$v_{avg} = \frac{1}{2}v_{max} \qquad \text{(Special case: } Re < 2000) \tag{6.11}$$

as shown in Figure 6.18(b). In calculating the Reynolds number for fluid flow in pipes, the average velocity v_{avg} and the pipe's diameter d should be used in Equation (6.6).

Combining Equations (6.8), (6.10), and (6.11), the volumetric flow rate in a pipe for steady, incompressible, laminar flow is

$$q = \frac{\pi d^4 \Delta p}{128 \mu L} \qquad \text{(Special case: } Re < 2000) \tag{6.12}$$

Table 6.3

Conversion Factors Between USCS and SI Units for Volumetric Flow Rate

m³/s	L/s	ft³/s	gal/s
1	1000	35.31	264.2
10^{-3}	1	3.531×10^{-2}	0.2642
2.832×10^{-2}	28.32	1	7.481
3.785×10^{-3}	3.785	0.1337	1

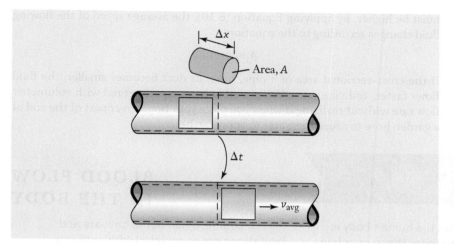

Figure 6.19

Volumetric flow rate in a pipe.

This equation is called *Poiseuille's law*, and like Equations (6.7), (6.8), and (6.11), it is limited to laminar flow conditions. As measured by volume, the rate of fluid flowing in a pipe grows with the fourth power of its diameter, is directly proportional to the pressure drop, and is inversely proportional to the pipe's length. Poiseuille's law can be used to determine the volumetric flow rate when the pipe's length, diameter, and pressure drop are known; to find the pressure drop; and to determine the necessary diameter for a pipe when q, L, and Δp are given.

Poiseuille's law

When a fluid's compressibility is insignificant, the volumetric flow rate remains constant even when there are changes in the pipe's diameter, as depicted in Figure 6.20. In essence, because fluid can't build up and become concentrated at some point in the pipe, the amount of fluid that flows into the pipe must also flow out of it. In Figure 6.20, the cross-sectional area of the pipe decreases between sections 1 and 2. For the same volume of fluid to flow out of the constriction per unit time as flows into it, the fluid's velocity in section 2

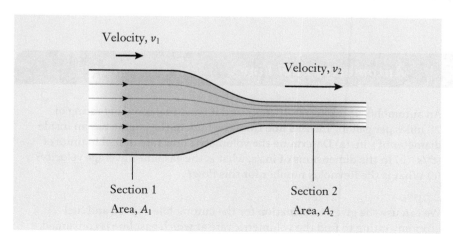

Figure 6.20

Fluid flow in a pipe having a constriction.

must be higher. By applying Equation (6.10), the average speed of the flowing fluid changes according to the equation

$$A_1v_1 = A_2v_2 \qquad (6.13)$$

If the cross-sectional area of a pipe, hose, or duct becomes smaller, the fluid flows faster, and vice versa. You probably have experimented with volumetric flow rate without realizing it when you place your finger over part of the end of a garden hose to cause the water to spray farther.

Focus On   BLOOD FLOW IN THE BODY

The flow of blood within the human body is an interesting application for these principles of fluids engineering. The circulatory system is regulated in part by muscles in the arterial walls that contract and expand to control how much blood will flow through different parts of the body. Blood pressure is determined by the output of the heart and by the extent of contraction and resistance in the capillary system. The diameter of a blood vessel can be an important factor in determining pressure, since Δp in Equation (6.12) scales with the fourth power of diameter. If the diameter of a blood vessel decreases by a factor of 2 and all other factors remain the same, the pressure must increase by a factor of 16 to maintain the same volumetric flow rate. Some medications for hypertension are designed around this principle, and they lower blood pressure by limiting the contraction of blood vessel walls.

Of course, a number of caveats and limitations are associated with applying Poiseuille's equation to describe blood flow in the human body. First of all, blood does not flow in a steady manner because it pulses with every beat of the heart. In addition, the analysis behind Poiseuille's law stipulates that the pipe is rigid, but blood vessels are flexible and compliant tissue. Blood is also not a homogeneous liquid. At very small scales, the diameter of a capillary is actually smaller than the blood cells themselves, and those cells must bend and fold in order to pass through the tiniest capillaries. While Poiseuille's equation may not be directly applicable, it nevertheless does provide a qualitative indication that blood-vessel diameter and the partial blockage of blood vessels are important factors in influencing blood pressure.

Example 6.6 | *Automotive Fuel Line*

An automobile is driving at 40 mph, and it has a fuel economy rating of 28 miles per gallon. The fuel line from the tank to the engine has an inside diameter of $\frac{3}{8}$ in. (a) Determine the volumetric flow rate of fuel in units of ft³/s. (b) In the dimensions of in./s, what is the gasoline's average velocity? (c) What is the Reynolds number for this flow?

Approach
We can use the given information for the automobile's speed and fuel economy rating to find the volumetric rate at which gasoline is consumed.

Example 6.6 | continued

Then, knowing the cross-sectional area of the fuel line [Equation (5.1)], we will apply Equation (6.10) to determine the flow's average velocity. Lastly, we calculate the Reynolds number using Equation (6.6), where the characteristic length is the fuel line's diameter. The density and viscosity of gasoline are listed in Table 6.1.

Solution

(a) The volumetric flow rate is the ratio of the vehicle's speed and the fuel economy rating:

$$q = \frac{40 \ \text{mi/h}}{28 \ \text{mi/gal}}$$

$$= 1.429 \left(\frac{\text{mi}}{\text{h}}\right)\left(\frac{\text{gal}}{\text{mi}}\right)$$

$$= 1.429 \frac{\text{gal}}{\text{h}}$$

Converting from a per-hour basis to a per-second basis, this rate is equivalent to

$$q = \left(1.429 \frac{\text{gal}}{\text{h}}\right)\left(\frac{1}{3600} \frac{\text{h}}{\text{s}}\right)$$

$$= 3.968 \times 10^{-4} \left(\frac{\text{gal}}{\text{h}}\right)\left(\frac{\text{h}}{\text{s}}\right)$$

$$= 3.968 \times 10^{-4} \frac{\text{gal}}{\text{s}}$$

By applying a factor from Table 6.3, we next convert this quantity to the dimensionally consistent basis of ft^3:

$$q = \left(3.968 \times 10^{-4} \frac{\text{gal}}{\text{s}}\right)\left(0.1337 \frac{\text{ft}^3/\text{s}}{\text{gal/s}}\right)$$

$$= 5.306 \times 10^{-5} \left(\frac{\text{gal}}{\text{s}}\right)\left(\frac{\text{ft}^3/\text{s}}{\text{gal/s}}\right)$$

$$= 5.306 \times 10^{-5} \frac{\text{ft}^3}{\text{s}}$$

(b) The cross-sectional area of the fuel line is

$$A = \frac{\pi}{4}(0.375 \ \text{in.})^2 \qquad \leftarrow \left[A = \frac{\pi d^2}{4}\right]$$

$$= 0.1104 \ \text{in}^2$$

Example 6.6 | continued

or $A = 7.670 \times 10^{-4}$ ft^2 since 1 ft = 12 in. The average velocity of the gasoline is

$$v_{\text{avg}} = \frac{5.306 \times 10^{-5} \text{ ft}^3/\text{s}}{7.670 \times 10^{-4} \text{ ft}^2} \quad \leftarrow [q = Av_{\text{avg}}]$$

$$= 6.917 \times 10^{-2} \left(\frac{\text{ft}^2 \cdot \text{ft}}{\text{s}}\right)\left(\frac{1}{\text{ft}^2}\right)$$

$$= 6.917 \times 10^{-2} \frac{\text{ft}}{\text{s}}$$

or $v_{\text{avg}} = 0.8301$ in./s

(c) Since the fuel line's diameter is $d = \frac{3}{8}$ in. $= 3.125 \times 10^{-2}$ ft, the Reynolds number for this flow is

$$Re = \frac{(1.32 \text{ slug/ft}^3)(6.917 \times 10^{-2} \text{ ft/s})(3.125 \times 10^{-2} \text{ ft})}{6.1 \times 10^{-6} \text{ slug/(ft} \cdot \text{s)}} \quad \leftarrow \left[Re = \frac{\rho v l}{\mu}\right]$$

$$= 467.8 \left(\frac{\text{slug}}{\text{ft}^3}\right)\left(\frac{\text{ft}}{\text{s}}\right)(\text{ft})\left(\frac{\text{ft} \cdot \text{s}}{\text{slug}}\right)$$

$$= 467.8$$

Discussion

Since $Re < 2000$, the flow is expected to be smooth and laminar. A higher fuel economy rating would result in a lower flow rate, velocity, and Reynolds number, as less fuel is needed to maintain the same vehicle speed.

$$q = 5.306 \times 10^{-5} \frac{\text{ft}^3}{\text{s}}$$

$$v_{\text{avg}} = 0.8301 \text{ in./s}$$

$$Re = 467.8$$

▶ 6.6 Drag Force

When mechanical engineers design automobiles, aircraft, rockets, and other vehicles, they generally need to know the drag force F_D that will resist high-speed motion through the air or water (Figure 6.21). In this section, we will discuss the drag force and a related quantity that is known as the *coefficient of drag*, denoted by C_D. That parameter quantifies the extent to which an object is streamlined, and it is used to calculate the amount of resistance that an object will experience as it moves through a fluid (or as fluid flows around it).

Coefficient of drag

Figure 6.21

Virgin Galactic's SpaceShip Two, shown in flight, was designed with adaptive drag capabilities. The twin tails pointing straight back off each wing are designed to pivot upright during descent to increase drag, slowing the vehicle without any risk of instability or overheating. While innovative, this adaptive drag system relies on accurate timing, as evidence by the 2014 crash of a SpaceShip Two vehicle.

Getty Images News/Getty Images

Fluid dynamics

Whereas buoyancy forces (Section 6.3) develop even in stationary fluids, the drag force and the lift force (discussed in Section 6.7) arise from the relative motion between a fluid and a solid object. The general behavior of moving fluids and the motions of objects through them define the field of mechanical engineering known as *fluid dynamics*.

For values of the Reynolds number that can encompass either laminar or turbulent flow, the magnitude of the drag force is determined by the equation

$$F_D = \frac{1}{2}\rho A v^2 C_D \qquad (6.14)$$

where ρ is the fluid's density, and the area A of the object facing the flowing fluid is called the *frontal area*. In general, the magnitude of the drag force increases with the area that impinges on the fluid. The drag force also increases with the density of the fluid (for instance, air versus water), and it grows with the square of the speed. If all other factors remain the same, the drag force exerted on an automobile going twice as fast as another would be four times greater.

Frontal area

The drag coefficient is a single numerical value that represents the complex dependency of the drag force on the shape of an object and its orientation relative to the flowing fluid. Equation (6.14) is valid for any object, regardless of whether the flow is laminar or turbulent, provided that one knows the numerical value for the coefficient of drag. However, mathematical equations for C_D are available only for idealized geometries (such as spheres, flat plates, and cylinders) and restricted conditions (such as a low Reynolds number). In many cases, mechanical engineers must still obtain practical results, even for situations where the coefficient of drag can't be described mathematically. In such cases, engineers rely on a combination of laboratory experiments and computer simulations. Through such methods, numerical values for the drag coefficient have been tabulated in the engineering literature for a wide range of applications. The representative data of Table 6.4 (see page 256) can help you to develop intuition for the relative magnitudes of the drag coefficient in different circumstances. For instance, a sport utility vehicle, which has a relatively broad

Table 6.4

Numerical Values of
the Drag Coefficient
and Frontal Area for
Different Systems

System	Frontal Area, A		Drag Coefficient, C_D
	ft²	m²	
Economy sedan (60 mph)	20.8	1.9	0.34
Sports car (60 mph)	22.4	2.1	0.29
Sport-utility vehicle (60 mph)	29.1	2.7	0.45
Bicycle and rider (racing)	4.0	0.37	0.9
Bicycle and rider (upright)	5.7	0.53	1.1
Person (standing)	6.7	0.62	1.2

and blunt front, has a larger coefficient of drag (and a larger frontal area as well) than a sports car. By using the values of C_D and A in this table, as well as other published data, the drag force can be calculated using Equation (6.14).

To illustrate, Figure 6.22 depicts the drag force that acts on a sphere as fluid flows around it (or the force that would develop as the sphere moves through the fluid). Regardless of whether the sphere or fluid is moving, the relative velocity v between the two is the same. The sphere's frontal area as seen by the fluid is $A = \pi d^2/4$. In fact, the interaction between a sphere and a fluid has important engineering applications to devices that deliver medicine through aerosol sprays, to the motion of pollutant particles in the atmosphere, and to the modeling of raindrops and hailstones in storms. Figure 6.23 shows how the drag coefficient for a smooth sphere changes as a function of the Reynolds number over the range $0.1 < Re < 100,000$. At the higher values, say $1000 < Re < 100,000$, the drag coefficient is nearly constant at the value $C_D \approx 0.5$.

When it is combined with Figure 6.23, Equation (6.14) can be used to calculate the drag force acting on a sphere. When Re is very low, so that the flow is smooth and laminar, the drag coefficient is given approximately by

$$C_D \approx \frac{24}{Re} \qquad \text{(Special case for a sphere: } Re < 1) \qquad (6.15)$$

This result is shown as the dotted line in the logarithmic representation of Figure 6.23. You can see that the result from Equation (6.15) agrees with the more general C_D curve only when the Reynolds number is less than one.

Figure 6.22

The drag force depends on the relative velocity between a fluid and an object. (a) Fluid flows past a stationary sphere and creates the drag force F_D. (b) The fluid is now stationary and the sphere moves through it.

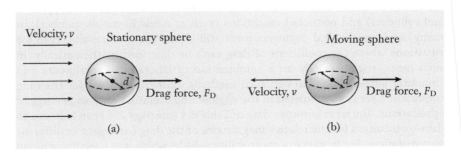

Velocity, v Stationary sphere

Drag force, F_D

(a)

Moving sphere

Velocity, v Drag force, F_D

(b)

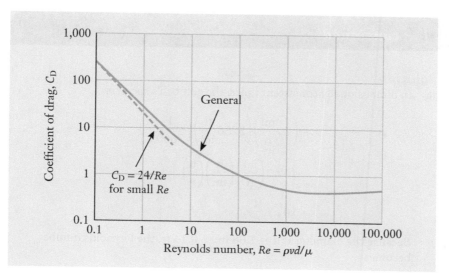

Figure 6.23

Dependence of the drag coefficient for a smooth sphere on the Reynolds number (solid line), and the value predicted for low Re from Equation (6.15) (dashed line).

The substitution of Equation (6.15) into Equation (6.14) gives the low-speed approximation for the sphere's drag force

$$F_D \approx 3\pi\mu dv \qquad \text{(Special case for a sphere: } Re < 1) \qquad (6.16)$$

Although this result is valid only for low speeds, you can see how the magnitude of F_D increases in relation to speed, the fluid's viscosity, and the sphere's diameter. Experiments show that Equation (6.16) starts to underestimate the drag force as the Reynolds number grows. Because the fundamental character of a fluid's flow pattern changes from laminar to turbulent with Re (Figure 6.15), Equations (6.15) and (6.16) are applicable only when Re is less than one and the flow is unmistakably laminar. When those equations are used in any calculation, you should be sure to verify that the condition $Re < 1$ is met.

Example 6.7 | *Golf Ball in Flight*

A 1.68 in. diameter golf ball is driven off a tee at 70 mph. Determine the drag force acting on the golf ball by (a) approximating it as a smooth sphere and (b) using the actual drag coefficient of 0.27.

Approach
To find the drag force in part (a), we will begin by calculating the Reynolds number [Equation (6.6)] with the density and viscosity of air listed in Table 6.1. If Re for this situation is less than one, then it is acceptable to apply Equation (6.16). On the other hand, if the Reynolds number is larger, that equation can't be used, and we will instead find the drag force from Equation (6.14) with C_D determined from Figure 6.23. We will use this latter approach to find the drag force in part (b).

Example 6.7 | *continued*

Solution

(a) In dimensionally consistent units, the golf ball's speed is

$$v = \left(70\,\frac{mi}{h}\right)\left(5280\,\frac{ft}{mi}\right)\left(\frac{1}{3600}\,\frac{h}{s}\right)$$

$$= 102.7\left(\frac{mi}{h}\right)\left(\frac{ft}{mi}\right)\left(\frac{h}{s}\right)$$

$$= 102.7\,\frac{ft}{s}$$

Because the diameter is $d = 1.68$ in. $= 0.14$ ft, the Reynolds number becomes

$$Re = \frac{(2.33 \times 10^{-3}\ \text{slug/ft}^3)(102.7\ \text{ft/s})(0.14\ \text{ft})}{3.8 \times 10^{-7}\ \text{slug/(ft} \cdot \text{s)}} \qquad \leftarrow \left[Re = \frac{\rho v l}{\mu}\right]$$

$$= 8.813 \times 10^4 \left(\frac{slug}{ft^3}\right)\left(\frac{ft}{s}\right)(ft)\left(\frac{ft \cdot s}{slug}\right)$$

$$= 8.813 \times 10^4$$

Since this value is much greater than one, we can't apply Equation (6.16) or Equation (6.15) for the drag coefficient. We refer to Figure 6.23 to see that this value for Re lies in the flat portion of the curve where $C_D \approx 0.5$. The frontal area of the ball is

$$A = \frac{\pi(0.14\ \text{ft})^2}{4} \qquad \leftarrow \left[A = \frac{\pi d^2}{4}\right]$$

$$= 1.539 \times 10^{-2}\ \text{ft}^2$$

The drag force then becomes

$$F_D = \frac{1}{2}\left(2.33 \times 10^{-3}\,\frac{slug}{ft^3}\right)(1.539 \times 10^{-2}\ \text{ft}^2)\left(102.7\,\frac{ft}{s}\right)^2 (0.5)$$

$$= 9.452 \times 10^{-2}\left(\frac{slug}{ft^3}\right)(ft^2)\left(\frac{ft^3}{s^2}\right) \qquad \leftarrow \left[F_D = \frac{1}{2}\rho A v^2 C_D\right]$$

$$= 9.452 \times 10^{-2}\,\frac{slug \cdot ft}{s^2}$$

$$= 9.452 \times 10^{-2}\ \text{lb}$$

where the final grouping of USCS dimensions is equivalent to the pound [Equation (3.3)].

Example 6.7 | *continued*

(b) With $C_D = 0.27$ instead, the drag force is reduced to

$$F_D = \frac{1}{2}\left(2.33 \times 10^{-3}\frac{\text{slug}}{\text{ft}^3}\right)(1.539 \times 10^{-2}\text{ ft}^2)\left(102.7\frac{\text{ft}}{\text{s}}\right)^2(0.27)$$

$$= 5.104 \times 10^{-2}\left(\frac{\text{slug}}{\text{ft}^3}\right)(\text{ft}^2)\left(\frac{\text{ft}^2}{\text{s}^2}\right) \quad \leftarrow \left[F_D = \frac{1}{2}\rho A v^2 C_D\right]$$

$$= 5.104 \times 10^{-2}\frac{\text{slug} \cdot \text{ft}}{\text{s}^2}$$

$$= 5.104 \times 10^{-2}\text{ lb}$$

Discussion

The simplification of treating the golf ball as a smooth sphere neglects the manner in which the dimples change the airflow around the ball, lowering its coefficient of drag. By reducing C_D, the ball will travel farther in flight. The aerodynamic behavior of golf balls is also significantly influenced by any spin that the ball might have when it is driven from the tee. Spin can provide extra lift force and enable the ball to travel farther than would otherwise be possible.

$$F_D = 9.452 \times 10^{-2}\text{ lb (smooth sphere)}$$
$$F_D = 5.104 \times 10^{-2}\text{ lb (actual)}$$

Example 6.8 | *Bicycle Rider's Air Resistance*

In Example 3.9, we made the order-of-magnitude approximation that a person can produce 100–200 W of power while exercising. Based on the upper value of 200 W, estimate the speed at which a person can ride a bicycle at that level of exertion and still overcome air resistance (Figure 6.24, see page 260). Express your answer in the dimensions of mph. In the calculation, neglect rolling resistance between the bicycle's tires and the road, as well as the friction in the bearings, chain, and sprockets. A mathematical expression for power is $P = Fv$, where F is a force's magnitude and v is the speed of the object to which the force is applied.

Approach

To find the speed, we assume that the only resistance that the rider encounters is air drag. The drag force is given by Equation (6.14), and Table 6.4 lists $C_D = 0.9$ with a frontal area of $A = 4.0$ ft^2 for a cyclist in

Example 6.8 | *continued*

Figure 6.24

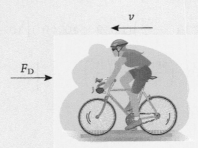

racing position. To calculate the drag force, we need the numerical value for the density of air, which is given as 2.33×10^{-3} slugs/ft^3 in Table 6.1.

Solution

We first will obtain a general symbolic equation for the cyclist's speed, and then we will substitute numerical values into it. The power produced by the rider balances the loss from air resistance

$$P = \left(\frac{1}{2}\rho A v^3 C_D\right) v \qquad \leftarrow [P = Fv]$$

$$= \frac{1}{2}\rho A v^3 C_D$$

The rider's velocity becomes

$$v = \sqrt[3]{\frac{2P}{\rho A C_D}}$$

Next, we will substitute numerical values into this equation. Converting the cyclist's power to dimensionally consistent units with the factor listed in Table 3.6,

$$P = (200\ \text{W})\left(0.7376\frac{(\text{ft} \cdot \text{lb})/\text{s}}{\text{W}}\right)$$

$$= 147.5\ (\text{W})\left(\frac{(\text{ft} \cdot \text{lb})/\text{s}}{\text{W}}\right)$$

$$= 147.5\frac{\text{ft} \cdot \text{lb}}{\text{s}}$$

Example 6.8 | *continued*

the cyclist's speed becomes

$$v = \sqrt[3]{\frac{2(147.5(\text{ft} \cdot \text{lb})/\text{s})}{(2.33 \times 10^{-3} \text{ slugs/ft}^3)(4.0 \text{ ft}^2)(0.9)}} \quad \leftarrow \left[v = \sqrt[3]{\frac{2P}{\rho A C_\text{D}}} \right]$$

$$= 32.77 \sqrt[3]{\frac{\text{lb} \cdot \text{ft}^2 \cdot \text{ft}^2}{\text{slug} \cdot \text{ft}^2 \cdot \text{s}}}$$

$$= 32.77 \sqrt[3]{\frac{((\text{slug} \cdot \text{ft})/\text{s}^2) \cdot \text{ft}^2}{\text{slug} \cdot \text{s}}}$$

$$= 32.77 \sqrt[3]{\frac{\text{ft}^3}{\text{s}^3}}$$

$$= 32.77 \frac{\text{ft}}{\text{s}}$$

Finally, we convert this value to the conventional units of mph:

$$v = \left(32.77 \frac{\text{ft}}{\text{s}} \right) \left(\frac{1 \text{ mi}}{5280 \text{ ft}} \right) \left(3600 \frac{\text{s}}{\text{h}} \right)$$

$$= 22.34 \left(\frac{\text{ft}}{\text{s}} \right) \left(\frac{\text{mi}}{\text{ft}} \right) \left(\frac{\text{s}}{\text{h}} \right)$$

$$= 22.34 \text{ mph}$$

Discussion

We recognize that this calculation overstates the rider's speed because other forms of friction have been neglected in our assumptions. Nevertheless, the estimate is quite reasonable, and, interestingly, the resistance that air drag provides is significant. The power required to overcome air drag increases with the cube of the cyclist's speed. If the rider exercises twice as hard, the speed will increase only by a factor of $\sqrt[3]{2} \approx 1.26$ or 26% faster.

$$v = 22.34 \text{ mph}$$

Example 6.9 | *Engine Oil's Viscosity*

An experimental engine oil with density of 900 kg/m³ is being tested in a laboratory to determine its viscosity. A 1 mm diameter steel sphere is released into a much larger, transparent tank of the oil (Figure 6.25, see page 262). After the sphere has fallen through the oil for a few seconds, it

Example 6.9 | *continued*

Figure 6.25

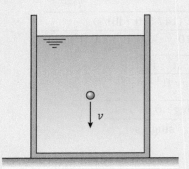

falls at a constant speed. A technician records that the sphere takes 9 s to pass marks on the container that are separated by 10 cm. Knowing that the density of steel is 7830 kg/m³, what is the oil's viscosity?

Approach

To calculate the oil's viscosity, we will use an equilibrium-force balance involving the drag force to determine the speed at which the steel sphere falls through the oil. When the sphere is initially dropped into the tank, it will accelerate downward with gravity. After a short distance, however, the sphere will reach a constant, or terminal, velocity. At that point, the drag F_D and buoyancy F_B forces that act upward in the free body diagram exactly balance the sphere's weight w (Figure 6.26). The viscosity can then be found from the drag force following Equation (6.16). Finally, we will double-check the solution by verifying that the Reynolds number is less than one, a requirement when Equation (6.16) is used.

Figure 6.26

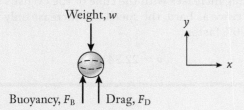

Solution

The terminal velocity of the sphere is $v = (0.10 \text{ m})/(9 \text{ s}) = 0.0111 \text{ m/s}$. By applying the equilibrium force balance in the y-direction,

$$F_D + F_B - w = 0 \quad \leftarrow \left[\sum_{i=1}^{N} F_{y,i} = 0 \right]$$

Example 6.9 | *continued*

and the drag force is $F_D = w - F_B$. The sphere's volume is

$$V = \frac{\pi(0.001 \text{ m})^3}{6} \quad \leftarrow \left[V = \frac{\pi d^3}{6} \right]$$

$$= 5.236 \times 10^{-10} \text{ m}^3$$

and its weight is

$$w = \left(7830 \frac{\text{kg}}{\text{m}^3} \right) \left(9.81 \frac{\text{m}}{\text{s}^2} \right) (5.236 \times 10^{-10} \text{ m}^3) \quad \leftarrow [w = \rho g V]$$

$$= 4.022 \times 10^{-5} \left(\frac{\text{kg}}{\text{m}^3} \right) \left(\frac{\text{m}}{\text{s}^2} \right) (\text{m}^3)$$

$$= 4.022 \times 10^{-5} \frac{\text{kg} \cdot \text{m}}{\text{s}^2}$$

$$= 4.022 \times 10^{-5} \text{ N}$$

When the sphere is immersed in the oil, the buoyancy force that develops is

$$F_B = \left(9000 \frac{\text{kg}}{\text{m}^3} \right) \left(9.81 \frac{\text{m}}{\text{s}^2} \right) (5.236 \times 10^{-10} \text{ m}^3) \quad \leftarrow [F_B = \rho_{\text{fluid}} g V_{\text{object}}]$$

$$= 4.623 \times 10^{-6} \left(\frac{\text{kg}}{\text{m}^3} \right) \left(\frac{\text{m}}{\text{s}^2} \right) (\text{m}^3)$$

$$= 4.623 \times 10^{-6} \frac{\text{kg} \cdot \text{m}}{\text{s}^2}$$

$$= 4.623 \times 10^{-6} \text{ N}$$

The drag force is therefore

$$F_D = (4.022 \times 10^{-5} \text{ N}) - (4.623 \times 10^{-6} \text{ N})$$

$$= 3.560 \times 10^{-5} \text{ N}$$

Following Equation (6.16) for the drag force on a sphere, the viscosity becomes

$$\mu = \frac{3.560 \times 10^{-5} \text{ N}}{3\pi (0.001 \text{ m}) (0.0111 \text{ m/s})} \quad \leftarrow [F_D = 3\pi \mu d v]$$

$$= 0.3403 \left(\frac{\text{kg} \cdot \text{m}}{\text{s}^2} \right) \left(\frac{1}{\text{m}} \right) \left(\frac{\text{s}}{\text{m}} \right)$$

$$= 0.3403 \frac{\text{kg}}{\text{m} \cdot \text{s}}$$

Example 6.9 | *continued*

Discussion

As a double check on the consistency of our solution, we will verify that the terminal velocity is low enough so that our assumption of $Re < 1$ and our use of Equation (6.16) is appropriate. By using the measured viscosity value, we calculate the Reynolds number to be

$$Re = \frac{(900 \text{ kg/m}^3)(0.011 \text{ m/s})(0.001 \text{ m})}{0.3403 \text{ kg/(m} \cdot \text{s)}} \quad \leftarrow \left[Re = \frac{\rho v l}{\mu} \right]$$

$$= 0.0291 \left(\frac{\text{kg}}{\text{m}^3} \right)\left(\frac{\text{m}}{\text{s}} \right)(\text{m})\left(\frac{\text{m} \cdot \text{s}}{\text{kg}} \right)$$

$$= 0.0291$$

Because this is less than one, we have confirmed that it was acceptable to apply Equation (6.16). Had we found otherwise, we would have discarded this prediction, and instead applied Equation (6.14) with the graph of Figure 6.23 for C_D.

$$\mu = 0.3403 \frac{\text{kg}}{\text{m} \cdot \text{s}}$$

▶ 6.7 Lift Force

Similar to drag, the lift force is also produced by the relative motion between a solid object and a fluid. While the drag force acts in parallel to the direction of the fluid's flow, the lift force acts in perpendicular to it. For instance, in the context of the airplane shown in Figure 6.27, the high-speed flow of air around the wings generates a vertical lift force F_L that balances the plane's weight. Four forces are shown acting on the aircraft in flight: the plane's weight w, the thrust F_T produced by its jet engines, the lift F_L produced by the wings, and the drag F_D that opposes the motion of the plane through the

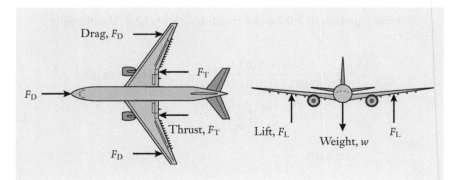

Figure 6.27

The weight, thrust, lift, and drag forces acting on an airplane.

air. In steady level flight, those forces balance to keep the plane in equilibrium: the engines' output overcomes wind resistance, and the wings' lift supports the weight of the aircraft. Lift force is important not only for aircraft wings and other flight control surfaces, but also for the design of propeller, compressor, and turbine blades; ship hydrofoils; and the body contours of commercial and racing automobiles.

The area of mechanical engineering that encompasses the interaction between structures and the air flowing around them is called *aerodynamics*. When engineers are performing aerodynamic analysis of drag and lift forces, invariably they make approximating assumptions with respect to geometry and the behavior of the fluid. For instance, neglecting a fluid's viscosity or compressibility can simplify an engineering analysis problem enough for an engineer to develop a preliminary design or interpret the results of measurements.

Aerodynamics

On the other hand, engineers are mindful of the fact that such assumptions, while meaningful in some applications, could be inappropriate in others. As in our usage of Equations (6.15) and (6.16), mechanical engineers are aware of the assumptions and restrictions involved when certain equations are applied. For example, in this chapter we assume that air is a continuous fluid, not a collection of discrete molecules that collide with one another. This assumption is effective for most applications involving the flow of air around automobiles and aircraft at low speeds and altitudes. However, for aircraft or space vehicles in the upper atmosphere, that assumption might not be appropriate, and engineers and scientists might instead examine fluid forces from the standpoint of the kinetic theory of gases.

Mechanical engineers often use *wind tunnels*, such as those shown in Figure 6.28, to conduct experiments to understand and measure the forces generated when air flows around a solid object. Wind tunnels enable engineers to optimize the performance of aircraft, spacecraft, missiles, and rockets at

Wind tunnel

Figure 6.28

An aerial view of several wind tunnels that are used for aircraft and flight research.

Courtesy of NASA.

Figure 6.29

During a test in a supersonic wind tunnel, shock waves propagate off the scale model of an upper-atmosphere research aircraft.

Courtesy of NASA.

Shock wave

Airfoil

Angle of attack

different speeds and flight conditions. In such a test, a scale model of the object is built and attached to a special fixture for measuring the drag and lift forces developed by the airstream (Figure 1.13). Wind tunnels can also be used to perform experiments related to high-altitude and supersonic flight. Figure 6.29 depicts the shock waves that propagate off the scale model of an upper-atmosphere research aircraft. *Shock waves* occur when the speed of air flowing around an aircraft exceeds the speed of sound, and they are responsible for the noise known as a sonic boom. Wind tunnels are also used to design automobile profiles and surfaces to reduce wind resistance and therefore increase fuel economy. Low-speed wind tunnels are even applied in the realm of Olympic sports to help ski jumpers improve their form, and to help engineers design bicycles, cycling helmets, and sporting apparel having improved aerodynamic performance.

In addition to speed, the magnitude of the lift force generated by an aircraft's wing (more generally known as an *airfoil*) depends on its shape and tilt relative to the air stream (Figure 6.30). The inclination of an airfoil is called its *angle of attack* α, and, up to a point known as the stall condition, the lift force will generally grow with α. In Figure 6.31, air flows past the wing and produces the vertical lift force F_L. Lift is associated with the pressure difference that exists between the airfoil's upper and lower surfaces. Since the force exerted by a fluid on an object can be interpreted as the product of pressure and area, the lift force develops because the pressure on the wing's lower surface is greater than on the upper surface.

In fact, airfoils are designed to take advantage of a trade-off between the pressure, velocity, and elevation of a flowing fluid, a result that is attributed to the eighteenth-century mathematician and physicist Daniel Bernoulli. This principle is based on the assumptions that no energy is dissipated because of the fluid's viscosity, no work is performed on the fluid or by it, and no heat

Figure 6.30

A high-performance military fighter climbs at a steep 55° angle of attack.

Reprinted with permission of Lockheed-Martin.

transfer takes place. Together, those restrictions frame the flowing fluid as a conservative energy system, and *Bernoulli's equation* becomes

Bernoulli's equation

$$\frac{p}{\rho} + \frac{v^2}{2} + gh = \text{constant} \qquad (6.17)$$

Here p and ρ are the pressure and density of the fluid, v is its speed, g is the gravitational acceleration constant, and h is the height of the fluid above some

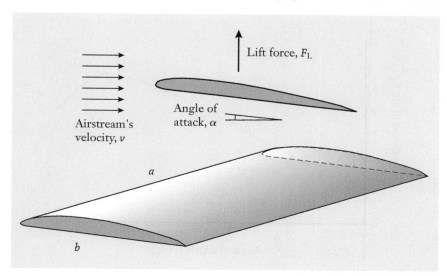

Figure 6.31

The lift force is created as fluid flows past an airfoil that is inclined by the angle of attack α.

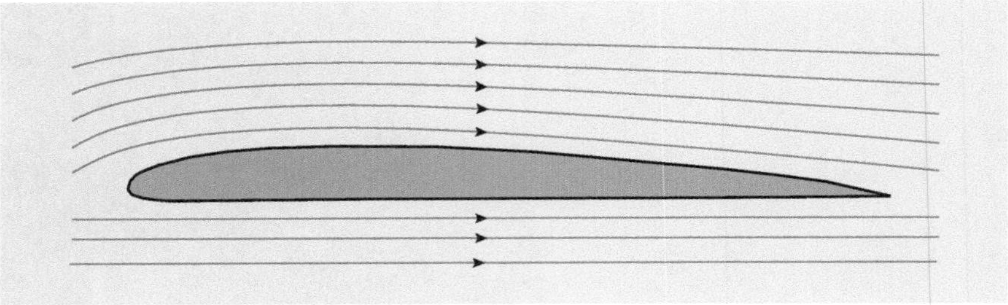

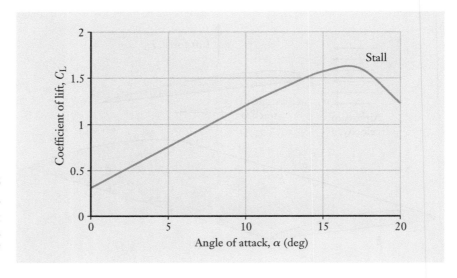

Figure 6.32

Flow pattern around the cross section of an airfoil.

reference point. The three terms on the left-hand side of the equality represent the work of pressure forces, the kinetic energy of the flowing fluid, and its gravitational potential energy. This equation is dimensionally consistent, and each quantity in it has the units of energy-per-unit-mass-of-fluid. For flow around the airfoil in Figure 6.32, the gravitational potential energy term gh can be neglected because the elevation changes are small when compared to the pressure and velocity. The quantity $(p/\rho) + (v^2/2)$ is therefore approximately constant as air moves along the airfoil's upper and lower surfaces. For a variety of reasons, air speeds up as it flows over the airfoil's upper surface, and its pressure therefore decreases by an amount corresponding to Equation (6.17). The airfoil's lift force is generated by the imbalance between the lower-pressure top surface and the higher-pressure bottom surface.

In a manner similar to our treatment of the drag force in Equation (6.14), the lift force created by the fluid acting on the airfoil is quantified by the *coefficient of lift* C_L and calculated from the expression

Coefficient of lift

$$F_L = \frac{1}{2}\rho A v^2 C_L \tag{6.18}$$

where the area is given by $A = ab$ in Figure 6.31. Numerical values for the coefficient of lift are available and tabulated in the engineering literature for various designs of airfoils. Figure 6.33 illustrates the dependence of C_L on the

Figure 6.33

Coefficient of lift for a type of airfoil that could be used in a small single-engine aircraft.

angle of attack for a type of airfoil that could be used in a small single-engine airplane. Aircraft wings generally have some amount of camber in which the airfoil's centerline is curved slightly with a concave downward shape. In this manner, the airfoil is able to develop a finite lift coefficient even with zero angle of attack. In Figure 6.33, for instance, $C_L \approx 0.3$ even with $\alpha = 0°$, enabling the aircraft to develop lift on a runway or during level flight. Aircraft wings are only slightly cambered for efficiency during steady cruising flight. During low-speed flight at takeoff and landing where stall may be a concern, additional camber can be created by extending flaps on the trailing edge of the wings. In addition, the lift coefficient decreases at large angles of attack, resulting in a flight phenomenon known as stall, where the airfoil's ability to develop lift rapidly diminishes.

Focus On   AERODYNAMICS IN SPORTS

When a ball is hit, thrown, or kicked in sports, the ball's trajectory can rapidly change direction. In baseball, this phenomenon is the strategy behind curve ball pitches; in golf, the phenomenon is responsible for slicing or hooking a ball driven off a tee, and in soccer, the ball can be "bent" during a free kick to outmaneuver the goalie. Cricket and tennis are other sports in which the ball can curve during play. In each case, the ball's trajectory depends on its complex interaction with the surrounding air and on the amount of spin imparted to the ball as it is thrown, hit, or kicked. During flight, a ball is subjected to lift and drag forces and to a sideways force commonly known as the Magnus effect and related to the ball's spin. As any ball spins, a thin layer of air is dragged alongside in rotation because of the air's viscosity. The roughness of the ball's surface, as well as seams and laces on the ball, are also important factors in causing air to rotate with the ball. On the side of the ball with rotation and airflow acting in the same direction, the air's speed increases, and the pressure drops following Bernoulli's principle. On the ball's other side, the rotation and airflow act in opposite directions, and the pressure is correspondingly higher. The imbalance in pressure between the ball's two sides produces a sideways force that causes the ball's trajectory to curve. This principle is used in the design of a wide range of sports equipment, including golf clubs and soccer cleats that effectively place a spin on golf and soccer balls upon impact.

Understanding the impact of flow, drag, buoyancy, and lift also helps engineers develop new technologies to improve the performance of athletes, including advanced materials for Olympic swimsuits; innovative bike, wheel, and helmet designs for cyclists and triathletes; and aerodynamic bodysuits for speed skaters.

Summary

I n this chapter, we introduced the physical properties of fluids, the distinction between laminar and turbulent flows, and the forces that are known as buoyancy, drag, and lift. Mechanical engineers categorize substances as being either solids or fluids, and the distinction between the two centers around how they respond to a shear stress. Although a solid material will deform by only a small amount and resist shear stress by virtue of its stiffness, a fluid will respond by flowing in a continuous steady motion.

Mechanical engineers apply the principles of fluids engineering to applications such as aerodynamics, biomedical engineering, microfluidics, and sports engineering. The flow of fluids through pipes, hoses, and ducts is an example of this diversity. In addition to the distribution of water, gasoline, natural gas, and air through piping systems, the principles behind fluid flow in pipes can be applied in studies of the human circulatory and respiratory systems. The primary variables, symbols, and conventional units that are used in this chapter are summarized in Table 6.5, and the key equations are listed in Table 6.6.

Quantity	Conventional Symbols	Conventional Units USCS	SI
Area	A	ft^2	m^2
Coefficient of drag	C_D	—	—
Coefficient of lift	C_L	—	—
Density	ρ	slug/ft^3	kg/m^3
Force			
Buoyancy	F_B	lb	N
Drag	F_D	lb	N
Lift	F_L	lb	N
Weight	w	lb	N
Length			
Characteristic length	l	ft	m
Pipe length	L	ft	m
Mach number	Ma	—	—
Pressure	p	psi, psf	Pa
Reynolds number	Re	—	—
Shear stress	τ	psi	Pa
Time interval	Δt	s	s
Velocity	v, v_{avg}, v_{max}	ft/s	m/s
Viscosity	μ	$\text{slug/(ft} \cdot \text{s)}$	$\text{kg/(m} \cdot \text{s)}$
Volume	$V, \Delta V$	gal, ft^3	L, m^3
Volumetric flow rate	q	$\text{gal/s, ft}^3\text{/s}$	$\text{L/s, m}^3\text{/s}$

Table 6.5

Quantities, Symbols, and Units that Arise in Fluids Engineering

Bernoulli's equation	$\dfrac{p}{\rho} + \dfrac{v^2}{2} + gh = \text{constant}$
Buoyancy force	$F_B = \rho_{\text{fluid}}\, g\, V_{\text{object}}$
Drag force	
General	$F_D = \dfrac{1}{2}\rho A v^2 C_D$
Special case: Sphere with $Re < 1$	$C_D = \dfrac{24}{Re}$
Lift force	$F_L = \dfrac{1}{2}\rho A v^2 C_L$
Pipe flow velocity	$v_{\max} = \dfrac{d^2 \Delta p}{16\mu L}$ $v_{avg} = \dfrac{1}{2} v_{\max}$ $v = v_{\max}\left(1 - \left(\dfrac{r}{R}\right)^2\right)$
Pressure	$p_1 = p_0 + \rho g h$
Reynolds number	$Re = \dfrac{\rho v l}{\mu}$
Shear stress	$\tau = \mu \dfrac{v}{h}$
Volumetric flow rate	$q = \dfrac{\Delta V}{\Delta t}$ $q = A v_{avg}$ $q = \dfrac{\pi d^4 \Delta p}{128\,\mu L}$ $A_1 v_1 = A_2 v_2$
Weight	$w = \rho g V$

Table 6.6

Key Equations that Arise in Fluids Engineering

The buoyancy force develops when an object is immersed in a fluid, and it is related to the weight of the displaced fluid. Drag and lift forces develop when there is relative motion between a fluid and a solid object, and they encompass situations where the fluid is stationary and the object is moving (as in the case of an automobile); the fluid is moving and the object is stationary (as in wind loading on a building); or some combination of the two. The magnitudes of drag and lift forces are generally calculated in terms of drag and lift coefficients, which are numerical quantities that capture the complex dependency of these forces on an object's shape and its orientation relative to the flowing fluid.

Self-Study and Review

6.1. What are the conventional USCS and SI dimensions for the density and viscosity of a fluid?

6.2. In what manner does the pressure within a fluid increase with depth?

6.3. Describe some of the differences between laminar and turbulent flows of a fluid.

6.4. What is the definition of the Reynolds number, and what is its significance?

6.5. Give examples of situations where fluids produce buoyancy, drag, and lift forces, and explain how those forces can be calculated.

6.6. What are the coefficients of drag and lift, and on what parameters do they depend?

6.7. What is Bernoulli's principle?

Problems

P6.1

Convert the viscosity of mercury (1.5×10^{-3} kg/(m · s)) to the dimensions of slug/(ft · s) and centipoise.

P6.2

Michael Phelps won a record-setting 8 gold medals at the 2008 Beijing Olympics. Now imagine if Phelps had competed in a pool filled with pancake syrup. Would you expect his race times to increase, decrease, or stay the same? Research the issue and prepare an approximately 250-word report supporting your answer. Cite at least two references for your information.

P6.3

The fuel tank on a sport-utility wagon holds 14 gal of gasoline. How much heavier is the automobile when the tank is full compared to when it is empty?

P6.4

The pressure at the bottom of an 18 ft deep storage tank for gasoline is how much greater than at the top? Express your answer in the units of psi.

P6.5

Blood pressure is conventionally measured in the dimensions of millimeters in a column of mercury, and the readings are expressed as two numbers, for example, 120 and 80. The first number is called the systolic value, and it is the maximum pressure developed as the heart contracts. The second number is called the diastolic reading, and it is the pressure when the heart is at rest. In the units of kPa and psi, what is the difference in pressure between the given systolic and diastolic readings? The density of mercury is 13.54 Mg/m³.

P6.6

A 6 m high, 4 m wide rectangular gate is located at the end of an open freshwater tank (Figure P6.6). The gate is hinged at the top and held in place by a force F. From Equation (6.4), the pressure is proportional to the depth of the water, and the average pressure p_{avg} exerted on the gate by the water is

$$p_{avg} = \frac{\Delta p}{2}$$

where Δp is the difference in pressure between the bottom of the gate (p_1) and the surface (p_0). The resulting force of the water on the gate is

$$F_{water} = p_{avg}A$$

where A is the area of the gate the water acts upon. The resulting force acts 2 m from the bottom of the gate because the pressure increases with depth. Determine the force required to hold the gate in place.

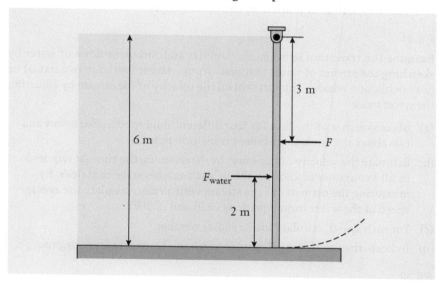

Figure P6.6

P6.7

For the system in P6.6, suppose your design requirement was to minimize the force necessary to hold the gate closed. Would you rather put the hinge at the top of the gate or at the bottom? Support your answer.

P6.8

Make an order-of-magnitude estimate for a safety floatation device that can be used by children who are playing in a swimming pool. The design concept is for one inflatable, annular, plastic balloon to slide over each of the child's arms and up to the shoulder. Take into account the child's weight and buoyancy force, and determine the dimensions that will be suitable for children weighing up to 50 lb.

P6.9

An ancient king's supposedly golden crown had a mass of 3 kg, but it was actually made by a dishonest metal smith from an equal mix of gold (1.93×10^4 kg/m^3) and silver (1.06×10^4 kg/m^3).

(a) Suppose that Archimedes suspended the crown from a string and lowered it into water until it was fully submerged. If the string was then connected to a balance scale, what tension would Archimedes have measured in the string?

(b) If the test was repeated, but this time with the crown replaced by a 3-kg block of pure gold, what tension would be measured?

P6.10

Scuba divers carry ballast weights to have neutral buoyancy. At that condition, the buoyancy force on the diver exactly balances weight, and there is no tendency either to float toward the surface or to sink. In freshwater, a certain diver carries 10 lb of lead alloy ballast of density 1.17×10^4 kg/m^3. During an excursion in seawater, the diver must carry 50% more ballast to remain neutrally buoyant. How much does this diver weigh?

P6.11

Examine the transition between the laminar and turbulent flows of water by sketching the stream of water that exits from a faucet (without an aerator) or hose (without a nozzle). You can control the velocity of the stream by adjusting the water valve.

(a) Make sketches of the flow for four different fluid speeds: two below and two above the laminar-turbulent transition point.

(b) Estimate the velocity of the water by determining the time Δt required to fill a container of known volume, such as a beverage container. By measuring the diameter of the stream with a ruler, calculate the average speed of the water from Equations (6.9) and (6.10). ·

(c) For each speed, calculate the Reynolds number.

(d) Indicate the value of the Reynolds number where turbulence begins.

P6.12

Water flows through a 5 cm diameter pipe at the average velocity of 1.25 m/s.

(a) In the dimensions of L/s, what is the volumetric flow rate?

(b) If the diameter of the pipe is reduced by 20% at a constriction, by what percentage does the water's velocity change?

P6.13

In a pipeline that connects a production oilfield to a tanker terminal, oil having density 1.85 slug/ft^3 and viscosity 6×10^{-3} slug/(ft · s) flows through a 48 in. diameter pipeline at 6 mph. What is the Reynolds number? Is the flow laminar or turbulent?

P6.14

For the parabolic pressure profile in Equation (6.7), show that the average flow velocity is half the maximum value [Equation (6.11)].

P6.15

The average velocity of blood flowing in a certain 4 mm diameter artery in the human body is 0.28 m/s. Calculate the Reynolds number and determine whether the flow is laminar or turbulent. The viscosity and density of blood are approximately 4 cP and 1.06 Mg/m³, respectively.

P6.16

(a) Determine the volumetric flow rate of blood in the artery of P6.15.

(b) Calculate the maximum velocity of blood across the artery's cross section.

(c) Determine the amount by which the blood pressure decreases along each 10 cm length of the artery.

P6.17

The Boeing 787 Dreamliner is designed to be 20% more fuel efficient than the comparable Boeing 767 and flies at an average cruise speed of Mach 0.85. The midsize Boeing 767 has a range of 12,000 km, a fuel capacity of 90,000 L, and flies at Mach 0.80. Assume the speed of sound is 700 mph, and calculate the projected volumetric flow rate of fuel for each of the two Dreamliner engines in m³/s.

P6.18

Assume that the fuel lines for each engine in the Boeing Dreamliner of P6.17 are 7/8 in. in diameter and that the density and viscosity of jet fuel are 800 kg/m³ and 8.0×10^{-3} kg/(m · s), respectively. Calculate the average velocity of the fuel in m/s and the Reynolds number of this flow. Also determine whether the flow is laminar or turbulent.

P6.19

(a) For a 1.25 in. diameter pipe, what is the maximum volumetric flow rate at which water can be pumped and the flow will remain laminar? Express your result in the dimensions of gallons per minute.

(b) What would be the maximum flow rate for SAE 30 oil?

P6.20

At any time, approximately 20 volcanoes are actively erupting on the Earth, and 50–70 volcanoes erupt each year. Over the past 100 years, an average of 850 people have died each year from volcano eruptions. As scientists and engineers study the mechanics of lava flow, accurately predicting the flow rate (velocity) of the lava is critical to saving lives after an eruption. Jeffrey's equation captures the relationship between flow rate and viscosity as:

$$V = \frac{\rho g t^2 \sin(\alpha)}{3\mu}$$

where ρ is the density of the lava, g is gravity, t is the flow thickness, α is the slope, and μ is the lava viscosity. Typical values for the viscosity and density of lava are 4.5×10^3 kg/(m · s) and 2.5 g/cm^3, respectively. Find the velocity of the flow in Figure P6.20 in cm/s and mph.

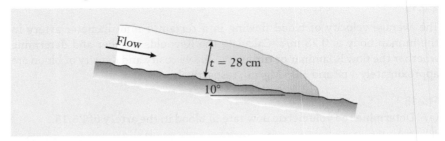

Figure P6.20

P6.21

Using the volcanic flow model and flow parameters from P6.20, prepare two charts.

(a) On one chart, plot the flow velocity in mph as a function of the slope, varying the slope from 0° to 90°.

(b) On the other chart, plot the flow velocity in mph as a function of the flow thickness, varying the thickness from 0 to 300 cm.

(c) Discuss and compare the influence of the slope and the thickness on the flow velocity.

P6.22

A steel storage tank is filled with gasoline. The tank has partially corroded on its inside, and small particles of rust have contaminated the fuel. The rust particles are approximately spherical, and they have a mean diameter of 25 μm and density of 5.3 g/cm^3.

(a) What is the terminal velocity of the particles as they fall through the gasoline?

(b) How long would it take the particles to fall 5 m and settle out of the liquid?

P6.23

A small water droplet in a mist of air is approximated as being a sphere of diameter 1.5 mil. Calculate the terminal velocity as it falls through still air to the ground. Is it reasonable to neglect the buoyancy force in this instance?

P6.24

In a production woodworking shop, 50 μm spherical dust particles were blown into the air while a piece of oak furniture was being sanded.

(a) What is the terminal velocity of the particles as they fall through the air?

(b) Neglecting air currents that are present, how long would it take the cloud of sawdust to settle out of the air and fall 2 m to the ground? The density of dry oak is approximately 750 kg/m^3.

P6.25

(a) A 1.5 mm diameter steel sphere (7830 kg/m^3) is dropped into a tank of SAE 30 oil. What is its terminal velocity?

(b) If the sphere is instead dropped into a different oil of the same density but develops a terminal speed of 1 cm/s, what is the oil's viscosity?

P6.26

A 175 lb skydiver reaches a terminal velocity of 150 mph during free fall. If the frontal area of the diver is 8 ft^2, what are:

(a) The magnitude of the drag force acting on the skydiver?

(b) The drag coefficient?

P6.27

A 14 mm diameter sphere is dropped into a beaker of SAE 30 oil. Over a portion of its downward motion, the sphere is measured to fall at 2 m/s. In the units of newtons, what is the drag force acting on the test sphere?

P6.28

A low-altitude meteorological research balloon, temperature sensor, and radio transmitter together weigh 2.5 lb. When inflated with helium, the balloon is spherical with a diameter of 4 ft. The volume of the transmitter can be neglected when compared to the balloon's size. The balloon is released from ground level and quickly reaches its terminal ascent velocity. Neglecting variations in the atmosphere's density, how long does it take the balloon to reach an altitude of 1000 ft?

P6.29

A submarine releases a spherical flotation buoy containing a radio beacon. The buoy has a diameter of 1 ft and weighs 22 lb. The coefficient of drag for the submerged buoy is $C_D = 0.45$. At what steady speed will the buoy rise to the surface?

P6.30

Place Equation (6.16) into the form of Equation (6.14) and show that the coefficient of drag for small Reynolds numbers is given by Equation (6.15).

P6.31

(a) A luxury sports car has a frontal area of 22.4 ft^2 and a 0.29 coefficient of drag at 60 mph. What is the drag force on the vehicle at this speed?

(b) A sport-utility vehicle has $C_D = 0.45$ at 60 mph and a slightly larger frontal area of 29.1 ft^2. What is the drag force in this case?

P6.32

A certain parachute has a drag coefficient of $C_D = 1.5$. If the parachute and skydiver together weigh 225 lb, what should the frontal area of the parachute be so that the skydiver's terminal velocity is 15 mph when approaching the ground? Is it reasonable to neglect the buoyant force that is present?

P6.33

Submarines dive by opening vents that allow air to escape from ballast tanks and water to flow in and fill them. In addition, diving planes located at the bow are angled downward to help push the boat below the surface. Calculate the diving force produced by a 20 ft^2 hydroplane having a lift coefficient of 0.11 as the boat cruises at 15 knots (1 knot = 1.152 mph).

P6.34

(a) Use the principle of dimensional consistency to show that when Bernoulli's equation is written in the form

$$p + \frac{1}{2}\rho v^2 + \rho g h = \text{constant}$$

each term has the dimension of pressure.

(b) When the equation is written alternatively as

$$\frac{p}{\rho g} + \frac{v^2}{2g} + h = \text{constant}$$

show that each term has the dimension of length.

P6.35

A device called a venturi flowmeter can be used to determine the velocity of a flowing fluid by measuring the change in its pressure between two points (Figure P6.35). Water flows through a pipe, and, at the constriction, the cross-sectional area of the pipe reduces from A_1 to the smaller value A_2. Two pressure sensors are located just before and after the constriction, and the change $p_1 - p_2$ that they measure is enough information to determine the water's velocity. By using Equations (6.13) and (6.17), show that the velocity downstream of the constriction is given by

$$v_2 = \sqrt{\frac{2(p_1 - p_2)}{\rho(1 - (A_2/A_1)^2)}}$$

This so-called venturi effect is the principle behind the operation of such hardware as automotive carburetors and the aspirators that deliver pharmaceutical products to medical patients by inhalation.

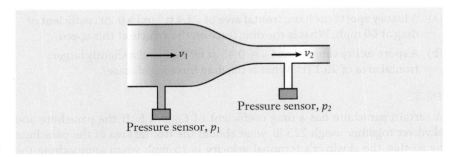

Figure P6.35

P6.36

Water flows through a circular pipe having a constriction in diameter from 1 in. to 0.5 in. The velocity of the water just upstream of the constriction is 4 ft/s. By using the result of P6.35, determine:

(a) The velocity of the water downstream from it.

(b) The pressure drop across the constriction.

P6.37*

An off-shore oil platform uses desalinated seawater as its source for a freshwater delivery system. The delivery system has four stages and starts with a large intake pipe for the seawater, eventually desalinating the flowing seawater and delivering freshwater to a precise location using a small nozzle. The intake pipe diameter in stage one is 0.9 m, and the seawater is flowing at 0.11 m/s. The pipe diameter in stage two is 0.5 m. Desalination occurs between stage two and three. The pipe diameter for the fresh water in stage three is 0.2 m, and the nozzle diameter for the final stage is 0.05 m.

(a) Assuming that the volumetric flow rate for each stage remains constant and neglecting any gravitational effects, determine whether the flow is laminar or turbulent at each stage of the process.

(b) If your group was now commissioned to design and deploy a similar four-stage delivery system in an environment where ensuring laminar flow at each stage was critical, what specific design recommendations would you have? Assume that the diameter of each subsequent stage is reduced, but that the overall size of the system can be changed and a different fluid can be used.

P6.38*

A recent study estimated that the energy consumed by the entire world over a calendar year was around 537 quadrillion BTU. Over the year, this equates to the following amount of power:

$$537 \times 10^{15} \frac{BTU}{year} \times 1055 \frac{J}{BTU} = 567 \times 10^{18} \frac{joules}{year}$$

$$567 \times 10^{18} \frac{joules}{year} \times \frac{1 \ year}{365 \ days} \times \frac{1 \ days}{24 \ hour} \times \frac{1 \ hour}{3600 \ sec} = 17 \times 10^{12} \frac{J}{s} = 17 \ TW$$

This staggering amount of power raises concerns about energy scarcity and has prompted the exploration for untapped energy sources. One major source of energy is stored in the world's dynamic fluid resources, whether in gas form (e.g., air and wind) or in liquid form (e.g., oceans and rivers). Engineers are creating innovative ways to leverage the dynamic properties of such fluids as energy sources. Your group's task is to research four innovative methods in which the movement of a fluid in gas or liquid form is being harnessed. Describe the primary mechanism for how the method generates energy, and how much energy each method could contribute annually to global production. Produce a professional technical report with clear assumptions and calculations.

6.39*

Your group has been hired to design a raft that can carry a payload of 275 kg. The basic design of the raft is pontoon style and will have 4 hollow pipes of the same diameter, two on each side, which hold up a platform that runs the length of the pipes (See Figure P6.39). The platform has a width of 2 meters and a thickness of 7.5 cm. Your group has 3 different size pipes and 3 different platform materials to choose from. Your pipe options are as follows:

- Pipe A: Inner Diameter: 0.2540 m, Outer Diameter: 0.2667 m, Cost: $7.10/pipe
- Pipe B: Inner Diameter: 0.2667 m, Outer Diameter: 0.2794 m, Cost: $14.50/pipe
- Pipe C: Inner Diameter: 0.2794 m, Outer Diameter: 0.2921 m, Cost: $21.80/pipe

Pipes can only be purchased in 3.048 m lengths, but can be cut into shorter lengths at no additional cost. Your platform options are as follows:

- Platform X: Density: 498 kg/m^3, Cost: $25.05/m
- Platform Y: Density: 560 kg/m^3, Cost: $16.10/m
- Platform Z: Density: 605 kg/m^3, Cost: $9.85/m

While the pipes can only be purchased in 3.048 m lengths, you can purchase platform lengths to the nearest centimeter.

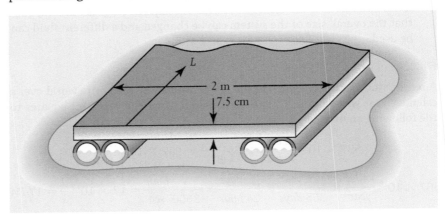

Figure P6.39

When considering your choice for the pipes and platform, you must make sure that your design is the most cost-effective option and that the cross sections of each pipe are half-way submerged in the water when supporting the platform and payload, as shown in Figure P6.39. Assume that the raft is used in fresh water at about 20 °C, that the pipes can be fixed to each other without gaining or losing length, that any added mass due to fastenings or other joining material is included in the platform material properties, that the collective mass of the pipes is negligible, and that the payload will be balanced in the middle of the platform.

References

Adian, R. J., "Closing in on Models of Wall Turbulence," *Science*, July 9, 2010, pp. 155–156.

Ehrenman, G., "Shrinking the Lab Down to Size," *Mechanical Engineering*, May 2004, pp. 26–32.

Jeffreys, H., "The Flow of Water in an Inclined Channel of Rectangular Section," *Philosophical Magazine*, XLIX, 1925.

Kuethe, A. M., and Chow, C. Y., *Foundations of Aerodynamics: Bases of Aerodynamic Design*, 5th ed. Hoboken, NJ: John Wiley and Sons, 1998.

Nichols, R. L., "Viscosity of Lava," *The Journal of Geology*, **47**(3), 1939, pp. 290–302.

Olson, R. M., and Wright, S. J., *Essentials of Engineering Fluid Mechanics*, 5th ed. New York: Harper and Row, 1990.

Ouellette, J., "A New Wave of Microfluidic Devices," *The Industrial Physicist*, August–September, 2003, pp. 14–17.

Thilmany, J., "How Does Beckham Bend It?" *Mechanical Engineering*, April 2004, p. 72.

CHAPTER 7

CHAPTER
OBJECTIVES

Thermal and Energy Systems

CHAPTER OBJECTIVES

- Calculate various energy, heat, work, and power quantities that are encountered in mechanical engineering, and express their numerical values in the SI and USCS.

- Describe how heat is transferred from one location to another by the processes of conduction, convection, and radiation.

- Apply the principle of energy conservation for a mechanical system.

- Explain how heat engines operate and understand the limitations on their efficiency.

- Outline the basic operating principles behind two-stroke and four-stroke internal-combustion engines and electric power plants.

▶ 7.1 Overview

Element 6:
Thermal and energy
systems

U p to this point, we have explored the first *five* elements of the mechanical engineering profession: mechanical design, professional practice, forces in structures and machines, materials and stresses, and fluids engineering. In Chapter 1, mechanical engineering was described in a nutshell as the process of developing machinery that either consumes or produces power. With that view, we now turn our attention to the sixth element of mechanical engineering and the practical topic of *thermal and energy systems* (Figure 7.1). This field encompasses such hardware as internal-combustion engines, aircraft propulsion, heating and cooling systems, and electrical power generation through both renewable (solar, wind, hydroelectric, geothermal, and biomass) and nonrenewable (oil, petroleum, natural gas, coal, and nuclear) sources. It will become more and more important for engineers to study and solve energy issues; three of the 14 NAE grand challenges (Chapter 2) are directly related to energy issues: making solar energy economical, providing energy from fusion, and developing carbon sequestration methods.

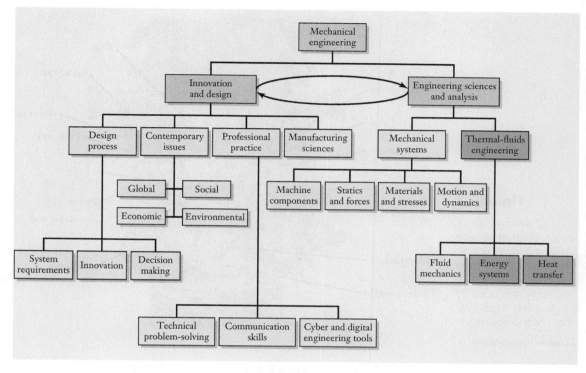

Figure 7.1

Relationship of the topics emphasized in this chapter (shaded blue boxes) relative to an overall program of study in mechanical engineering.

Although you can neither see energy nor hold it in your hand, energy is needed to accelerate an object, stretch it, heat it, and elevate it. The characteristics of energy, the different forms that energy can take, and the methods by which energy can be converted from one form to another lie at the very heart of mechanical engineering. In the internal-combustion engine of Figure 7.2 (see page 284), for instance, diesel fuel is burned to release thermal energy. In turn, the engine converts thermal energy into the rotation of its crankshaft and ultimately into the motion of a vehicle. Machinery that consumes or produces power often involves processes for converting the chemical energy that is stored in a fuel into thermal energy, converting thermal energy into mechanical work and shaft rotation, and moving energy between locations for the purposes of heating or refrigeration.

In the first portion of this chapter, we will introduce the physical principles and terminology needed to understand the operation and efficiency of such thermal and energy systems. Later, in Sections 7.6–7.7, we will apply those ideas to two-stroke and four-stroke engines, and electrical power generation.

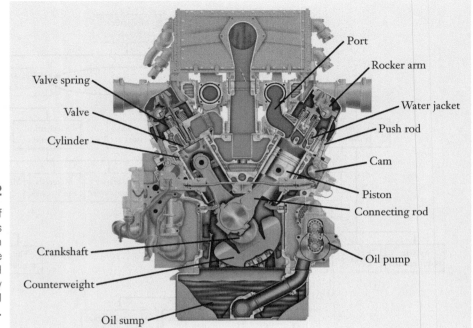

Figure 7.2

Cutaway view of an engine that powers off-road construction equipment. The engine converts the thermal energy produced by burning diesel fuel into mechanical work.

Reprinted courtesy of Caterpillar, Inc.

▶ 7.2 Mechanical Energy, Work, and Power

Gravitational Potential Energy

Gravitational acceleration

Near the surface of the Earth, the *acceleration of gravity* is given by the standard acceleration values

$$g = 32.174 \, \frac{\text{ft}}{\text{s}^2} \approx 32.2 \, \frac{\text{ft}}{\text{s}^2}$$

$$g = 9.8067 \, \frac{\text{m}}{\text{s}^2} \approx 9.81 \, \frac{\text{m}}{\text{s}^2}$$

that have been adopted by international agreement for sea level and a latitude of 45°. *Gravitational potential energy* is associated with changing the elevation of an object within a gravitational field, and it is measured relative to a reference height, for instance the ground or the top of a workbench. The change U_g in gravitational potential energy as an object moves through the vertical distance Δh is given by

$$U_g = mg\Delta h \tag{7.1}$$

where m is the object's mass. When $\Delta h > 0$, the gravitational potential energy increases ($U_g > 0$) as the object is elevated. Conversely, the gravitational potential energy decreases ($U_g < 0$) as the object is lowered and $\Delta h < 0$. Gravitational potential energy is stored by virtue of vertical position.

ft · lb	J	Btu	kW · h
1	1.356	1.285×10^{-3}	3.766×10^{-7}
0.7376	1	9.478×10^{-4}	2.778×10^{-7}
778.2	1055	1	2.930×10^{-4}
2.655×10^{6}	3.600×10^{6}	3413	1

Table 7.1

Conversion Factors between Various Units for Energy and Work in the USCS and SI

For calculations involving energy and work, it is conventional to use the unit joule in the SI, and ft · lb in the USCS. In Table 3.2, the joule is listed as a derived unit having the definition $1\,J = 1\,N \cdot m$. A prefix can also be applied in the SI to represent either small or large quantities. As examples, $1\,kJ = 10^{3}\,J$ and $1\,MJ = 10^{6}\,J$. In some circumstances, the *British thermal unit* (Btu) and *kilowatt-hour* (kW · h) are also used as dimensions for energy and work, and those unit choices will be discussed later in this chapter. The conversion factors among four commonly used units for energy and work are listed in Table 7.1. Reading off the table's first row, for instance, we see that

British thermal unit
Kilowatt-hour

$$1\,\text{ft} \cdot \text{lb} = 1.356\,\text{J} = 1.285 \times 10^{-3}\,\text{Btu} = 3.766 \times 10^{-7}\,\text{kW} \cdot \text{h}$$

Elastic Potential Energy

Elastic potential energy is stored by an object when it is stretched or bent in the manner described by Hooke's law (Section 5.3). For a spring having stiffness k, the elastic potential energy stored within it is given by

$$U_e = \frac{1}{2}k\Delta L^2 \tag{7.2}$$

where ΔL is the spring's elongation, defined as the distance that it has been stretched or compressed. If the spring had an original length of L_0 and if it was then stretched to the new length L after a force had been applied, the elongation would be $\Delta L = L - L_0$. Although ΔL can be positive when the spring is stretched and negative when it is compressed, the elastic potential energy is always positive. As discussed in Chapter 5, k has the units of force-per-unit length, and engineers generally use N/m and lb/in. in the SI and USCS, respectively, as the dimensions for stiffness. You should note that Equation (7.2) can be applied whether or not a machine component actually looks like a coiled spring. In Chapter 5, the force–deflection behavior of a rod in tension and compression was discussed, and in that case the stiffness was given by Equation (5.7).

Kinetic Energy

Kinetic energy is associated with an object's motion. As forces or moments act on a machine, they cause its components to move and store kinetic energy by virtue of velocity. The motion can be in the form of vibration (for instance, the cone of a stereo speaker), rotation (the flywheel attached to an engine's

Figure 7.3

Work of a force that is applied to a piston sliding within a cylinder. (a) The force reinforces displacement as the gas in the cylinder is compressed ($\Delta d > 0$). (b) The force opposes displacement as the gas expands ($\Delta d < 0$).

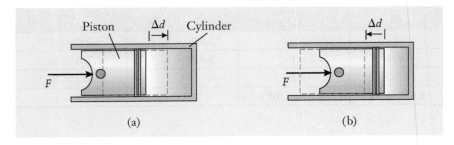

crankshaft), or translation (the straight-line motion of the piston in an engine or compressor). For an object of mass m that moves in a straight line with velocity v, the kinetic energy is defined by

$$U_k = \frac{1}{2}mv^2 \tag{7.3}$$

You can verify that J and ft $\cdot$ lb are appropriate units for kinetic energy, just as they are for gravitational and elastic potential energy.

Work of a Force

The *work of a force* is illustrated in Figure 7.3 in the context of a piston that slides horizontally in its cylinder. This situation arises in internal-combustion engines, air compressors, and pneumatic and hydraulic actuators. In Figure 7.3(a), the force F is applied to the piston to compress the gas in the cylinder while the piston moves to the right. On the other hand, if the gas already has been compressed to a high pressure and the piston moves to the left [Figure 7.3(b)], the force F can be applied to resist that expansion. Those two situations are analogous to the compression and power strokes occurring in an automotive engine. The work W of the force as the piston moves through the distance Δd is defined by

$$W = F\Delta d \tag{7.4}$$

In Figure 7.3(a), the work of the force is positive ($\Delta d > 0$) because the force acts in the same direction as the piston's motion. Conversely, the work is negative if the force opposes motion with $\Delta d < 0$ [Figure 7.3(b)].

Power

Power, the last of the quantities introduced in this section, is defined as the rate at which work is performed. When a force performs work during the interval of time Δt, the average power is

$$P_{\text{avg}} = \frac{W}{\Delta t} \tag{7.5}$$

As work is performed more rapidly, Δt becomes smaller, and the average power increases accordingly. Engineers conventionally express power in the units of watt (1 W = 1 J/s) in SI, and either (ft $\cdot$ lb)/s or horsepower (hp) in USCS.

(ft · lb)/s	W	hp
1	1.356	1.818×10^{-3}
0.7376	1	1.341×10^{-3}
550	745.7	1

Table 7.2

Conversion Factors between Various Units for Power in USCS and SI

Table 7.2 lists conversion factors among those choices of units. Reading the first row, for instance, we see that

$$1\frac{\text{ft} \cdot \text{lb}}{\text{s}} = 1.356\,\text{W} = 1.818 \times 10^{-3}\,\text{hp}$$

In particular, the horsepower is equivalent to 550 ft · lb/s.

Example 7.1 | *Power Conversion Factor*

Beginning with the definitions of the derived units (ft · lb)/s and W in the USCS and SI, verify the conversion factors between them in Table 7.2.

Approach
The watt is defined by work being performed at the rate of one joule per second. To relate the two units for power, we must convert the dimensions for force and length appearing in the definition of the joule (1 J = 1 N · m). Referring to Table 3.6, 1 ft = 0.3048 m and 1 lb = 4.448 N.

Solution
The conversion factor is given by

$$1\frac{\text{ft} \cdot \text{lb}}{\text{s}} = \left(1\frac{\text{ft} \cdot \text{lb}}{\text{s}}\right)\left(0.3048\frac{\text{m}}{\text{ft}}\right)\left(4.448\frac{\text{N}}{\text{lb}}\right)$$

$$= 1.356\left(\frac{\text{ft} \cdot \text{lb}}{\text{s}}\right)\left(\frac{\text{m}}{\text{ft}}\right)\left(\frac{\text{N}}{\text{lb}}\right)$$

$$= 1.356\frac{\text{N} \cdot \text{m}}{\text{s}}$$

$$= 1.356\frac{\text{J}}{\text{s}}$$

$$= 1.356 \text{ W}$$

where we have used the definition of the derived unit joule from Table 3.2.

Discussion
The conversion factor between the dimensions watt and (ft · lb)/s is the reciprocal of the conversion factor from (ft · lb)/s to watts. Thus, 1 W = $(1.356)^{-1}$ (ft · lb)/s = 0.7376 (ft · lb)/s.

$$1\frac{\text{ft} \cdot \text{lb}}{\text{s}} = 1.356 \text{ W}$$

$$1 \text{ W} = 0.7376\frac{\text{ft} \cdot \text{lb}}{\text{s}}$$

Example 7.2 | *Potential Energy Stored in a U-Bolt*

Calculate the elastic potential energy stored in the two straight sections of the U-bolt examined in Examples 5.2 and 5.3.

Figure 7.4

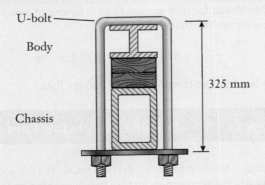

U-bolt

Body

Chassis

325 mm

Approach

We are tasked with finding the potential energy stored in the straight sections of the U-bolt as a result of the elongation of these sections. With the dimensions and material properties given in those examples, the spring stiffness for one straight section can be found by applying Equation (5.7), which is repeated here

$$k = \frac{EA}{L} \qquad (5.7)$$

Each straight section of the U-bolt, having length $L = 325$ mm and cross-sectional area $A = 7.854 \times 10^{-5}$ m^2, was found to stretch by $\Delta L = 78.82$ μm $= 7.882 \times 10^{-5}$ m. The elastic potential energy is found by using Equation (7.2).

Solution

By using the rule-of-thumb value for the elastic modulus of steel (page 184), the stiffness of one of the U-bolt's straight sections is

$$k = \frac{(210 \times 10^9 \text{ Pa})(7.854 \times 10^{-5} \text{ m}^2)}{0.325 \text{ m}} \qquad \leftarrow \left[k = \frac{EA}{L} \right]$$

$$= 5.075 \times 10^7 \left(\frac{\text{N}}{\text{m}^2} \right)(\text{m}^2)\left(\frac{1}{\text{m}} \right)$$

$$= 5.075 \times 10^7 \frac{\text{N}}{\text{m}}$$

Here we used the definition of the derived unit pascal (1 Pa = 1 N/m^2) in manipulating the dimensions for the U-bolt's elastic modulus. The elastic potential energy becomes

$$U_e = \frac{1}{2}\left(5.075 \times 10^7 \frac{\text{N}}{\text{m}} \right)(7.882 \times 10^{-5} \text{ m})^2 \qquad \leftarrow \left[U_e = \frac{1}{2}k\Delta L^2 \right]$$

Example 7.2 | continued

$$= 0.1576 \left(\frac{N}{m} \right) (m)^2$$

$$= 0.1576 \ N \cdot m$$

$$= 0.1576 \ J$$

Discussion

Because the U-bolt has two identical straight sections, the amount of stored potential energy is $2(0.1576 \ J) = 0.3152 \ J$, a relatively modest amount. The U-bolt could also have kinetic energy if it is moving and gravitational potential energy if it is elevated.

$$\boxed{U_e = 0.3152 \ J}$$

Example 7.3 | Jet Aircraft's Kinetic Energy

Calculate the kinetic energy of a Boeing 767 airliner that is loaded to its maximum weight of 350,000 lb and travels at 400 mph. Express the kinetic energy in both USCS and SI dimensions.

Approach

We are tasked with finding the instantaneous kinetic energy of an airliner travelling at a given velocity and having a given weight. To apply Equation (7.3), we must first determine the aircraft's mass based on its weight (which is given) and convert the speed to dimensionally consistent units.

Solution

The aircraft's mass is

$$m = \frac{3.5 \times 10^5 \ lb}{32.2 \ ft/s^2} \quad \leftarrow [w = mg]$$

$$= 1.087 \times 10^4 \frac{lb \cdot s^2}{ft}$$

$$= 1.087 \times 10^4 \ slugs$$

where we applied the definition of the derived unit slug from Equation (3.2). By using the definition of the derived unit mile from Table 3.5, the aircraft's speed in dimensionally consistent units is

$$v = \left(400 \frac{mi}{h} \right) \left(5280 \frac{ft}{mi} \right) \left(\frac{1}{3600} \frac{h}{s} \right)$$

$$= 586.7 \left(\frac{mi}{h} \right) \left(\frac{ft}{mi} \right) \left(\frac{h}{s} \right)$$

$$= 586.7 \frac{ft}{s}$$

Example 7.3 | *continued*

The kinetic energy then becomes

$$U_k = \frac{1}{2}(1.087 \times 10^4 \text{ slugs})\left(586.7 \frac{\text{ft}}{\text{s}}\right)^2 \quad \leftarrow \left[U_k = \frac{1}{2}mv^2\right]$$

$$= 1.871 \times 10^9 \left(\frac{\text{slug} \cdot \text{ft}}{\text{s}^2}\right)(\text{ft})$$

$$= 1.871 \times 10^9 \text{ ft} \cdot \text{lb}$$

Referring to Table 7.1, this amount of energy is equivalent to

$$U_k = (1.871 \times 10^9 \text{ ft} \cdot \text{lb})\left(1.356 \frac{\text{J}}{\text{ft} \cdot \text{lb}}\right)$$

$$= 2.537 \times 10^9 \text{ (ft} \cdot \text{lb})\left(\frac{\text{J}}{\text{ft} \cdot \text{lb}}\right)$$

$$= 2.537 \times 10^9 \text{ J}$$

$$= 2.537 \text{ GJ}$$

in the SI.

Discussion
While manipulating the dimensions in this example, we used the definition of the derived unit slug in the USCS (1 slug = 1 (lb $\cdot$ s^2)/ft), and the SI prefix "giga" for the factor of 1 billion. Also, recognize that an airliner's gravitational potential energy would be significant during flight.

$$U_k = 1.871 \times 10^9 \text{ ft} \cdot \text{lb}$$
$$U_k = 2.537 \text{ GJ}$$

Example 7.4 | *Elevator's Power Requirement*

Use an order-of-magnitude calculation to estimate the capacity of an electric motor that will power a freight elevator in a four-story building. Express your estimate in the units of horsepower. The elevator car weighs 500 lb, and it should be able to carry an additional 2500 lb of freight.

Approach
To make this calculation of the elevator's power capacity, we will neglect air resistance, friction in the elevator's drive mechanism, and all other sources of inefficiency. We also need to make reasonable estimates for information that is not explicitly known at this point in the elevator's design but that is necessary to determine the motor's power rating. We estimate that it will take the elevator 20 s to travel from ground level to the building's top floor and that the total elevation change is 50 ft. The amount of work that the motor must perform is the product of the total weight (3000 lb) and the change in

Example 7.4 | *continued*

height. We can then apply Equation (7.5) to determine the average power produced by the motor.

Solution

The work performed in raising the fully loaded elevator's car is

$$W = (3000 \text{ lb})(50 \text{ ft}) \qquad \leftarrow [W = F\Delta d]$$
$$= 1.5 \times 10^5 \text{ ft} \cdot \text{lb}$$

The average power is

$$P_{avg} = \frac{1.5 \times 10^5 \text{ ft} \cdot \text{lb}}{20 \text{ s}} \qquad \leftarrow \left[P_{avg} = \frac{W}{\Delta t} \right]$$

$$= 7500 \frac{\text{ft} \cdot \text{lb}}{\text{s}}$$

Finally, we will convert this quantity to the dimensions of horsepower by using the factor listed in Table 7.2:

$$P_{avg} = 7500 \left(\frac{\text{ft} \cdot \text{lb}}{\text{s}} \right) \left(1.818 \times 10^{-3} \frac{\text{hp}}{(\text{ft} \cdot \text{lb})/\text{s}} \right)$$

$$= 13.64 \left(\frac{\text{ft} \cdot \text{lb}}{\text{s}} \right) \left(\frac{\text{hp}}{(\text{ft} \cdot \text{lb})/\text{s}} \right)$$

$$= 13.64 \text{ hp}$$

Discussion

As a designer would later account for such other factors as the motor's inefficiency, the possibility of the elevator being overloaded, and a factor of safety, the calculation would become more accurate. From a preliminary standpoint, however, a motor rated for a few tens of horsepower, rather than one with a few horsepower or several hundred horsepower, would seem sufficient for the task at hand.

$$P_{avg} = 13.64 \text{ hp}$$

▶ 7.3 Heat as Energy in Transit

In the previous section, we saw several different forms of energy that can be stored in a mechanical system. In addition to being stored, energy can also be converted from one form to another—for instance, when the potential energy of an elevated object is transferred to kinetic energy as it falls. In a similar manner, the energy stored in a fuel by virtue of its chemical structure is released when the fuel is burned. We view heat as energy that is in transit (or being moved) from one location to another because of a temperature difference.

When they design machinery that consumes and produces energy, mechanical engineers exploit the properties of heat to manage temperature and to move energy

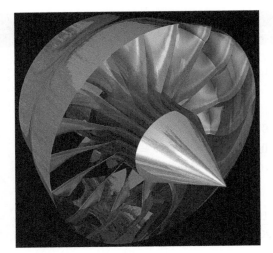

Figure 7.5

Mechanical engineers
use computer-aided
engineering software
to calculate the
temperature of turbine
blades in a jet engine.

Paul Nylander, http://bugman123.com

between locations (Figure 7.5). In this section, we will explore several engineering concepts related to heat, its release as a fuel is burned, and its transfer through the processes known as conduction, convection, and radiation.

Heating Value

When a fuel is burned, the chemical reactions that take place release thermal energy and waste products including water vapor, carbon monoxide, and particulate material. The fuel can be a liquid (such as oil for a furnace), a solid (such as coal in an electrical power plant), or a gas (such as propane in a mass-transit bus). In each case, chemical energy is stored in the fuel, and that energy is liberated during combustion. Mechanical engineers design machinery that manages the release of energy stored in chemical form, and that energy is subsequently converted to more useful forms.

In a furnace, power plant, or gasoline engine, the energy output of the combustion process is measured by a quantity called the *heating value H*. As listed in Table 7.3, the magnitude of a heating value describes the ability of a fuel to release heat. Because heat is defined as energy that is in motion between two locations, heating values are the amount of energy released per unit mass

Type	Fuel	Heating Value, H	
		MJ/kg	Btu/lbm
Gas	Natural gas	47	20.2×10^3
	Propane	46	19.8×10^3
Liquid	Gasoline	45	19.3×10^3
	Diesel	43	18.5×10^3
	Fuel oil	42	18.0×10^3
Solid	Coal	30	12.9×10^3
	Wood	20	8.6×10^3

Table 7.3

Heating Values for
Certain Fuels*

*The numerical values are representative, and values for specific fuels could vary with their chemical composition.

of fuel burned. The greater the amount of fuel that is burned, the more heat is released. To express the energy content of a fuel, the joule is the standard dimension in the SI, and in the USCS the *British thermal unit* (abbreviated Btu) is the most commonly used dimension. Historically, one Btu was defined as the quantity of heat necessary to raise the temperature of one pound of water by one degree on the Fahrenheit scale. In the modern definition, the Btu is equivalent to 778.2 ft · lb or 1055 J, as listed in the energy conversion factors of Table 7.1.

British thermal unit

As discussed in Chapter 3, the slug is the preferred mass unit in mechanical engineering for USCS calculations involving gravitation, motion, momentum, kinetic energy, acceleration, and other mechanics quantities. For engineering calculations related to the thermal and combustion properties of materials, however, it is conventional to use the pound-mass as the derived unit for mass in the USCS. This usage in the USCS is convenient because a quantity of matter that has a mass of 1 lbm also weighs 1 lb. The pound-mass is defined in Equation (3.4), and it is related to the slug by Equation (3.5). As outlined in Table 3.6, the kg and lbm are related through

$$1\,\text{lbm} = 0.4536\,\text{kg} \qquad 1\,\text{kg} = 2.205\,\text{lbm}$$

The conventional dimensions for a fuel's heating value are therefore MJ/kg in the SI and Btu/lbm in the USCS.

In calculations involving the combustion of a fuel, the heat Q that is released by burning mass m is given by

$$Q = mH \qquad (7.6)$$

Referring to the heating values in Table 7.3, when 1 kg of gasoline is burned, 45 MJ of heat are released. Equivalently in the USCS, 19,800 Btu are released when 1 lbm of gasoline is consumed. If we could design an automobile engine that perfectly converts that quantity of heat into kinetic energy, a 1000 kg vehicle could be accelerated to a speed of 300 m/s—roughly the speed of sound at sea level. Of course, that view is idealistic to the extent that no engine can operate at 100% efficiency and convert all of a fuel's stored chemical energy into useful mechanical work. The heating value simply tells us the amount of heat supplied by a fuel, and it is up to the furnace, power plant, or engine to use that heat as efficiently as possible.

The heating values of fuels can vary widely based on their chemical composition and how they are burned. The heating value of coal, for instance, can lie between 15 and 35 MJ/kg depending on the geographical location where it was mined. In addition, when a fuel is burned, water will be present in the by-products of combustion, either in the form of vapor or liquid. A fuel's heating value in a specific application depends on whether the water leaves the combustion process as vapor or is condensed into liquid. Approximately 10% additional heat can be extracted from a fuel by condensing the water vapor produced during combustion and recapturing the heat associated with the vapor-to-liquid phase change. In an automotive engine, the water produced as a combustion by-product is exhausted in vapor form, and the thermal energy contained in it is therefore lost to the environment along with the combustion gases. Likewise, most residential natural gas furnaces are noncondensing; the water vapor produced is exhausted up the furnace's chimney.

Specific Heat

One example of heat flow and temperature change occurs in the commercial production of steel. During processing in a mill, hot steel is rapidly cooled by

Quenching   immersing it in a bath of oil or water. The purpose of this step, known as *quenching*, is to harden the steel by modifying its internal structure. The material's ductility (Section 5.5) can be improved subsequently by a reheating operation called

Tempering   *tempering*. When a steel rod is held at, say, 800 °C and then quenched in oil, heat flows from the steel, and the oil bath becomes warmer. The thermal energy that was stored in the steel decreases, and that loss of energy is manifested by a change in its temperature. Likewise, the temperature of the oil bath and the energy stored in it increase. The ability of a material to accept heat and store it as internal energy depends on the amount of material, its physical properties, and the temperature change. The specific heat is defined as the amount of heat necessary to change the temperature of a unit mass of material by 1°.

The flow of heat between the steel and oil is intangible in the sense that we cannot see it take place. However, we can measure the effect of the heat flow by the temperature changes that occur. Although heat is not the same as temperature, changes in temperature indicate that heat has been transferred. For instance, when the sun heats an asphalt driveway during the afternoon, it remains warm well into the night. The large and massive driveway is able to store more energy than a pot of water that was warmed on a stove to the same temperature.

As heat flows into an object, its temperature rises from an initial value T_0 to T according to

$$Q = mc\,(T - T_0) \tag{7.7}$$

where m is the object's mass. Known as the *specific heat*, the parameter c is a property that captures how materials differ with respect to the amount of heat they must absorb to raise their temperature. Therefore, the specific heat has the units of energy per unit mass per degree of temperature change. The conventional dimensions in the SI and USCS are kJ/(kg · °C) and Btu/(lbm · °F), respectively. You can convert temperature values between the Celsius and Fahrenheit scales using

$$°F = \left(\frac{9}{5}°C\right) + 32$$

$$°C = \left(\frac{5}{9}\right)(°F - 32) \tag{7.8}$$

Table 7.4 lists numerical values of the specific heat for several materials. To raise the temperature of a 1-kg sample of steel by 1°C, 0.50 kJ of heat

| Type | Substance | Specific Heat, c | |
		kJ/(kg · °C)	Btu/(lbm · °F)
Liquid	Oil	1.9	0.45
	Water	4.2	1.0
Solid	Aluminum	0.90	0.21
	Copper	0.39	0.093
	Steel	0.50	0.11
	Glass	0.84	0.20

Table 7.4

Specific Heat of Certain Materials

must be added to it. As you can see from the table, the specific heat of water is higher than the values for oil and metals. That is one reason why water plays an important role in regulating temperature; it can be an efficient means for storing and transporting thermal energy.

In Equation (7.7), if $T > T_0$, then Q is positive and heat flows into the object. Conversely, Q is negative when $T < T_0$; in that case, the negative value implies that the direction of heat flow has reversed so that heat flows out of the object. As we have encountered in other elements of mechanical engineering—forces, moments, angular velocity, and mechanical work—sign conventions are useful to keep track of the direction of certain physical quantities.

For changes in temperature that are not too large, it is acceptable to treat c as being constant. Equation (7.7) does not apply if the material changes phase, say from a solid to liquid or from a liquid to vapor, because the heat added (or removed) during a phase change does not alter the material's temperature. The physical property that quantifies the amount of heat that must flow into or out of a material to produce a phase change is called the *latent heat*.

Latent heat

Transfer of Heat

We have described heat as energy that is being transferred from one location to another because of a temperature difference. The three mechanisms for *heat transfer* are known as conduction, convection, and radiation, and they arise in different mechanical engineering technologies. When you grip the hot handle of a pan on a stove, you feel *conduction* in action. Heat flows from the pan and along the length of the handle to its cooler free end. That process is illustrated by the metallic rod in Figure 7.6. One end of the rod is held at the high temperature T_h, and the other end stays at the lower temperature T_l. Although the rod itself doesn't move, heat flows down it like a current because points along the rod have different temperatures. Just as a change in voltage produces a current in an electrical circuit, the change in temperature $T_h - T_l$ causes heat to be conducted along the rod.

Conduction

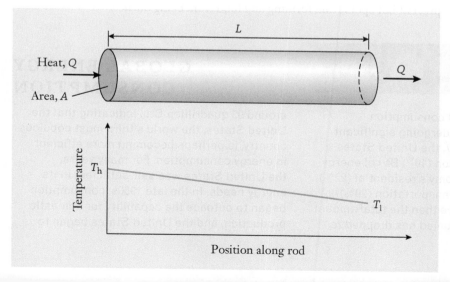

Figure 7.6

Heat conduction along the length of a metal rod.

Table 7.5

Thermal Conductivity
of Certain Materials

Material	Thermal Conductivity, κ	
	W/(m · °C)	(Btu/h)/(ft · °F)
Steel	45	26
Copper	390	220
Aluminum	200	120
Glass	0.85	0.50
Wood	0.3	0.17

The quantity of heat that flows along the rod during a time interval Δt is given by

$$Q = \frac{\kappa A \Delta t}{L}(T_h - T_l) \qquad (7.9)$$

Fourier's law

Thermal
conductivity

This principle is known as *Fourier's law* of heat conduction, and it is named in honor of the French scientist Jean Baptiste Joseph Fourier (1768–1830). Heat conduction occurs in proportion to the cross-sectional area A, and it is inversely proportional to the rod's length L. The material property κ (the lowercase Greek character kappa) is called the *thermal conductivity*, and values for various materials are listed in Table 7.5.

When the thermal conductivity is large, heat flows through the material quickly. Metals generally have large values of κ, and insulators such as fiberglass have low values. With reference to Table 7.5, 200 J of heat will flow each second through a square panel of aluminum that is 1 m on a side and 1 m thick, when the temperature difference across the panel's two faces is 1°C. Even among metals, the numerical values for κ vary significantly. The thermal conductivity of aluminum is over 400% greater than the value for steel, and κ for copper is nearly twice as large again. Aluminum and copper are preferred metals for use in cookware for precisely that reason; heat can more easily flow around a pan to prevent hot spots from forming and food from being burned.

Focus On

GLOBAL ENERGY CONSUMPTION

Energy production and consumption around the world is undergoing significant transformation. In 2007, the United States peaked at 101 quadrillion (10^{15}) Btu of energy consumed by its economy's residential (22%), business (49%), and transportation (29%) sectors. However, since then the total amount of annual energy consumed has dropped to

around 95 quadrillion Btu, indicating that the United States, the world's third most populous country, is perhaps becoming more efficient in energy consumption. For many years, the United States was self-sufficient in its energy needs. In the late 1950s, consumption began to outpace the capability for domestic production, and the United States began to

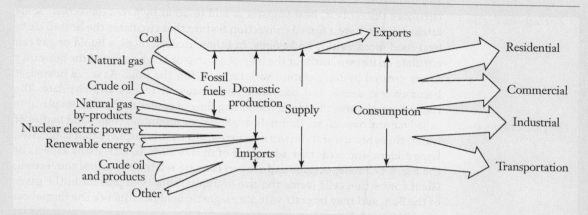

Figure 7.7

A profile of energy production and consumption in the United States.

United States Department of Energy, Energy Information Administration.

import energy to fill the gap between its supply of and demand for energy. Between 2000 and 2010, the United States imported 25–35% of the net energy that it consumed, with crude oil accounting for the bulk of the imports. See Figure 7.7.

While the United States was the world's largest energy consumer in 2000, China, the world's most populous country, is now the world's largest consumer of energy. Within the same time frame, China also went from being the world's largest exporter of coal to the leading importer. Chinese families are plugging in more electronic consumer products and driving more vehicles than ever before. In fact, in 2009, China surpassed the United States as the country that sold the most new cars. Although some see China's sudden rise on the energy consumption curve worrying, it is making significant green efforts, such as nationwide renewable energy targets and the pursuit of greener technologies. Projects

such as the Three Gorges Dam, the largest hydroelectric power plant in the world, highlight China's attempt to shift to cleaner energy sources.

The energy consumption needs in India, the world's second most populous country, are also changing rapidly. In 2012, India consumed 32 quadrillion Btu of energy, placing it as the fourth-largest energy consumer in the world after China, the United States, and Russia. India's transportation, business, and residential infrastructures are growing quickly. As a result, India's energy consumption is projected to increase by almost 3% per year through 2040. Approximately 25% of the population in India does not have access to electricity. However, this number should decrease not only in India but in other places around the world as engineers develop effective and efficient means to harness, store, and distribute renewable and nonrenewable sources of energy.

In addition to conduction, heat can also be transferred by a fluid that is in motion; that process is known as *convection*. The cooling system of an automobile engine, for instance, operates by pumping a mixture of water and antifreeze through passageways inside the engine's block. Excess heat is removed from the engine, transferred temporarily to the coolant by convection,

Convection

Forced convection

and ultimately released into the air by the vehicle's radiator. Because a pump circulates the coolant, heat transfer is said to occur by *forced convection*. Some kitchen ovens have a forced convection feature that circulates the heated air to heat food more quickly and evenly. In other circumstances, a liquid or gas can circulate on its own, without the help of a pump or fan, because of the buoyancy forces created by temperature variations within the fluid. As air is heated, it becomes less dense, and buoyancy forces cause it to rise and circulate. The rising flow of warm fluid (and the falling flow of cooler fluid to fill its place) is

Natural convection

called *natural convection*. Thermals develop near mountain ridges and bodies of water; these are natural convection currents in the atmosphere that sailplanes, hang gliders, and birds take advantage of to stay aloft. In fact, many aspects of the Earth's climate, oceans, and molten core are related to natural convection. Giant convection cells (some the size of Jupiter) are even present in the gases of the Sun, and they interact with its magnetic field to influence the formation of sunspots.

Radiation

The third mechanism of heat transfer is *radiation*, which refers to the emission and absorption of heat without direct physical contact. It is unrelated to radiation in the context of nuclear processes or power generation. Radiation occurs when heat is transmitted by the long infrared waves of the electromagnetic spectrum. Those waves are able to propagate through air and even through the vacuum of space. Energy from the sun reaches the Earth by means of radiation. As the electromagnetic waves are absorbed by air, land, and water, they are converted to heat. The radiators in a home heating system comprise winding metal pipes through which steam or hot water circulates. If you place your hand directly on the radiator, you feel heat flow directly into your hand by conduction. However, even if you stand some distance away and do not touch the radiator, you are still warmed by it through radiation.

Example 7.5 | *Household Energy Consumption*

The average single-family household in the United States consumes 98 million Btu of energy each year. How many tons of coal must be burned to produce that amount of energy?

Approach

To calculate the amount of coal necessary to produce this energy, we will use the heating value of coal listed in Table 7.3 as 12,900 Btu/lbm and apply Equation (7.6) to determine the coal's mass. In Table 3.5, a ton is equivalent to 2000 lb.

Solution

The mass of coal is

$$m = \frac{98 \times 10^6 \text{ Btu}}{12.9 \times 10^3 \text{ Btu/lbm}} \qquad \leftarrow [Q = mH]$$

$$= 7.597 \times 10^3 \text{ (Btu)}\left(\frac{\text{lbm}}{\text{Btu}}\right)$$

$$= 7.597 \times 10^3 \text{ lbm}$$

Example 7.5 | *continued*

Since a 1-lbm object also weighs 1 lb, the weight is $w = 7.597 \times 10^3$ lb. By using the definition of the derived unit ton in the USCS,

$$w = (7.597 \times 10^3 \text{ lb})\left(\frac{1}{2000}\frac{\text{ton}}{\text{lb}}\right)$$

$$= 3.798 \text{ (lb)}\left(\frac{\text{ton}}{\text{lb}}\right)$$

$$= 3.798 \text{ tons}$$

Discussion

Compared to a sedan automobile that weighs 2500 lb, this amount of coal has the equivalent weight of three vehicles. Since coal is a nonrenewable source of energy, mechanical engineers must continue to develop other renewable sources of energy to meet the needs of households around the world.

$$w = 3.798 \text{ tons}$$

Example 7.6 | *Engine's Fuel Consumption*

A gasoline-powered engine generates an average power output of 50 kW. Neglecting any inefficiency that may be present, calculate the volume of fuel consumed each hour. Express your result in the dimensions of liters and gallons.

Approach

To find the volumetric use of fuel, we use the heating value of gasoline, 45 MJ/kg, from Table 7.3. Therefore, following Equation (7.6), 45 MJ of heat are released for each kilogram of gasoline burned. To convert the fuel usage into a volume measure, we will use Table 6.1, which lists the density of gasoline as 680 kg/m³ from Table 6.1.. The volume can be converted between the SI and USCS by using the factor 1 L = 0.2642 gal from Table 3.6.

Solution

In terms of the definition for the derived unit kW, the engine produces

$$W = \left(50\frac{\text{kJ}}{\text{s}}\right)(3600 \text{ s}) \quad \leftarrow \left[P_{avg} = \frac{W}{\Delta t}\right]$$

$$= 1.8 \times 10^5\left(\frac{\text{kJ}}{\text{s}}\right)(\text{s})$$

$$= 1.8 \times 10^5 \text{ kJ}$$

Example 7.6 | *continued*

in 1 h. This energy output is equivalent to 180 MJ. The mass of gasoline that must be burned to release that amount of heat is

$$m = \frac{180 \text{ MJ}}{45 \text{ MJ/kg}} \qquad \leftarrow [Q = mH]$$

$$= 4 \,(\text{MJ}) \left(\frac{\text{kg}}{\text{MJ}} \right)$$

$$= 4 \text{ kg}$$

We next determine the volume of the fuel:

$$V = \frac{4 \text{ kg}}{680 \text{ kg}/m^3} \qquad \leftarrow [m = \rho V]$$

$$= 5.882 \times 10^{-3} \,(\text{kg}) \left(\frac{m^3}{\text{kg}} \right)$$

$$= 5.882 \times 10^{-3} \text{ m}^3$$

By using the definition of the derived unit liter from Table 3.2,

$$V = (5.882 \times 10^{-3} \text{ m}^3) \left(1000 \frac{\text{L}}{\text{m}^3} \right)$$

$$= 5.882 \,(m^3) \left(\frac{\text{L}}{m^3} \right)$$

$$= 5.882 \text{ L}$$

In the USCS, the quantity of fuel is

$$V = (5.882 \text{ L}) \left(0.2642 \frac{\text{gal}}{\text{L}} \right)$$

$$= 1.554 \,(\text{L}) \left(\frac{\text{gal}}{\text{L}} \right)$$

$$= 1.554 \text{ gal}$$

Discussion
By disregarding the engine's inefficiency when converting the heat of combustion into mechanical energy, we recognize that our calculation will underestimate the actual rate of fuel consumption.

$$V = 5.882 \text{ L}$$
$$V = 1.554 \text{ gal}$$

Example 7.7 | *Drill Rod Quenching*

A steel drill rod that is 8 mm in diameter and 15 cm long is being heat treated in an oil bath. The 850 °C rod is quenched and held at 600 °C and then quenched a second time to 20 °C. Calculate the quantities of heat that must be removed at the two quenching stages.

Approach

We will calculate the quantities of heat that flow from the rod during the two quenching stages by applying Equation (7.7). Table 5.2 lists the weight density of steel as $\rho_w = 76$ kN/m³, which we will use in calculating the rod's mass. The specific heat of steel is listed in Table 7.4 as 0.50 kJ/(kg · °C).

Solution

The volume V of the rod is calculated based on its length L and cross-sectional area A

$$A = \pi \frac{(0.008\ \text{m})^2}{4} \qquad \leftarrow \left[A = \pi \frac{d^2}{4} \right]$$

$$= 5.027 \times 10^{-5}\ \text{m}^2$$

The volume becomes

$$V = (5.027 \times 10^{-5}\ \text{m}^2)(0.15\ \text{m}) \qquad \leftarrow [V = AL]$$

$$= 7.540 \times 10^{-6}\ \text{m}^3$$

The rod's weight is therefore

$$w = \left(76 \times 10^3 \frac{\text{N}}{\text{m}^3} \right)(7.540 \times 10^{-6}\ \text{m}^3) \qquad \leftarrow [w = \rho_w V]$$

$$= 0.5730 \left(\frac{\text{N}}{\text{m}^3} \right)(\text{m}^3)$$

$$= 0.5730\ \text{N}$$

and its mass is

$$m = \frac{0.5730\ \text{N}}{9.81\ \text{m/s}^2} \qquad \leftarrow [w = mg]$$

$$= 5.841 \times 10^{-2}\ (\text{N}) \left(\frac{\text{s}^2}{\text{m}} \right)$$

$$= 5.841 \times 10^{-2} \left(\frac{\text{kg} \cdot \text{m}}{\text{s}^2} \right) \left(\frac{\text{s}^2}{\text{m}} \right)$$

$$= 5.841 \times 10^{-2}\ \text{kg}$$

Example 7.7 | continued

In the last step, we expanded the derived unit newton in terms of the base units meter, kilogram, and second. The quantities of heat removed during the two quenching stages become

$$Q_1 = (5.841 \times 10^{-2}\,\text{kg})\left(0.50\,\frac{\text{kJ}}{\text{kg} \cdot {}^{\circ}\text{C}}\right)(850\,{}^{\circ}\text{C} - 600\,{}^{\circ}\text{C}) \leftarrow [Q = mc(T - T_0)]$$

$$= 7.301\,(\text{kg})\left(\frac{\text{kJ}}{\text{kg} \cdot {}^{\circ}\text{C}}\right)({}^{\circ}\text{C})$$

$$= 7.301\ \text{kJ}$$

and

$$Q_2 = (5.841 \times 10^{-3}\,\text{kg})\left(0.50\,\frac{\text{kJ}}{\text{kg} \cdot {}^{\circ}\text{C}}\right)(600\,{}^{\circ}\text{C} - 20\,{}^{\circ}\text{C}) \leftarrow [Q = mc(T - T_0)]$$

$$= 16.94\,(\text{kg})\left(\frac{\text{kJ}}{\text{kg} \cdot {}^{\circ}\text{C}}\right)({}^{\circ}\text{C})$$

$$= 16.94\ \text{kJ}$$

Discussion
The heat flows from the drill rod into the oil bath. Following the two quenching stages, the temperature of the oil bath rises by amounts that can be calculated by using Equation (7.7). Since the second temperature reduction is significantly more than the reduction in the first quenching, it makes sense that the second quenching removes more heat.

$$Q_1 = 7.301\ \text{kJ}$$
$$Q_2 = 16.94\ \text{kJ}$$

Example 7.8 | Heat Loss through a Window

A small office has a 3×4 ft window on one wall. The window is made from single-pane glass that is $\frac{1}{8}$ in. thick. While evaluating the building's heating and ventilation system, an engineer needs to calculate the heat loss through the window on a winter day. Although the temperature difference of the air inside and outside the office is much larger, the two surfaces of the glass differ by only 3 °F. In units of watts, what quantity of heat is lost through the window each hour?

Approach
To calculate the conductive flow of heat through the window, we apply Equation (7.9). The thermal conductivity of glass is listed in Table 7.5 as 0.50 (Btu/h)/(ft · °F). After calculating the energy loss in the USCS dimensions of Btu, we will

Example 7.8 | *continued*

convert to the SI unit of watt by applying the conversion factor 1 Btu = 1055 J from Table 7.1.

Solution

To keep the dimensions in Equation (7.9) consistent, we first convert the window's thickness as follows:

$$L = (0.125 \text{ in.})\left(\frac{1 \text{ ft}}{12 \text{ in.}}\right)$$

$$= 1.042 \times 10^{-2} \text{ ft}$$

The heat loss in 1 h becomes

$$Q = \frac{(0.50(\text{Btu/h})/(\text{ft} \cdot {}^{\circ}\text{F}))(3 \text{ ft})(4 \text{ ft})(1 \text{ h})}{1.042 \times 10^{-2} \text{ ft}}(3\,{}^{\circ}\text{F}) \quad \leftarrow \left[Q = \frac{\kappa A \Delta t}{L}(T_{\text{h}} - T_{\text{l}})\right]$$

$$= 1728 \left(\frac{\text{Btu/h}}{\text{ft} \cdot {}^{\circ}\text{F}}\right)(\text{ft}^2)(\text{h})\left(\frac{1}{\text{ft}}\right)({}^{\circ}\text{F})$$

$$= 1728 \text{ Btu}$$

Expressed in the SI, this quantity of heat is equivalent to

$$Q = (1728 \text{ Btu})\left(1055 \frac{\text{J}}{\text{Btu}}\right)$$

$$= 1.823 \times 10^6 \text{ (Btu)}\left(\frac{\text{J}}{\text{Btu}}\right)$$

$$= 1.823 \times 10^6 \text{ J}$$

Because the heat flows continuously during a 1 h period, the average rate at which heat is lost becomes

$$\frac{Q}{\Delta t} = \left(1.823 \times 10^6 \frac{\text{J}}{\text{h}}\right)\left(\frac{1 \text{ h}}{3600 \text{ s}}\right)$$

$$= 506.4 \left(\frac{\text{J}}{\text{h}}\right)\left(\frac{\text{h}}{\text{s}}\right)$$

$$= 506.4 \frac{\text{J}}{\text{s}}$$

$$= 506.4 \text{ W}$$

where we have used the definition of the derived SI unit watt.

Discussion

In practice, the temperature of the window's surfaces is difficult to measure because of the convection taking place between the window and the surrounding air. A small electric heater in the 500 W range would be sufficient to compensate for this rate of heat loss.

$$\frac{Q}{\Delta t} = 506.4 \text{ W}$$

▶ 7.4 **Energy Conservation and Conversion**

With the concepts of energy, work, and heat in mind, we will now explore the conversion of energy from one form to another. Automobile engines, jet engines, and electrical power plants are three examples of systems that convert energy and produce power by burning fuel as efficiently as possible. In particular, the chemical energy stored in the fuel (be it gasoline, jet fuel, or natural gas) is released as heat, which in turn is converted into mechanical work. In the case of the automobile's engine, the mechanical work takes the form of the crankshaft's rotation; in the jet engine, the output is primarily thrust for powering an aircraft; and in the power plant, the end product is electrical energy.

System   The principles of energy conservation and conversion are built on the concept of a *system*, which is a collection of materials and components that are grouped together with respect to their thermal and energy behavior. The system is imagined to be isolated from its surroundings so that it is separated from extraneous effects that are not important for the problem at hand. We took a very similar viewpoint in Chapter 4 when we used free body diagrams to examine forces acting on structures and machines. In that case, all forces that crossed an imaginary boundary drawn around the body were included on the diagram, and other effects were disregarded. In much the same way, engineers analyze thermal and energy systems by isolating a system from its surroundings and identifying heat that flows into or out of the system, work that is performed on the surroundings (or vice versa), and potential or kinetic energy levels that change within the system.

At a high level of abstraction, consider the thermal and energy system sketched in Figure 7.8. The quantity of heat Q, which may have been produced by burning a fuel, flows into the system. The heat can be transferred by the processes of conduction, convection, or radiation. At the same time, the system performs mechanical work W as an output. In addition, it is possible that the internal energy of the system changes by the amount labeled as ΔU in Figure 7.8. The internal energy change could correspond to the temperature of the system being raised (in which case thermal energy is stored), to its kinetic energy (U_k) changing, or

First law of thermodynamics   to changes in its gravitational (U_g) or elastic (U_e) potential energy. The *first law of thermodynamics* states that these three quantities balance according to

$$Q = W + \Delta U \qquad (7.10)$$

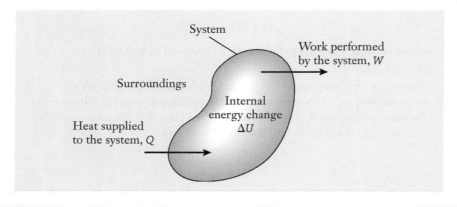

Figure 7.8

Schematic of the first law for energy balance in a thermal and energy system.

As we have found in other aspects of mechanical engineering, a *sign convention* is useful when applying this equation in order to keep track of the direction in which heat flows, whether the system performs work on the surroundings or vice versa, and whether the internal energy increases or decreases. Heat Q is positive when it flows into the system, W is positive when the system performs work on its surroundings, and ΔU is positive when the system's internal energy increases. If one of these quantities is negative, then the converse is true; for instance, when the surroundings perform work on the system, $W < 0$. In the event that the system's internal energy remains constant ($\Delta U = 0$), the heat supplied precisely balances the work that the system performs on its surroundings. Likewise, if the system performs no work ($W = 0$) while heat flows into it ($Q > 0$), the system's internal energy must increase ($\Delta U > 0$) accordingly.

Sign convention

In addition to the requirement of energy conservation, Equation (7.10) also demonstrates that heat, work, and energy are equivalent and that it is possible to design machinery that exchanges one form for another. For instance, the first law describes how an internal-combustion engine operates; the heat Q that is released by burning gasoline is converted into mechanical work W. The first law establishes the opportunity for many other devices in mechanical engineering, ranging from air conditioners to jet engines to the power plants used for generating electricity. In Section 7.5, we discuss the practical limitations that restrict how efficiently those devices are able to exchange heat, work, and energy.

Example 7.9 | *Hydroelectric Power Plant*

The vertical drop of water in a hydroelectric power plant is 100 m (Figure 7.9). Water flows through the plant and into the lower river at the rate of 500 m³/s. Discounting viscous losses in the flowing water and the inefficiency of the turbines and generators, how much electrical energy can be produced each second?

Approach
To calculate the amount of electrical energy produced, we recognize that the potential energy of water in the reservoir is converted into kinetic energy as it falls, and, in turn, the water's kinetic energy is transferred to the rotation of

Figure 7.9

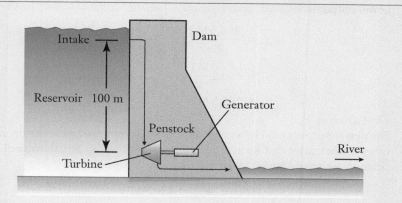

Example 7.9 | *continued*

the turbines and generators. We will neglect the water's relatively small velocity as the level of the reservoir falls and as the water exits the turbines. Those kinetic energy components are small when compared to the overall change in gravitational potential energy of the flowing water. Since no heat is involved, the change in gravitational potential energy balances the work output following Equation (7.10). As listed in Table 6.1, the density of freshwater is 1000 kg/m³.

Solution

The mass of water that exits the reservoir, flows through the penstocks and turbines, and discharges into the lower river each second is

$$m = (500 \text{ m}^3)\left(1000\frac{\text{kg}}{\text{m}^3}\right)$$

$$= 5 \times 10^5 \, (\cancel{\text{m}^3})\left(\frac{\text{kg}}{\cancel{\text{m}^3}}\right)$$

$$= 5 \times 10^5 \text{ kg}$$

Each second, the reservoir's gravitational potential energy changes by the amount

$$\Delta U_g = (5 \times 10^5 \text{ kg})\left(9.81\frac{\text{m}}{\text{s}^2}\right)(-100 \text{ m}) \quad \leftarrow [U_g = mg\Delta h]$$

$$= -4.905 \times 10^8\left(\frac{\text{kg} \cdot \text{m}}{\text{s}^2}\right)(\text{m})$$

$$= -4.905 \times 10^8 \text{ N} \cdot \text{m}$$

$$= -4.905 \times 10^8 \text{J}$$

$$= -490.5 \text{ MJ}$$

Since the reservoir's potential energy decreases, the internal energy change is negative. Neglecting friction and with ideal efficiencies for the water turbines and generators, the first law energy balance becomes

$$W - (490.5 \text{ MJ}) = 0 \quad \leftarrow [Q = W + \Delta U]$$

and $W = 490.5$ MJ. Therefore, 490.5 MJ of electrical energy can be produced each second. By using the definition of average power in Equation (7.5), the power output is

$$P_{\text{avg}} = \frac{490.5 \text{ MJ}}{1 \text{ s}} \quad \leftarrow \left[P_{\text{avg}} = \frac{W}{\Delta t}\right]$$

$$= 490.5\frac{\text{MJ}}{\text{s}}$$

$$= 490.5 \text{ MW}$$

Discussion

The power output could be as high as 490.5 MW, but owing to friction and other inefficiencies that we assumed were not present, an actual hydroelectric power plant would have a lower capacity.

$$P_{\text{avg}} = 490.5 \text{ MW}$$

Example 7.10 | *Automotive Disk Brakes*

The driver of a 1200 kg automobile traveling at 100 km/h applies the brakes and brings the vehicle to a complete stop. The vehicle has front and rear disk brakes, and the braking system is balanced so that the front set provides 75% of the total braking capacity. Through friction between the brake pads and the rotors, the brakes convert the automobile's kinetic energy into heat (Figure 7.10). If the two front 7 kg brake rotors are initially at a temperature of 25 °C, how hot are they after the vehicle stops? The specific heat of cast iron is $c = 0.43$ kJ/(kg · °C).

Approach

As the vehicle comes to a stop, a portion of its initial kinetic energy is lost through air drag, the rolling resistance of the tires, and the wear of the brake pads, but we will neglect those extraneous factors at the first level of approximation. The automobile's kinetic energy [Equation (7.3)] decreases as work is performed on it, and the heat produced will cause the temperature of the brake rotors to increase. That temperature rise can be found by applying Equation (7.7).

Solution

In dimensionally consistent units, the initial velocity is

$$v = \left(100 \frac{km}{h}\right)\left(1000 \frac{m}{km}\right)\left(\frac{1}{3600} \frac{h}{s}\right)$$

$$= 27.78\left(\frac{km}{h}\right)\left(\frac{m}{km}\right)\left(\frac{h}{s}\right)$$

$$= 27.78 \frac{m}{s}$$

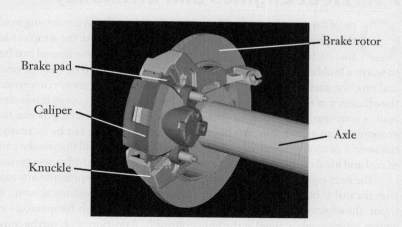

Brake rotor

Brake pad

Caliper

Axle

Knuckle

Figure 7.10

Reprinted with permission by Mechanical Dynamics, Inc.

Example 7.10 | *continued*

The automobile's kinetic energy decreases by the amount

$$\Delta U_k = \frac{1}{2}(1200 \text{ kg})\left(27.78\frac{\text{m}}{\text{s}}\right)^2 \quad \leftarrow \left[U_k = \frac{1}{2}mv^2\right]$$

$$= 4.630 \times 10^5 \left(\frac{\text{kg} \cdot \text{m}}{\text{s}^2}\right)(\text{m})$$

$$= 4.630 \times 10^5 \text{ N} \cdot \text{m}$$

$$= 4.630 \times 10^5 \text{ J}$$

$$= 463.0 \text{ kJ}$$

Because the front brakes provide three-quarters of the braking capacity, the quantity

$$Q = (0.75)(463.0 \text{ kJ})$$

$$= 347.3 \text{ kJ}$$

of heat flows into the front rotors. Their temperature rises according to

$$347.3 \text{ kJ} = 2(7 \text{ kg})\left(0.43\frac{\text{kJ}}{\text{kg} \cdot {}^\circ\text{C}}\right)(T - 25\,{}^\circ\text{C}) \quad \leftarrow [Q = mc(T - T_0)]$$

where the factor of 2 accounts for both front rotors. The equation is dimensionally consistent, and the final temperature becomes $T = 82.69\,{}^\circ\text{C}$.

Discussion

The brakes convert the automobile's kinetic energy to heat, which in turn is stored as thermal energy by virtue of the rotors' temperature rise. This temperature rise is an upper bound because we assumed that all the kinetic energy is converted to heat and not lost in other forms.

$$T = 82.69\,{}^\circ\text{C}$$

▶ 7.5 Heat Engines and Efficiency

One of the most important functions of engineering is developing machines that produce mechanical work by burning a fuel. At the simplest level, a fuel such as natural gas can be burned, and the heat released can be used to warm a building. Of equal practical importance is the need to take the next step and produce useful work from that heat. Mechanical engineers are concerned with the efficiency of machines in which a fuel is burned, thermal energy is released, and heat is converted into work. By increasing the efficiency of that process, the fuel economy of an automobile can be raised, its power output can be increased, and the weight of its engine can be reduced. In this section, we will discuss the concepts of real and ideal efficiencies as applied to energy conversion and power generation.

Heat engine
  The *heat engine* sketched schematically in Figure 7.11 represents any machine that is capable of converting the heat supplied to it into mechanical work. As the input, the engine absorbs a quantity of heat Q_h from the high-temperature energy source, which is maintained at the temperature T_h. A portion of Q_h can be converted into mechanical work W by the engine. The remainder of the heat, however, is

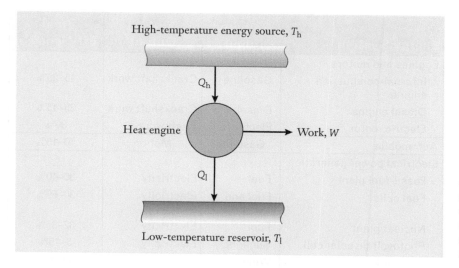

High-temperature energy source, T_h

Q_h

Heat engine

Work, W

Q_l

Low-temperature reservoir, T_l

Figure 7.11

Conceptual view of
a heat engine that
operates between
sources of energy that
are maintained at high
and low temperatures.

exhausted from the engine as a waste product. The lost heat Q_l is released back into the low-temperature reservoir, which is maintained at the constant value $T_l < T_h$. From this conceptual viewpoint, the high temperature source of energy and the low temperature reservoir are large enough so that their temperatures do not change as heat is removed or added.

Heat reservoir

In the context of an automotive engine, Q_h represents the heat released by burning fuel in the engine's combustion chamber, W is the mechanical work associated with the rotation and torque of the engine's crankshaft, and Q_l is the heat that warms the engine block and is expelled from the exhaust pipe. Our everyday experience is that the engine is not able to convert all of the supplied heat into mechanical work, and some fraction of it is wasted and released into the environment. Heat is transferred to the engine at the gasoline's combustion temperature, labeled as T_h in Figure 7.11, and the wasted heat Q_l is released at the lower temperature T_l. Alternatively, as applied to an electrical power plant, Q_h would represent the heat produced by burning a fuel such as coal, oil, or natural gas. The energy released by combustion is used to produce electricity, but the plant also returns some waste heat into the atmosphere through cooling towers or into a nearby river or lake. With W being regarded as the heat engine's useful output, all thermal and energy systems such as these have the characteristic that the supplied heat cannot be converted entirely into useful work.

Referring to Equation (7.10), the energy balance for the heat engine is

$$Q_h - Q_l = W \qquad (7.11)$$

because there is no change in its internal energy. The *real efficiency* η (the lowercase Greek character eta) of the heat engine is defined as the ratio of the work output to the amount of heat supplied to it:

Real efficiency

$$\eta = \frac{W}{Q_h} \qquad (7.12)$$

In the light of Equation (7.11), we see that $\eta = 1 - Q_l/Q_h$ for the heat engine. Since the amount of heat supplied to the engine is greater than the amount wasted ($Q_h > Q_l$), the efficiency always lies between zero and one.

Energy System	Required Input	Desired Output	Real Efficiency, η
Engines and motors			
Internal-combustion engine	Gasoline	Crankshaft work	15–25%
Diesel engine	Diesel	Crankshaft work	35–45%
Electric motor	Electricity	Shaft work	80%
Automobile	Gasoline	Motion	10–15%
Electrical power generation			
Fossil-fuel plant	Fuel	Electricity	30–40%
Fuel cells	Fuel and oxidant	Electricity	30–60%
Nuclear plant	Fuel	Electricity	32–35%
Photovoltaic solar cell	Sunlight	Electricity	5–15%
Wind turbine	Wind	Shaft work	30–50%
Wave energy convertor	Waves and currents	Electricity	10–20%
Hydroelectric plant	Water flow	Electricity	70–90%
Residential			
Forced-air furnace	Natural gas	Heat	80–95%
Hot water heater	Natural gas	Heat	60–65%

Table 7.6

Typical Efficiencies of Thermal and Energy Systems

The real efficiency is sometimes described as the ratio of "what you get" (the work produced by the engine) to "what you paid" (the heat input), and it is often expressed as a percentage. If an automobile's engine has an overall efficiency of 20%, then for every 5 gal of fuel consumed, only 1 gal's worth of energy is converted for the useful purpose of powering the vehicle. Such is the case even for a heat engine that has been optimized by millions of hours of research and development effort by skilled mechanical engineers. That seemingly low level of efficiency, however, is not as bad as you might think at first glance, given the limitations that physical laws place on our ability to convert heat into work. Table 7.6 lists real efficiencies for a variety of thermal and energy systems encountered in mechanical engineering.

Focus On

RENEWABLE ENERGY

Renewable energy sources are those that are naturally replenished at a rate that outpaces consumption. Nonrenewable energy sources, such as oil, petroleum, natural gas, and uranium, either exist in a fixed amount or cannot be produced or created quickly enough to meet current consumption rates. Mechanical engineers will play significant roles in the harvesting, storing, and efficient use of both renewable and nonrenewable energy sources as the global use of all types of energy continues to increase (Figure 7.12). However, as nonrenewable sources diminish, it is critical that engineers design and develop innovative systems to capture and use renewable energy sources.

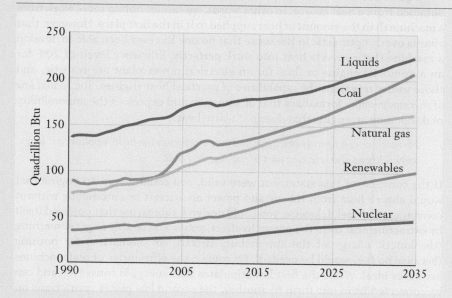

Figure 7.12

World energy use by fuel type.

Based on data from United States Energy Information Administration, International Energy Outlook 2014 report #:DOE/EIA-0383(2014).

Hydropower is the world's leading renewable source of electricity. This source includes hydroelectric power plants; tidal barrages, fences, and turbines; devices to capture wave energy on the surface of oceans and seas; and systems that convert the thermal energy of the ocean.

Wind energy is the world's second largest renewable source of electricity. Various types of wind turbines currently exist, including the traditional horizontal type and the less common vertical (e.g., Darrieus) type.

Every minute, the sun provides the Earth with more energy than the entire world consumes in a year. *Solar energy* can be collected and converted into electricity using photovoltaic solar cells. Also, solar thermal power plants use sunlight to turn a fluid into steam to power a generator. These power plants use solar concentrators such as parabolic troughs, solar dishes, and solar power towers to efficiently heat the fluid. Energy from the sun is also stored in *biomass,* which can be burned for fuel. Biomass is organic material from living or recently living

organisms including wood, crops, manure, alcohol fuels such as ethanol and biodiesel, and certain types of garbage.

Geothermal energy is heat from within the Earth that can be recovered as steam or hot water and used to heat buildings or generate electricity. Unlike solar and wind energy, geothermal energy is always available, 365 days a year.

Fuel cells are devices that use a fuel and an oxidant to create electricity from an electrochemical process. The fuel can be from a renewable source such as alcohol, methane from waste digestion, or hydrogen generated by the conversion of water using wind or solar energy. The oxidant can also be from a renewable source such as oxygen (from air) or chlorine.

So whether it is developing the necessary mechanical hardware for these energy systems, making appropriate thermal energy calculations, or designing new energy solutions, mechanical engineers will continue to be leaders in addressing global energy challenges.

Equation (7.10) sets an upper limit on the amount of work that can be obtained from a heat source. In other words, we cannot obtain more work from a machine than the amount of heat supplied to it in the first place. However, that view is overly optimistic in the sense that no one has ever been able to develop a machine that converts heat into work perfectly. Efficiency levels of 20% for an automobile engine or 35% for an electrical power plant are realistic, and those values represent the capabilities of practical heat engines. The *second law of thermodynamics* formalizes that observation and expresses the impossibility of developing an engine that does not waste heat:

Second law of thermodynamics

> *No machine can operate in a cycle and only transform the heat supplied to it into work without also rejecting part of the supplied heat.*

If the converse of this statement were valid, you could design an engine that would absorb heat from the air and power an aircraft or automobile without consuming any fuel. Likewise, you could design a submarine that powered itself by extracting heat from the ocean. In effect, you would be able to get something (the kinetic energy of the automobile, aircraft, or submarine) for nothing (because no fuel would be needed). Of course, the efficiencies of real machines are never ideal. While the first law stipulates that energy is conserved and can be converted from one form to another, the second law places restrictions on the manner in which that energy can be used.

Given that heat cannot be converted into work ideally, what is the maximum efficiency that a heat engine could ever reach? That upper theoretical bound was established by the French engineer Sadi Carnot (1796–1832), who was interested in building engines that produced the greatest amount of work within the constraints set by physical laws. Although an engine that operates precisely on Carnot's cycle cannot be built, it nevertheless provides a useful point of comparison when evaluating and designing real engines. The cycle is based on expanding and compressing an ideal gas in a cylinder and piston mechanism. At various stages of the cycle, heat is supplied to the gas, work is performed by it, and waste heat is expelled. The gas, cylinder, and piston form a heat engine that operates between energy reservoirs at the high and low temperatures of T_h and T_l, as in Figure 7.11. The *ideal Carnot efficiency* of the heat engine is given by

Ideal Carnot efficiency

$$\eta_C = 1 - \frac{T_l}{T_h} \qquad (7.13)$$

In calculating the ratio of temperatures, T_l and T_h must be expressed on the same absolute scale using either the dimensions *Kelvin* (K) in the SI or degrees *Rankine* (°R) in the USCS. Referring to Tables 3.2 and 3.5, absolute temperatures are found from values on the familiar Celsius and Fahrenheit scales by using the conversion expressions

Kelvin Rankine

$$K = °C + 273.15$$
$$°R = °F + 459.67 \qquad (7.14)$$

From the theoretical standpoint, the efficiency calculated from Equation (7.13) doesn't depend on the details of the engine's construction.

Performance is determined entirely by the temperatures of the two heat reservoirs. Efficiency can be increased by lowering the temperature at which heat is exhausted or by raising the temperature at which heat is supplied to the engine. However, the temperature at which fuel burns, the melting point of the metals making up the engine, and the temperature of our environment all restrict the ability to raise T_h or to lower T_1.

Further, the realistic inefficiencies associated with friction, fluid viscosity, and other losses are not considered by η_C. For those reasons, you should view Carnot's efficiency as a limit based on the physical principles of energy conversion; as such, it is an upper bound for the real efficiency η of Equation (7.12) and Table 7.6. The real efficiency of an engine will always be lower than Carnot's efficiency, and usually it will be much lower. Real efficiency values, such as those listed in Table 7.6, should be used whenever possible in calculations involving thermal and energy systems.

▶ 7.6 Internal-Combustion Engines

The internal-combustion engine is a heat engine that converts the chemical energy stored in gasoline, diesel, ethanol (derived primarily from corn), or propane into mechanical work. The heat generated by rapidly burning a mixture of fuel and air in the engine's combustion chamber is transformed into the rotation of the engine's crankshaft at a certain speed and torque. As you know, the applications of internal-combustion engines are wide and varied, and they include automobiles, motorcycles, rickshaws, aircraft, ships, pumps, power tools, recreational equipment, and electrical generators. When designing these engines, mechanical engineers must develop simulation models to predict a number of critical factors, including fuel efficiency, power-to-weight ratio, noise levels, emissions, and cost. In this section, we discuss some of the design, terminology, and energy principles behind four-stroke and two-stroke engines.

The main elements of the single-cylinder engine (shown in Figure 7.13) are the piston, cylinder, connecting rod, and crankshaft. Those components convert the back-and-forth motion of the piston into the crankshaft's rotation. As fuel burns, the high pressure that develops in the cylinder pushes the piston, moves the connecting rod, and rotates the crankshaft. The engine also contains a means for fuel and fresh air to be drawn into the cylinder, and for the exhaust

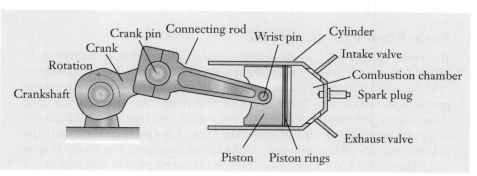

Figure 7.13

Layout of a single-cylinder internal-combustion engine.

Figure 7.14

This 1.4-L four-cylinder engine produces a peak power output of 55 kW. This engine is the range extender for the plug-in hybrid electric Chevrolet Volt.

gases to be vented away. We will discuss those processes separately in the context of four-stroke and two-stroke engines.

While the single-cylinder configuration may be relatively simple, its power output is limited by the small size. In multicylinder engines, the pistons and cylinders can be placed in the vee, in-line, or radial orientations. For instance, four-cylinder engines with the cylinders arranged in a single straight line are common in automobiles (Figure 7.14). In a V-6 or V-8 engine, the engine block is made short and compact by setting the cylinders into two banks of three or four cylinders each (Figures 1.9(c) and 7.2). The angle between the banks is usually between 60° and 90°, and, in the limit of 180°, the cylinders are said to horizontally oppose one another. Engines as large as V-12s and V-16s are found in heavy-duty trucks and luxury vehicles, and 54-cylinder engines comprising six banks of nine cylinders each have been used in some naval applications.

The power that can be produced by an internal-combustion engine depends not only on the number of cylinders, but also on its throttle setting and speed. Each automotive engine, for instance, has a speed at which the output power is the greatest. An engine that is advertised as generating 200 hp does not do so under all operating conditions. As an example, Figure 7.15 shows the *power curve* of a V-6 automobile engine. This engine was tested at full throttle, and it reached peak power output near a speed of 6000 rpm.

Power curve

Four-Stroke Engine Cycle

Engine value

The cross section of a single-cylinder four-stroke engine is illustrated in Figure 7.16 at each major stage of its operation. This engine has two *valves* per cylinder: one for drawing in fresh fuel and air and one for exhausting the by-products of combustion. The mechanism that causes the valves to open and close is an important aspect of this type of engine, and Figure 7.17 depicts one design for an intake or exhaust valve. The valve closes as its head contacts the polished surface of the cylinder's intake or exhaust port. A specially shaped metal lobe, called a *cam*, rotates and controls the opening and closing motions of

Cam

140
120
100
80
60
40
20
0
Power, hp
2000 2500 3000 3500 4000 4500 5000 5500 6000 6500
Engine speed, rpm

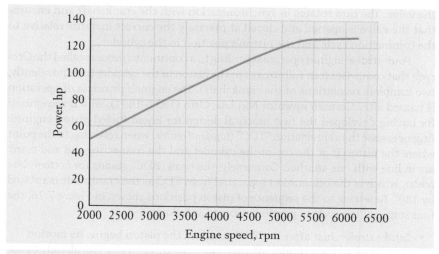

Figure 7.15

Measured power output of an automobile's 2.5-L engine as a function of speed.

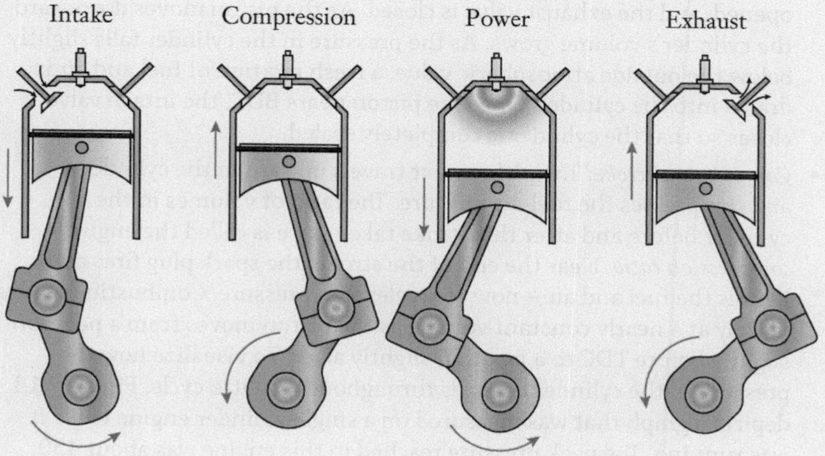

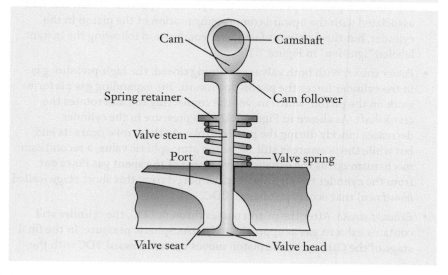

Intake
Compression
Power
Exhaust

Figure 7.16

Major stages of the four-stroke engine's cycle.

Cam
Camshaft
Spring retainer
Cam follower
Valve stem
Port
Valve spring
Valve seat
Valve head

Figure 7.17

One type of cam and valve mechanism that is used in a four-stroke internal-combustion engine.

the valve. The cam rotates in synchronization with the crankshaft and ensures that the valve is opened and closed at precisely the correct instants relative to the combustion cycle and the piston's position in the cylinder.

Four-stroke engines operate according to a continuous process called the *Otto cycle* that comprises four full strokes of the piston in the cylinder (or equivalently, two complete revolutions of the crankshaft). The engine's principle of operation is named after German inventor Nicolaus Otto (1832–1891), who is recognized for having developed the first practical design for liquid-fueled piston engines. Engineers use the abbreviation "TDC" (*top dead center*) when referring to the point where the piston is at the top of the cylinder and the connecting rod and crank are in line with one another. Conversely, the term "BDC" stands for *bottom dead center,* which is the orientation separated from TDC as the crankshaft is rotated by 180°. Referring to the sequence of piston positions shown in Figure 7.16, the four stages of the cycle's operation proceed as follows:

Otto cycle

Top dead center

Bottom dead center

- *Intake stroke*: Just after the TDC position, the piston begins its motion downward in the cylinder. At this stage, the intake valve has already been opened, and the exhaust valve is closed. As the piston moves downward, the cylinder's volume grows. As the pressure in the cylinder falls slightly below the outside atmospheric value, a fresh mixture of fuel and air is drawn into the cylinder. Once the piston nears BDC, the intake valve closes so that the cylinder is completely sealed.

- *Compression stroke*: The piston next travels upward in the cylinder and compresses the fuel-air mixture. The ratio of volumes in the cylinder before and after this stroke takes place is called the engine's *compression ratio*. Near the end of the stroke, the spark plug fires and ignites the fuel and air—now at an elevated pressure. Combustion occurs rapidly at a nearly constant volume as the piston moves from a position slightly before TDC to a position slightly after. To visualize how the pressure in the cylinder changes throughout the Otto cycle, Figure 7.18 depicts a graph that was measured on a single-cylinder engine while it was running. The peak pressure reached in this engine was about 450 psi. For each pulse of pressure in the cylinder, a portion of that rise was associated with the upward compressing motion of the piston in the cylinder, but the dominant factor was combustion following the instant labeled "ignition" in Figure 7.18.

Compression ratio

- *Power stroke*: With both valves remaining closed, the high-pressure gas in the cylinder forces the piston downward. The expanding gas performs work on the piston, which moves the connecting rod and rotates the crankshaft. As shown in Figure 7.18, the pressure in the cylinder decreases quickly during the power stroke. As the stroke nears its end, but while the pressure is still above the atmospheric value, a second cam mechanism opens the exhaust valve. Some of the spent gas flows out from the cylinder through the exhaust port during this short stage (called *blowdown*) that occurs just before BDC.

Blowdown

- *Exhaust stroke*: After the piston passes through BDC, the cylinder still contains exhaust gas at approximately atmospheric pressure. In the final stage of the Otto cycle, the piston moves upward toward TDC with the

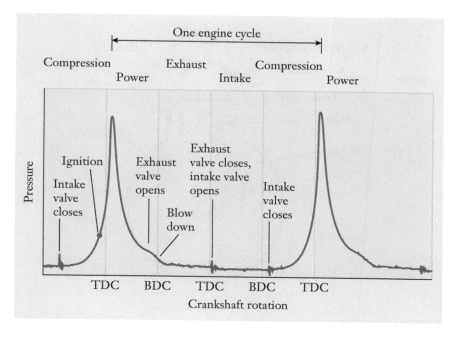

Figure 7.18

Measured pressure curve for a four-stroke engine operating at 900 rpm.

exhaust valve open to force the spent gas from the cylinder. Near the end of the exhaust stroke and just before TDC is reached, the exhaust valve closes, and the intake valve begins to open in preparation for the next repetition of the cycle.

To place this orchestration of valve and piston motions in perspective for a single-cylinder engine, consider that, at a speed of only 900 rpm, the crankshaft completes 15 revolutions each second, and each of the four strokes occurs in only 33 ms. In fact, automobile engines often operate several times faster, and the short time intervals between valves opening and closing highlight the need for accurate timing and sequencing of the four stages. In the Otto cycle, only one stroke out of every four produces power, and in that sense, the crankshaft is being driven only 25% of the time. However, the engine continues to rotate during the other three strokes because of the momentum stored in the engine's flywheel or, for a multicylinder engine, because of the overlapping power strokes from the other cylinders.

Two-Stroke Engine Cycle

The second common type of internal-combustion engine operates on the two-stroke cycle, which was invented in 1880 by British engineer Dugald Clerk (1854–1932). Figure 7.19 (see page 318) depicts the cross section of an engine that operates on this cycle. In contrast to its four-stroke cousin, this engine has no valves and hence no need for springs, camshafts, cams, or other elements of a valve train. Instead, a two-stroke engine has a passageway, called the *transfer port*, that allows fresh fuel and air to flow from the crankcase, through the transfer port, and into the cylinder. As it moves within the cylinder, the piston itself acts as the valve by uncovering or sealing the exhaust port, intake port,

Transfer port

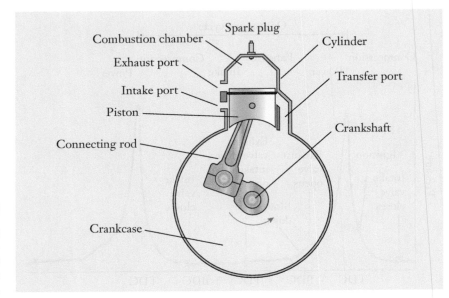

Figure 7.19

Cross section of a two-stroke internal-combustion engine using crankcase compression.

and transfer port in the proper sequence. This type of engine operates by using the principle of crankcase compression.

A two-stroke engine completes a full cycle for each revolution of the crankshaft, and power is therefore produced during every other stroke of the piston. As shown by the sequence of events in Figure 7.20, the *Clerk cycle* operates as follows:

Clerk engine cycle

- *Downstroke*: The piston begins near TDC with a compressed mixture of fuel and air in the combustion chamber. After the spark plug fires, the piston is driven downward in its power stroke, and torque and rotation are transferred to the crankshaft. When the piston has moved only partially down the cylinder, the exhaust port labeled in Figure 7.19 is uncovered. Because the gas in the cylinder is still at a relatively high pressure, it begins to vent outward through the newly opened port. This allows a fresh charge of fuel and air to be already waiting in the crankcase for the upcoming upstroke stage. The piston continues downward, and as the volume in the crankcase decreases, the pressure grows in the fuel-air mixture stored there. Eventually, the transfer port becomes uncovered as the upper edge of the piston moves past it, and the fuel and air in the crankcase flow through the transfer port to fill the cylinder. During this process, the piston continues to block the intake port.

- *Upstroke*: Once the piston passes the BDC position, most of the exhaust gas has been expelled from the cylinder. As the piston continues moving upward, the transfer port and exhaust port are both covered, and the pressure in the crankcase decreases because its volume is expanding. The intake port becomes uncovered as the lower edge of the piston moves past it, and fresh fuel and air flow into and fill the crankcase. That mixture will be stored for use during the engine's next combustion cycle. Slightly before the piston reaches TDC, the spark plug fires, and the power cycle begins anew.

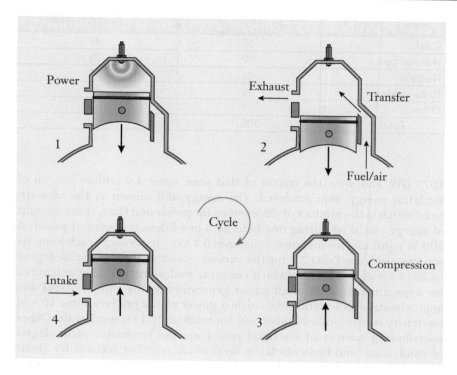

Figure 7.20

Sequence of stages in the operation of a two-stroke internal-combustion engine.

Each type of engine, two-stroke and four-stroke, has its own advantages. Relative to their four-stroke counterparts, two-stroke engines are simpler, lighter, and less expensive. Whenever possible, mechanical engineers attempt to keep matters simple, and because two-stroke engines have few moving parts, very little can go wrong with them. On the other hand, the intake, compression, power, and exhaust stages in a two-stroke engine are not as well separated from one another as they are in a four-stroke engine. In a two-stroke engine, the spent exhaust and fresh fuel-air gases mix unavoidably as the new charge flows from the crankcase into the cylinder. For that reason, a portion of unburned fuel is present in a two-stroke engine's exhaust, and that leakage contributes both to environmental pollution and to reduced fuel economy. In addition, because the crankcase is used to store fuel and air between cycles, it cannot be used as an oil sump, as is the case in a four-stroke engine. Lubrication in a two-stroke engine is instead provided by oil that has been premixed with the fuel, a factor that further contributes to its emissions and environmental problems.

▶ 7.7 **Electrical Power Generation**

The vast majority of electrical energy used around the world is produced by power plants that employ an energy cycle based on driving turbines and generators using high-temperature and high-pressure steam. To place the technology of electrical power into perspective, in 2013, the United States had an electrical power generation capacity of

Power Plant Type	US Contribution (%)	World Contribution (%)
Coal	38	40
Natural gas	30	23
Nuclear	19	11
Renewables	12	22
Other	1	4
Total	100	100

Kilowatt-hour

1075 GW, and, over the course of that year, some 4.0 trillion kW · h of electrical energy were produced. The energy unit known as the *kilowatt-hour*, which is the product of dimensions for power and time, is the amount of energy produced during one hour by a one-kilowatt source of power. A 100-W lightbulb, for instance, consumes 0.1 kW · h of energy each hour. As you can see from Table 7.7 for the various power generation technologies, in 2012 fossil fuel plants (which consume coal or natural gas) accounted for approximately 68% of all power generation in the United States and approximately 63% worldwide. Nuclear power plants produce some 19% of electricity in the United States and approximately 11% worldwide. Other contributing sources of electrical power are the renewable technologies of wind, solar, and hydroelectric power, which together account for about 12% of the United States' electrical power generation and approximately 22% worldwide.

In the macroscopic view, an electrical power plant receives fuel and air as its inputs. In turn, the plant produces electricity, along with two side effects that are released into the environment: the by-products of combustion and waste heat. As we saw in Section 7.5, heat engines are not able to convert all of the heat supplied to them into useful mechanical work, and some amount of heat must be rejected. To disperse the wasted heat, electrical power plants are often located near large bodies of water, or they employ cooling towers. This

Thermal pollution

by-product is sometimes called *thermal pollution*, and it can disturb the habitats of wildlife and the growth of vegetation.

As shown in Figure 7.21, the cycle for a fossil fuel power plant comprises

Primary loop

two loops. The *primary loop* consists of the steam generator, turbine, electrical generator, condenser, and pump. Water circulates in that closed loop, and it is

Secondary loop

continuously cycled between liquid and steam. The purpose of the *secondary loop* is to condense the low-pressure steam in the primary loop back into liquid water after it exits the turbine. Cold water is drawn from a lake, river, or ocean, and it is pumped through a bank of many tubes. When the steam exiting the turbine makes contact with those tubes, it cools and condenses into liquid water. As a result, the water in the secondary loop is heated slightly before it is returned to the original source. In areas without a large natural body of water, cooling towers are used. In that case, the water in the secondary loop is drawn from a pond, heated in the condenser, and then sprayed around the base of the tower. Natural convection currents draw the water up the tower to cool it. Much of the water can be recaptured, but a small amount evaporates into the atmosphere. Note that the water circulating in the primary loop is

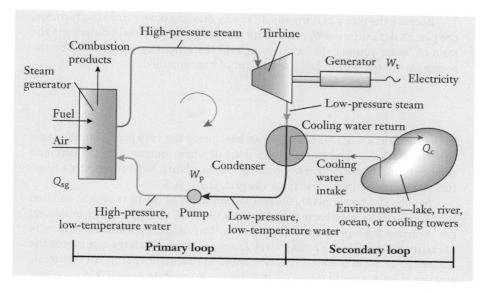

Figure 7.21

The energy cycle used in an electrical power generation plant.

entirely separate from the cooling water in the secondary loop; no mixing occurs between the two loops.

Continuously converting water between liquid and steam, the power plant operates on a *cycle* named in honor of the Scottish engineer and physicist William John Rankine (1820–1872). Water and steam are used to physically move energy from one location in the power plant (for instance, from the steam generator) to another (the turbines). Some 90% of the electrical power in the United States is produced through this process in one variation or another.

Beginning our description of the power plant at its *pump*, liquid water is pressurized and pumped to the *steam generator*. Mechanical work, which we label as W_p, is drawn from the power plant to drive the pump. As fuel is burned, the combustion gases heat a network of tubes in the steam generator, and the water in the primary loop turns into steam as it is pumped through those tubes. In the steam generator, the quantity of heat Q_{sg} is transferred to the water in the primary loop to create steam at high temperature and high pressure. The steam then flows to the *turbine* and performs mechanical work by causing the turbine's shaft to rotate. Each stage of the turbine is analogous to a water wheel, and as jets of high-pressure steam strike the blades in the turbine, its shaft is forced to rotate. The turbine is connected to a generator that ultimately produces the electricity that leaves the power plant. The turbine's work output is denoted by W_t. The spent low-pressure steam leaves the turbine and next enters the *condenser*, where the quantity of waste heat Q_c is removed from it and released into the environment. The steam is condensed to low-pressure and low-temperature water so that it can be pumped and recirculated through the system; water is more easily pumped than steam. The cycle repeats as the water exits the pump and is fed into the steam generator.

Rankine cycle

Pump
Steam generator

Turbine

Condenser

Because the pump is normally driven by drawing power from the turbines, the plant's net output is $W_t - W_p$. With the adage that the real efficiency is the ratio of "what you get" to "what you paid," the efficiency is measured by the power plant's net output and the amount of heat supplied:

$$\eta = \frac{W_t - W_p}{Q_{sg}} \approx \frac{W_t}{Q_{sg}} \tag{7.15}$$

In the last portion of this equation, we have made the very good simplification that the work supplied to the pump is small when compared to the turbine's total output. The real efficiency of most power plants, beginning with a fossil fuel and ending with electricity on the grid, is 30–40%.

In a nuclear power plant, the reactor functions as the heat source, and fuel rods, made of a radioactive material, replace the fossil fuels of coal, oil, or natural gas. Reactors operate on the principle of nuclear fission by which the structure of matter is altered at the atomic level. Energy is released in large quantity as the nucleus of an atom is split. Fuels consumed in this manner store an enormous amount of energy per unit mass. Just 1 g of the uranium isotope U-235 is capable of releasing the same amount of heat as some 3000 kg of coal.

The layout of a nuclear power plant is shown in Figure 7.22. It differs from a fossil fuel plant in that it requires two distinct internal loops. Water flowing in the primary loop comes into direct contact with the reactor's core, and its purpose is to transfer heat from the reactor to the steam generator. For safety reasons, the primary loop and the water in it do not pass outside a hardened containment wall. The steam generator functions as a means to transfer heat from the water in the primary loop to the water in the secondary loop; in short, it keeps the two loops entirely isolated from one another. The steam in the secondary loop drives the turbines and electrical generators in the same manner as in a conventional fossil fuel cycle. At the outermost

Figure 7.22

The primary, secondary, and tertiary loops in a nuclear power plant.

Reprinted with permission by Westinghouse Electric Company.

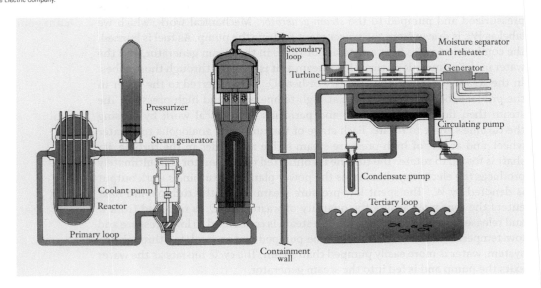

level, the tertiary loop draws water for cooling purposes from a lake, river, or ocean, and passes it through the condenser. In that manner, the low-pressure steam leaving the turbines is condensed back into liquid water so that it can be pumped back into the steam generator. The water supplies in the three loops do not mix, but heat is exchanged between them via the steam generator and the condenser.

Example 7.11 | *Design of a Solar-Power Generator*

A solar power system capable of heating water into steam to power a turbine/generator needs to be designed. The problem definition includes the following information and system requirements:

- Each second, 0.9 kJ of solar energy strikes a square meter of ground.
- The target value for the real efficiency of this system is 26%.
- The system needs to produce an average of 1 MW of power during a 24-h period.
- The heat engine releases its waste heat into the surrounding air.

(a) Design a solar power system to fulfill these requirements. (b) Assuming the resulting steam powers an ideal Carnot heat engine that operates between $T_l = 25°C$ and $T_h = 400°C$, calculate the upper bound on efficiency for the engine.

Approach
Following the general design process introduced in Chapter 2, we will first develop any necessary additional system requirements. Then, we will develop conceptual alternatives and converge on the most effective option. We will then determine the more detailed geometric specifications of the system. Lastly, we will use Equation (7.13) to calculate the efficiency of the proposed engine.

Solution
(a) *System requirements*: In addition to the requirements given in the problem, we need to make other assumptions about the system performance.

- We recognize that the solar power plant can operate only during the peak daylight hours, which is roughly one-third of a day. Therefore, to have an average output of 1 MW over the course of a full day, the plant must be sized to produce 3 MW during the daylight hours.
- A method to store the excess power during the day and retrieve it at night has to be devised, but we will not consider that aspect of the power plant's design in this example.

Example 7.11 | *continued*

Conceptual design: At this stage, we would develop a number of conceptual alternatives for the solar power system. This could include a system of solar cell panels to generate electricity to heat the water, a passive solar system to store heat in a thermal mass to heat the water, or a solar thermal energy system using mirrors to directly heat the water. Ideally, we would generate many alternatives here, potentially even developing ideas for new technologies. Simulation models can then be used to help mechanical engineers select the most effective option. We will assume that the solar thermal energy using trough-shaped mirrors has been determined to be the most effective option through preliminary cost, efficiency, and manufacturing analysis. See Figure 7.23.

Detailed design: We determine that parabolic trough-shaped mirrors will be used to collect the sunlight to heat water into steam. We choose to use standard 12 m² mirrors and then need to determine the number of mirrors needed to meet the 3 MW power requirement. By using the target real efficiency of $\eta = 0.26$, we will calculate the heat input and energy output of the power plant during a 1 s time interval. The heat Q_h supplied to the system by sunlight is related to the plant's output by Equation (7.12).

Using Equation (7.5), the plant produces

$$W = (3 \text{ MW})(1 \text{ s}) \quad \leftarrow \left[P_{avg} = \frac{W}{\Delta t} \right]$$

$$= 3 \text{ MW} \cdot \text{s}$$

$$= 3 \left(\frac{\text{MJ}}{\text{s}} \right) (\text{s})$$

$$= 3 \text{ MJ}$$

of electrical energy each second. Given the assumed level of plant efficiency, the quantity of heat that must be supplied each second is

$$Q_h = \frac{3 \text{ MJ}}{0.26} \quad \leftarrow \left[\eta = \frac{W}{Q_h} \right]$$

$$= 11.54 \text{ MJ}$$

Figure 7.23

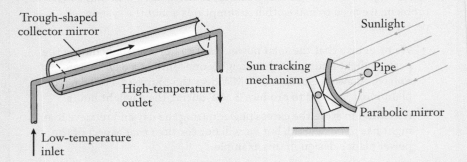

Trough-shaped collector mirror

High-temperature outlet

Low-temperature inlet

Sunlight

Sun tracking mechanism

Pipe

Parabolic mirror

Example 7.11 | continued

Sunlight will therefore need to be collected over the area

$$A = \frac{1.154 \times 10^4 \text{ kJ}}{0.9 \text{ kJ/m}^2}$$

$$= 1.282 \times 10^4 \text{ (kJ)}\left(\frac{\text{m}^2}{\text{kJ}}\right)$$

$$= 1.282 \times 10^4 \text{ m}^2$$

This area is equivalent to a square parcel of land over 100 m on a side. The power plant would therefore require

$$N = \frac{1.282 \times 10^4 \text{ m}^2}{12 \text{ m}^2}$$

$$= 1068 \frac{\text{m}^2}{\text{m}^2}$$

$$= 1068$$

individual collector mirrors.

(b) The ideal Carnot efficiency of a heat engine operating between the given low temperature (surrounding air) and high temperature (heated by sunlight) is

$$\eta_c = 1 - \frac{(25 + 273.15) \text{ K}}{(400 + 273.15) \text{ K}} \qquad \leftarrow \left[\eta_c = 1 - \frac{T_l}{T_h}\right]$$

$$= 1 - 0.4429\left(\frac{\text{K}}{\text{K}}\right)$$

$$= 0.5571$$

or roughly 56%. Here we have converted temperatures to the absolute Kelvin scale by using Equation (7.14) as required in Equation (7.13).

Discussion

In our design, we have neglected the factors of cloud cover, atmospheric humidity, and dirt on the mirrors, each of which would decrease the amount of solar radiation available for power conversion. The power plant's footprint would therefore need to be larger than our estimate suggests, but from a preliminary design standpoint, this calculation begins to firm up the plant's size. The ideal engine efficiency of 56% is the maximum value that physical laws allow, but it can never be achieved in practice. The power plant's real efficiency is about half of this theoretically possible value. In that sense, the efficiency of 26% in this application is not as low as you might think at first glance.

A parabolic trough-shaped mirror system with
$$N = 1068 \text{ mirrors}$$
$$\eta_c = 0.5571$$

Example 7.12 | *Power Plant's Emissions*

A coal-fired power plant produces a net electricity output of 1 GW. A key problem faced by the power plant is the emission of sulfur as a combustion by-product. The coal contains 1% sulfur, and it is known to react with rainwater in the atmosphere to produce sulfuric acid—the main ingredient of acid rain. The power plant is equipped with a scrubber system that removes 96% of the sulfur from the exhaust gas. See Figure 7.24. (a) By using the plant's real efficiency of 32%, calculate the plant's daily fuel consumption. (b) How much sulfur escapes the scrubber and is released into the environment each day?

Approach

We will consider a 1 s time interval during which the power plant draws heat, produces work, and exhausts some sulfur into the atmosphere. The plant's real efficiency relates "what we get" relative to "what we paid," and by using Equation (7.15), the heat supplied to the plant by the fuel can be determined. Table 7.3 lists the heating value of coal as 30 MJ/kg. The given information for the sulfur percentage and the scrubber's efficiency can be used to determine the amount of sulfur released each day.

Solution

(a)   By using Equation (7.5), the plant produces

$$W_t = (1 \text{ GW})(1 \text{ s}) \quad \leftarrow \left[P_{avg} = \frac{W}{\Delta t} \right]$$

$$= 1 \text{ GW} \cdot \text{s}$$

$$= 1 \left(\frac{\text{GJ}}{\cancel{\text{s}}} \right)(\cancel{\text{s}})$$

$$= 1 \text{ GJ}$$

of electrical energy each second. The amount of heat that must be supplied to the steam generator by burning coal is

$$Q_{sg} = \frac{1 \text{ GJ}}{0.32} \quad \leftarrow \left[\eta = \frac{W_t}{Q_{sg}} \right]$$

$$= 3.125 \text{ GJ}$$

Figure 7.24

Example 7.12 | *continued*

where we have neglected the small amount of pump work, W_p, in Equation (7.15). By using Equation (7.6), the mass of fuel consumed each second is

$$m_{coal} = \frac{3125 \text{ MJ}}{30 \text{ MJ/kg}} \quad \leftarrow [Q = mH]$$

$$= 104.2(\text{MJ})\frac{\text{kg}}{\text{MJ}}$$

$$= 104.2 \text{ kg}$$

where we have converted the SI prefix on the numerical value for Q_{sg} so that the numerator and denominator are dimensionally consistent. Over the course of a day, the plant must therefore use

$$m_{coal} = \left(104.2\frac{\text{kg}}{\text{s}}\right)\left(60\frac{\text{s}}{\text{min}}\right)\left(60\frac{\text{min}}{\text{h}}\right)\left(24\frac{\text{h}}{\text{day}}\right)$$

$$= 9.000 \times 10^6 \left(\frac{\text{kg}}{\text{s}}\right)\left(\frac{\text{s}}{\text{min}}\right)\left(\frac{\text{min}}{\text{h}}\right)\left(\frac{\text{h}}{\text{day}}\right)$$

$$= 9000 \times 10^6 \frac{\text{kg}}{\text{day}}$$

(b) With the coal containing 1% sulfur and a scrubber efficiency of 96%,

$$m_{sulfur} = (0.01)(1 - 0.96)\left(9.000 \times 10^6 \frac{\text{kg}}{\text{day}}\right)$$

$$= 3.600 \times 10^3 \frac{\text{kg}}{\text{day}}$$

is the daily release of sulfur.

Discussion

Because each quantity contains a large power-of-ten exponent, we apply SI prefixes from Table 3.3 to represent the final results more concisely. Coal typically contains about 0.5–4% sulfur by weight; so the emission from a power plant could be less or more depending on the type of coal used. Although power plants in the United States are equipped with scrubber systems that remove a large fraction of the sulfur (but not all of it) from the exhaust stack, not all countries use such scrubber systems. As a result, the large-scale pollution effects of fossil-fuel consumption can indeed be high. Mechanical engineers have an active role to play in the public policy arena by balancing environmental issues with the need for an abundant and inexpensive supply of electricity.

$$m_{coal} = 9.000 \frac{\text{Gg}}{\text{day}}$$

$$m_{sulfur} = 3.600 \frac{\text{Mg}}{\text{day}}$$

Focus On

CROWDSOURCING INNOVATIVE ENERGY SOLUTIONS

Engineers working at large companies are not the only ones developing innovative solutions to global energy problems. Through crowdsourcing competitions such as the XPRIZE and OpenIDEO, student teams, private engineers, and innovators from all around the world can compete for millions of dollars in prizes and a chance to have their ideas brought to a global stage. The Energy and Environment Prize Group at XPRIZE is developing challenges to stimulate innovative solutions in clean energy, energy distribution, energy storage, energy efficiency, and energy use, among other areas. A number of challenge areas have been formulated or already awarded including:

- A revolutionary battery technology capable of powering a wide range of systems from small and wearable consumer electronics to electric vehicles and aircraft
- Electrification of vehicle transportation infrastructures by embedding equipment in roadways in order to charge vehicles while driving

- Wireless power transmission technologies to provide carbon-free energy, including powering aircraft using ground-based power
- Technology to capture, store, and distribute solar energy at large scales to provide power to the nearly 2 billion people throughout world who live off-grid
- A home-based energy storage appliance to provide balanced power supply, renewable power generation, and greater resiliency from blackouts

Through these and other similar competitions in a wide range of fields, mechanical engineers can have substantial and widespread impact on technical, global, social, environmental, and economic challenges. A number of university teams from all over the world have competed for these prizes, many times in partnerships with small startups, large companies, and government organizations. Mechanical engineers will continue to play important roles in these innovative partnerships and transformational opportunities.

Summary

I n this chapter, we introduced several principles behind the thermal and energy systems encountered in mechanical engineering. Internal-combustion engines, electrical power plants, and jet aircraft engines figured prominently in Chapter 1 during our discussion of the mechanical engineering profession's top ten achievements, and each technology is based on converting heat into work. The innovative production, storage, and use of energy from renewable and nonrenewable sources will also play a significant role in meeting global energy challenges in the coming years. The different forms that energy can take and the methods by which energy can be efficiently exchanged from one form to another are, in short, central to mechanical engineering.

The main quantities introduced in this chapter, common symbols representing them, and their units are summarized in Table 7.8. The equations used to analyze thermal and energy systems are listed in Table 7.9 (see page 330). Heat engines are often powered by burning a fuel, and its heating value represents the quantity of heat released during that combustion process. In accordance with the first law, the heat can flow into a material and cause its temperature to rise by an amount depending on its specific heat, or the heat can be transformed into mechanical work. We view heat as energy that is in transit through the modes of conduction, convection, and radiation. Mechanical engineers design engines that convert heat into work with a view toward increasing their efficiency. In an ideal Carnot cycle, the efficiency is limited by the temperatures at which heat is supplied to the engine and released to its environment. In practice, efficiencies of real heat engines are substantially lower, with some typical values being listed in Table 7.6.

Table 7.8

Quantities, Symbols, and Units That Arise When Analyzing Thermal and Energy Systems

Quantity	Conventional Symbols	Conventional Units	
		USCS	SI
Energy	$U_g, U_e, U_k, \Delta U$	ft · lb, Btu	J, kW · h
Work	W, W_t, W_p	ft · lb	J
Heat	Q, Q_h, Q_l, Q_{sg}, Q_c	Btu	J
Average power	P_{avg}	(ft · lb)/s, hp	W
Heating value	H	Btu/slug	MJ/kg
Specific heat	c	Btu/(slug · °F)	kJ/(kg · °C)
Thermal conductivity	κ	(Btu/h)/(ft · °F)	W/(m · °C)
Time interval	Δt	s, min, h	s, min, h
Temperature	T, T_0, T_h, T_l	°F, °R	°C, K
Real efficiency	η	—	—
Ideal Carnot efficiency	η_C	—	—

Energy	
Gravitational potential	$U_g = mg\,\Delta h$
Elastic potential	$U_e = \dfrac{1}{2}k\Delta L^2$
Kinetic	$U_k = \dfrac{1}{2}mv^2$
Work of a force	$W = F\Delta d$
Average power	$P_{avg} = \dfrac{W}{\Delta t}$
Heating value	$Q = mH$
Specific heat	$Q = mc(T - T_0)$
Heat conduction	$Q = \dfrac{\kappa A\Delta t}{L}(T_h - T_l)$
Energy conversion/conservation	$Q = W + \Delta U$
Efficiency	
Real	$\eta = \dfrac{W}{Q_h}$
Ideal Carnot	$\eta_c = 1 - \dfrac{T_l}{T_h}$
Power plant	$\eta \approx \dfrac{W_t}{Q_{sg}}$

Table 7.9

Key Equations That Arise When Analyzing Thermal and Energy Systems

Self-Study and Review

7.1. How are gravitational and elastic potential energy, and kinetic energy calculated?

7.2. What is the difference between work and power? What are their conventional units in the SI and USCS?

7.3. What is the heating value of a fuel?

7.4. What is the specific heat of a material?

7.5. Give examples of situations where heat is transferred through conduction, convection, and radiation.

7.6. Define the term "thermal conductivity."

7.7. What is a heat engine?

7.8. How is efficiency defined?

7.9. What are some of the differences between real and ideal Carnot efficiencies? Which is always higher?

7.10. How are absolute temperatures calculated on the Rankine and Kelvin scales?

7.11. Sketch the piston, connecting rod, and crankshaft mechanism in an internal-combustion engine.

7.12. Explain how the four-stroke and two-stroke engine cycles operate.

7.13. What are the relative advantages and disadvantages of four-stroke and two-stroke engines?

7.14. Sketch a diagram of an electric power plant and briefly explain how it works.

Problems

P7.1

Beginning from their definitions, determine the conversion factor between ft · lb and kW · h in Table 7.1.

P7.2

A remote-controlled toy car weighs 3 lb and moves at 15 ft/s. What is its kinetic energy?

P7.3

A lawn mower engine is started by pulling a cord wrapped around a hub of radius 6.0 cm. If a constant tension of 80 N is maintained in the cord and the hub makes three revolutions before the motor starts, how much work is done?

P7.4

In the movie *Back to the Future*, Doc Brown and the young Marty McFly need 1.21 GW of power for their time machine.

(a) Convert that power requirement to horsepower.

(b) If a stock DeLorean sports car produces 145 hp, how many times more power does the time machine need?

P7.5

A sprinter runs using a force of 200 N and a power output of 600 W. Calculate how many minutes it takes for the runner to run 1 km.

P7.6

A baseball catcher stops a 98 mph fastball over a distance of 0.1 m. What is the force necessary to stop the 0.14 kg baseball?

P7.7

For the two automobiles of P6.31 in Chapter 6, how much power must the engines produce just to overcome air drag at 60 mph?

P7.8

A light truck weighs 3100 lb and is rated at 30 miles per gallon for 60 mph highway driving on level ground. Under those conditions, the engine must overcome air resistance, rolling resistance, and other sources of friction. Give your answers in the units shown.

(a) The coefficient of drag is 0.6 at 60 mph, and the truck's frontal area is 32 ft^2. What is the drag force on the truck?

(b) How much power must the engine produce at 60 mph just to overcome air resistance?

(c) In part (b) how much gasoline would be consumed each hour (neglecting other frictional effects)?

P7.9

Suppose that the truck in P7.8 was going up a hill with a grade of 2%. How much additional power must the engine produce to climb the hill, neglecting various frictional effects?

P7.10

The heating value of agricultural residue biomass (e.g., crop residues, animal manure and bedding, and organic material from food production) can range from 4300 to 7300 Btu/lbm. How much heat is released when 500 kg of biomass is burned?

P7.11

During processing in a steel mill, a 750 lb steel casting at 800 °F is quenched by plunging it into a 500 gal oil bath, which is initially at a temperature of 100 °F. After the casting cools and the oil bath warms, what is the final temperature of the two? The weight per unit volume of the oil is 7.5 lb/gal.

P7.12

The interior contents and materials of a small building weigh 25 tons, and together they have an average specific heat of 0.25 Btu/(lbm · °F). Neglecting any inefficiency in the furnace, what amount of natural gas must be burned to raise the building's temperature from freezing to 70 °F?

P7.13

A 5.0 kg steel gear is heated to 150 °C and then placed into a 0.5 gal container of water at 10 °C. What is the final temperature of the metal and water?

P7.14

Give two examples each of engineering applications where heat would be transferred primarily through conduction, convection, and radiation.

P7.15

A hollow square box is made from 1 ft^2 sheets of a prototype insulating material that is 1 in. thick. Engineers are performing a test to measure the new material's thermal conductivity. A 100 W electrical heater is placed inside the box. Over time, thermocouples attached to the box show that the interior and exterior surfaces of one face have reached the constant temperatures of 150 °F and 90 °F. What is the thermal conductivity? Express your result in both the SI and USCS.

P7.16

A welding rod with $\kappa = 30$ (Btu/h)/(ft · °F) is 20 cm long and has a diameter of 4 mm. The two ends of the rod are held at 500 °C and 50 °C.

(a) In the units of Btu and J, how much heat flows along the rod each second?

(b) What is the temperature of the welding rod at its midpoint?

P7.17

A brick wall 3 m high, 7.5 m wide, and 200 mm thick has a thermal conductivity of 0.7 W/(m · °C). The temperature on the inner face is 25 °C, and the temperature on the outer face is 0 °C. How much heat is lost per day through the wall?

P7.18

A 2500 lb automobile comes to a complete stop from 65 mph. If 60% of the braking capacity is provided by the front disk brake rotors, determine their temperature rise. Each of the two cast-iron rotors weighs 15 lb and has a specific heat of 0.14 Btu/(lbm · °F).

P7.19

A small hydroelectric power plant operates with 500 gal of water passing through the system each second. The water falls through a vertical distance of 150 ft from a reservoir to the turbines. Calculate the power output, and express it in the units of both hp and kW. The density of water is listed in Table 6.1.

P7.20

As part of a packing and distribution system, boxes are dropped onto a spring and pushed onto a conveyor belt. The boxes are originally at a height h above the uncompressed spring [Figure P7.20(a)]. Once dropped, the box of mass m compresses the spring a distance ΔL [Figure P7.20(b)]. If all the potential energy of the box is converted into elastic energy in the spring, find an expression for ΔL.

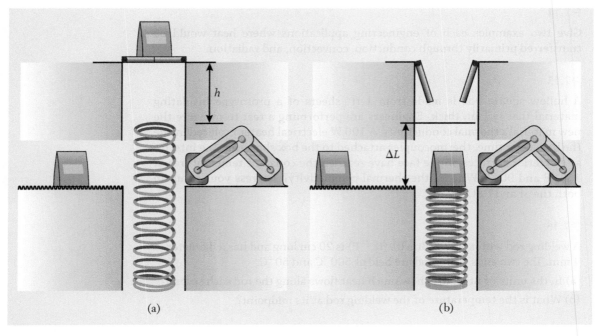

(a)   (b)

Figure P7.20

P7.21

Wind turbines convert the kinetic energy of wind to mechanical or electrical power. The mass of air that hits a wind turbine each second is given by:

$$\frac{\text{mass}}{\text{sec}} = \text{velocity} \cdot \text{area} \cdot \text{density}$$

where the density of air is 1.23 kg/m³, and the area is the area swept by the turbine rotor blades. This mass flow rate can be used to calculate the amount of kinetic energy per second that the air generates. One of the largest wind turbines in the world is in Norway and is projected to generate 10 MW of power with winds of 35 mph. The diameter of the rotor blades is 145 m. How much power is generated by the wind? Recall that power is the amount of energy per unit time.

P7.22

Neglecting the presence of friction, air drag, and other inefficiencies, how much gasoline is consumed when a 1300 kg automobile accelerates from rest to 80 km/h? Express your answer in the units of mL. The density of gasoline is listed in Table 6.1.

P7.23

In the summer of 2002, a group of miners in Quecreek, Pennsylvania, became trapped 240 ft underground when a section of the coal seam they were working in collapsed into an adjacent, but abandoned, mine that was not shown on their map. The area became flooded with water, and the miners huddled in an air pocket at the end of a passageway until they were safely rescued. As the first step in the rescue operation, holes were drilled into the mine to provide the miners with warm fresh air and to pump out the underground water. Neglecting friction in the pipes and the inefficiency of the pumps themselves, what average power would be required to remove 20,000 gal of water from the mine each minute? Express your answer in the units of horsepower. The density of water is listed in Table 6.1.

P7.24

Geothermal energy systems extract heat stored below the Earth's crust. For every 300 ft below the surface, the temperature of groundwater increases by about 5 °F. Heat can be brought to the surface by steam or hot water to warm homes and buildings, and it also can be processed by a heat engine to produce electricity. Using the real efficiency value of 8%, calculate the output of a geothermal power plant that processes 50 lb/s of groundwater at 180 °F and discharges it on the surface at 70 °F.

P7.25

A heat engine idealized as operating on the Carnot cycle is supplied with heat at the boiling point of water (212 °F), and it rejects heat at the freezing point of water (32 °F). If the engine produces 100 hp of mechanical work, calculate in units of Btu the amount of heat that must be supplied to the engine each hour.

P7.26

An inventor claims to have designed a heat engine that receives 120 Btu of heat and generates 30 Btu of useful work when operating between a high-temperature energy reservoir at 140 °F and a low-temperature energy reservoir at 20 °F. Is the claim valid?

P7.27

A person can blink an eye in approximately 7 ms. At what speed (in revolutions per minute) would a four-stroke engine be operating if its power stroke took place literally in the blink of an eye? Is that a reasonable speed for an automobile engine?

P7.28

A four-stroke gasoline engine produces an output of 35 kW. Using the density of gasoline listed in Table 6.1, the heating value for gasoline in Table 7.3, and a typical efficiency listed in Table 7.6, estimate the engine's rate of fuel consumption. Express your answer in the units of liters per hour.

P7.29

An automobile's engine produces 30 hp while being driven at 50 mph on a level highway. In those circumstances, the engine's power is used to overcome air drag, rolling resistance between the tires and the road, and friction in the drivetrain. Estimate the vehicle's fuel economy rating in the units of miles per gallon. Use a typical engine efficiency from Table 7.6 and the density of gasoline in Table 6.1.

P7.30

A university's campus has 20,000 computers with cathode-ray tube monitors that are powered up even when the computer is not being used. This type of monitor is relatively inefficient, and it draws more power than a flat panel display.

(a) If each cathode-ray tube monitor draws 0.1 kW of power over the course of a year, how much energy has been consumed? Express your answer in the conventional units of kW · h for electricity.

(b) At the cost of 12¢ per kW · h, how much does it cost the university each year to keep these monitors powered up?

(c) On average, a computer monitor that has the automatic sleep feature enabled will consume 72% less energy than one that is continually powered. What is the cost savings associated with enabling this feature on all of the university's computers?

P7.31

When a desktop personal computer is operating, its power supply is able to convert only about 65% of the supplied electrical power into the direct-current electricity that the computer's internal electronic components require. The remainder of the energy is mostly lost as heat. On average, each of the estimated 233 million personal computers in the United States consumes 300 kW · h of energy each year.

(a) If the efficiency of the power supply could be increased to 80% by using a new type of power supply that is under development (a so-called resonance-based switching-mode system), how much energy could be saved each year? Express your answer in the conventional units of kW · h for electricity.

(b) The United States produced 4.1 trillion kW · h of electrical energy in 2014. By what percentage would the nation's electricity needs decrease?

(c) At the cost of 12¢ per kW · h, what would the cost savings be?

P7.32

Suppose that the new type of computer power supply described in P7.31 costs an additional $5.

(a) At the cost of 12¢ per kW · h, after what period of time would the cost savings in electricity offset the power supply's added cost?

(b) How often do you estimate that individuals and companies generally upgrade their desktop personal computers?

(c) From an economic standpoint, what recommendation would you make regarding the new type of power supply if you worked for a computer manufacturer?

P7.33

A natural-gas-fired electrical power plant produces an output of 750 MW. By using a typical efficiency from Table 7.6 and neglecting the small amount of power drawn by the pump, calculate the rates at which:

(a) Heat is supplied to the water/steam in the steam generator.

(b) Waste heat is spent into the river adjacent to the power plant.

P7.34

For the plant in P7.33, 25,000 gal/s of water flow in the river adjacent to the power plant. The river is used as the source of cooling water for the condenser. Considering the heat transferred from the power plant to the river each second and the specific heat of water, by what amount does the temperature of the river rise as it passes the power plant? The density and specific heat of water are listed in Tables 6.1 and 7.4.

P7.35

Develop a novel design application for some form of renewable energy. Through research, describe the technical, social, environmental, and economic requirements involved in the application. You should take a global perspective in your research, but describe the requirements based on a specific geographic region, people group, or demographic.

P7.36*

Your school's engineering library just received a set of 1000 new books, and they have been delivered to the first floor of the library. Your group has been commissioned to place the books on the appropriate shelves in the library. Estimate how much work is done by your group to move the books to the proper shelf location. Next, estimate how much internal energy was used by your group members to accomplish this task (i.e., how many calories were burned). Finally, using the first law of thermodynamics and assuming other sources of energy loss are negligible, calculate the average change in internal temperature experienced by your group members.

P7.37*

Your group has been tasked by university administration to develop upper- and lower-bound estimates of how much energy could be stored by taking advantage of the foot traffic on your main campus on a weekday during the academic year. Provide clear support for your answer, including:

- The assumptions made regarding the foot traffic (e.g., location, volume of traffic, etc.)

- A description of what method(s) you plan to use to harness the foot traffic energy

- A clear set of calculations and a discussion of the reasonableness of your estimated bounds.

P7.38*

Your group has been contracted to perform a feasibility study on the use of solar energy to power appliances or other personal electronics. You are asked to choose one commonly used appliance or electronic consumer product (excluding a light bulb or smart phone) that is used frequently or runs continuously throughout an entire day. You are then asked to design a power supply system that would allow for continuous use of your appliance or product using the minimum number of solar panels and batteries. Determine the specifications for your solar panels, batteries, and any other components you will need. Provide clear support for any assumptions you are making about the power consumption of your appliance or product, including what kind of current it requires. Show all calculations that you make to support and validate your conclusions. As part of your discussion, also address the feasibility of scaling your solution up to providing power for 1000 of the same appliances or products for commercial or industrial purposes, including what barriers may exist for such scaling.

Your group should start with the following assumptions:

- This system will be used year-round within 100 km of your university and will have to accommodate the appropriate exposure to the sun in this region.
- Any excess power that is generated and not able to be stored in the batteries or used immediately by your appliance or product will be lost in the form of heat.
- Any battery used in your design will supply its specified voltage for the specified number of amp-hours.
- Any connections between components are appropriately rated and have zero energy loss.

References

Ferguson, C. R., and Kirkpatrick, A. T., *Internal Combustion Engines: Applied Thermosciences*, 2nd ed. Hoboken, NJ: Wiley, 2000.

Holman, J. P., *Heat Transfer*, 9th ed. New York: McGraw-Hill, 2002.

International Energy Outlook 2014. United States Department of Energy, Energy Information Administration, report DOE/EIA-0383(2014).

Sonntag, R. E., Borgnakke, C., and Van Wylen, G. J., *Fundamentals of Thermodynamics*, 6th ed. Hoboken, NJ: Wiley, 2002.

CHAPTER 8

Motion and Power Transmission

CHAPTER OBJECTIVES

- Perform calculations involving rotational velocity, work, and power.
- Discuss the circumstances in which one type of gear would be selected for use over another in terms of design.
- Explain some of the design characteristics of V-belts and timing belts.

- Calculate shaft speeds and torques and calculate the amount of power that is transferred in simple and compound greatrains and belt drives.
- Sketch a planetary geartrain, identify its input and output connection points, and explain how it works.

▶ 8.1 Overview

In this chapter, we turn our attention to the design and operation of *power-transmission* equipment as the seventh element of mechanical engineering. Machinery generally comprises gears, shafts, bearings, cams, linkages, and other building-block components. These mechanisms are capable of transmitting power from one location to another—for instance, from the engine in an automobile to the drive wheels. Another function of a mechanism might be to transform one type of motion into another. An application in that regard, which we encountered in Section 7.6, is the mechanism for converting the back-and-forth motion of the piston in an internal-combustion engine into the rotational motion of the crankshaft. The robotic manipulator arms shown in Figures 8.1 and 8.2 (see page 340) are other examples of mechanisms. Each arm is a chain of interconnected links in which position is controlled by motors at the joints.

Mechanical engineers evaluate the position, velocity, and acceleration of machines such as these, as well as the forces and torques that make them move. Overall, the topics discussed in this chapter fit into the hierarchy of mechanical engineering disciplines as shown in Figure 8.3 (see page 341). The analysis and design of machinery is, in part, an extension of the topics of the force systems and energy systems that we encountered in previous chapters.

Element 7:
Motion and power
transmission

Figure 8.1

Mechanical engineers design the links and joints that form this robot's manipulator arm in order to precisely move and place cargo in a warehouse.

Reprinted with permission by FANUC Robotics North America Inc.

Figure 8.2

The Canadarm2 at the International Space Station positions the unpiloted Japan Aerospace Exploration Agency (JAXA) HTV-3 Vehicle in preparation for its release. Astronauts from JAXA and NASA controlled the multiple links of the robotic arm to orient and release the HTV-3.

Courtesy of NASA.

A particularly common type of mechanism in mechanical engineering comprises individual gears assembled into geartrains and transmissions for the purposes of transmitting power, changing the rotation speed of a shaft, or modifying the torque applied to a shaft. A closely related type of drivetrain uses belts or chains to accomplish the same functions. After introducing the basic concepts of rotational motion in the first portion of this chapter, we will discuss various types of gears and belts and the circumstances in which an engineer would select one type over another in a particular design application.

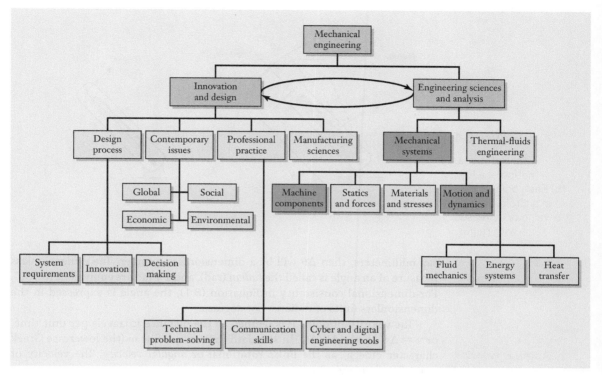

Figure 8.3

Relationship of the
topics emphasized in
this chapter (shaded
blue boxes) relative to
an overall program of
study in mechanical
engineering.

We will also explore some of the methods that can be used to analyze simple,
compound, and planetary geartrains.

▶ 8.2 Rotational Motion

Angular Velocity

When a gear (or, for that matter, any object) rotates, each point on it moves
in a circle about the center of rotation. The straight link in Figure 8.4 (see
page 342) could represent a component of a robotic arm such as those shown in
Figures 8.1 and 8.2. The link turns in a bearing about the center of its shaft. All
points on the link move along concentric circles, each having the same center
point O, as the angle θ (the lowercase Greek character theta) increases. The
velocity of any point P on the link is determined by its change in position as
the rotation angle grows. During the time interval Δt, the link moves from the
initial angle θ to the final angle $\theta + \Delta\theta$; the latter position is shown by dashed
lines in Figure 8.4(a). As point P moves along a circle of radius r, the distance
that it travels is the geometric arc length

$$\Delta s = r\Delta\theta \tag{8.1}$$

The angle $\Delta\theta = \Delta s/r$ is the ratio of two lengths: circumference along the
circle and the circle's radius r. When Δs and r are expressed in the same units,

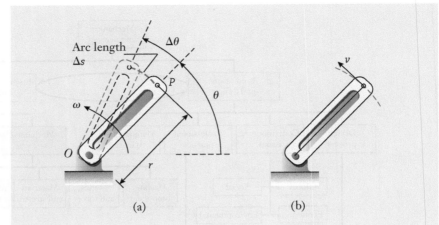

Figure 8.4

(a) Rotational motion of a link. (b) Each point moves in a circle, and the velocity increases with radius.

say millimeters, then $\Delta\theta$ will be a dimensionless number. The dimensionless measure of an angle is called the *radian* (rad), and 2π rad are equivalent to 360°. For dimensional consistency in Equation (8.1), the angle is expressed in the dimensionless units of radians, not degrees.

Radian

The velocity of point P is defined as the distance it travels per unit time, or $v = \Delta s/\Delta t = r(\Delta\theta/\Delta t)$. In standard form, we define ω (the lowercase Greek character omega) as the link's rotational or *angular velocity*. The velocity of point P is then given by

Angular velocity

$$v = r\omega \qquad (8.2)$$

When ω is given in radians per second (rad/s) and r in millimeters, for instance, then v will have the units of millimeters per second. The appropriate units for the angular velocity are chosen based on the context in which ω is being used. When a mechanical engineer refers to the speed of an engine, shaft, or gear, it is customary to use the units of revolutions per minute (rpm), as might be measured by a tachometer. If the rotational speed is very high, the angular velocity might instead be expressed in the dimensions of revolutions per second (rps), which is smaller by a factor of 60. It would be inaccurate, however, to express ω in the dimensions of either rpm or rps in Equation (8.2). To see why, recall that in the derivation for v, ω was defined by the angle $\Delta\theta$, which must have been measured in dimensionless radians for the arc length Δs to be calculated properly. In short, whenever you apply Equation (8.2), express ω in the dimensions of radians per unit time (for instance, rad/s).

Table 8.1 lists the conversion factors between the four common choices of units for angular velocity: rpm (revolutions per minute), rps (revolutions per second), degrees per second, and radians per second. Reading off the first row, for instance, we see that

$$1 \text{ rpm} = 1.667 \times 10^{-2} \text{ rps} = 6\frac{\text{deg}}{\text{s}} = 0.1047\frac{\text{rad}}{\text{s}}$$

rpm	rps	deg/s	rad/s
1	1.667×10^{-2}	6	0.1047
60	1	360	6.283
0.1667	2.777×10^{-3}	1	1.745×10^{-2}
9.549	0.1592	57.30	1

Table 8.1

Conversion Factors between Various Units for Angular Velocity

Rotational Work and Power

In addition to specifying the speeds at which shafts rotate, mechanical engineers also determine the amount of power that machinery draws, transfers, or produces. As expressed by Equation (7.5), power is defined as the rate at which work is performed over a certain interval of time. The mechanical work itself can be associated with forces moving through a distance (which we considered in Chapter 7) or, by analogy, with torques rotating through an angle. The definition of a torque in Chapter 4 as a moment that acts about the axis of a shaft makes the latter case particularly relevant for machinery, geartrains, and belt drives.

Figure 8.5 illustrates the torque T that a motor applies to a gear. The gear, in turn, may be connected to other gears in a transmission that is in the process of transmitting power to a machine. The motor applies torque to the rotating gear, and work is therefore being performed. Analogous to Equation (7.4), the *work of a torque* is calculated from the expression

Work of a torque

$$W = T \, \Delta\theta \qquad (8.3)$$

where the angle $\Delta\theta$ again has the dimensions of radians. As is the case for the work of a force, the sign of W depends on whether the torque tends to reinforce the rotation (W is positive) or oppose it (W is negative). Those two situations are shown in Figure 8.5. Engineering units for work are summarized in Table 7.1. Two particularly useful conversion factors for calculations involving geartrains and belt drives are

$$1 \text{ ft} \cdot \text{lb} = 1.356 \text{ J} \qquad 1 \text{ J} = 0.7376 \text{ ft} \cdot \text{lb}$$

Mechanical power has been defined as the rate at which the work of a force or torque is performed over an interval of time. In machinery applications, power is generally expressed in the units of kW in the SI and hp in the USCS (Table 7.2). In Equation (7.5), we defined the average power, but the instantaneous power is often more useful when machinery is being analyzed. *Instantaneous power* is the product of force and velocity in translational systems and of torque and angular velocity in rotational ones.

Instantaneous power

$$P = Fv \text{ (force)} \qquad (8.4)$$

$$P = T\omega \text{ (torque)} \qquad (8.5)$$

Figure 8.5

Work performed by a torque acting on a rotating gear. (a) The torque supplied by the motor reinforces the rotation and W is positive. (b) The torque opposes the rotation, and W is negative.

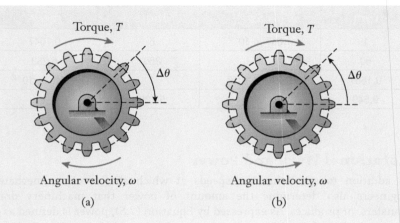

Torque, T $\Delta\theta$ Torque, T $\Delta\theta$

Angular velocity, ω Angular velocity, ω

(a)   (b)

Example 8.1 | *Angular Velocity Conversions*

Verify the conversion factor between rpm and rad/s that is listed in Table 8.1.

Approach
To convert from the dimensions of revolutions per minute to radians per second, we need conversions for both time and an angle. There are 60 s in 1 min, and 2π rad in one revolution.

Solution
We convert dimensions as follows:

$$1\frac{\text{rev}}{\text{min}} = \left(\frac{\text{rev}}{\text{min}}\right)\left(\frac{1}{60}\frac{\text{min}}{\text{s}}\right)\left(2\pi\frac{\text{rad}}{\text{rev}}\right)$$

$$= 0.1047\left(\frac{\text{rev}}{\text{min}}\right)\left(\frac{\text{min}}{\text{s}}\right)\left(\frac{\text{rad}}{\text{rev}}\right)$$

$$= 0.1047\frac{\text{rad}}{\text{s}}$$

Discussion
We can obtain the inverse conversion factor between rad/s and rpm as the reciprocal of this quantity, so that 1 rad/s = $(0.1047)^{-1}$ rpm = 9.549 rpm.

$$1\text{ rpm} = 0.1047\frac{\text{rad}}{\text{s}}$$

Example 8.2 | *Cooling Fan for Electronics*

The small fan spins at 1800 rpm and cools a rack of electronic circuit boards (See Figure 8.6). (a) Express the motor's angular velocity in the units of deg/s and rad/s. (b) Determine the velocity of the 4 cm tip of the fan blade.

Figure 8.6

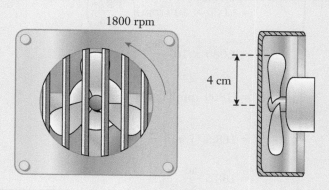

1800 rpm

4 cm

Approach

The conversion factors between different dimensions for angular velocity are listed in Table 8.1, and we see from the first row that 1 rpm = 6 deg/s = 0.1047 rad/s. In part (b), the velocity of the fan blade's tip is perpendicular to the line from the tip to the shaft's center (Figure 8.7). The velocity is calculated by applying Equation (8.2) with angular velocity in the consistent dimensions of rad/s.

Figure 8.7

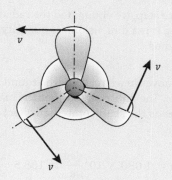

v

v

v

Example 8.2 | *continued*

Solution

(a) In the two different dimensions, the angular velocity is

$$\omega = (1800 \text{ rpm})\left(6\frac{\text{deg/s}}{\text{rpm}}\right)$$

$$= 1.080 \times 10^4 (\text{rpm})\left(\frac{\text{deg /s}}{\text{rpm}}\right)$$

$$= 1.080 \times 10^4 \frac{\text{deg}}{\text{s}}$$

$$\omega = (1800 \text{ rpm})\left(0.1047\frac{\text{rad/s}}{\text{rpm}}\right)$$

$$= 188.5 \ (\text{rpm})\left(\frac{\text{rad/s}}{\text{rpm}}\right)$$

$$= 188.5\frac{\text{rad}}{\text{s}}$$

(b) The velocity of a blade's tip is directed perpendicularly to the radius line drawn back to the shaft's center, and it has the magnitude

$$v = (0.04 \text{ m})\left(188.5\frac{\text{rad}}{\text{s}}\right) \qquad \leftarrow [v = rw]$$

$$= 7.540(\text{m})\left(\frac{\text{rad}}{\text{s}}\right)$$

$$= 7.540\frac{\text{m}}{\text{s}}$$

In the intermediate step, we eliminated the radian unit in the calculation because it is a dimensionless quantity, defined as the ratio of arc length to radius.

Discussion

The tip of a fan blade moves in a circular path around the shaft's center at a speed of 7.540 m/s. The direction (but not the magnitude) of each blade's velocity changes as the shaft rotates, an effect that is the basis of centripetal acceleration.

$$\omega = 1.080 \times 10^4 \frac{\text{deg}}{\text{s}} = 188.5\frac{\text{rad}}{\text{s}}$$

$$v = 7.540\frac{\text{m}}{\text{s}}$$

Example 8.3 | *Automotive Engine's Power*

At full throttle, a four-cylinder automobile engine produces 149 ft · lb of torque at a speed of 3600 rpm. In the units of hp, how much power does the engine produce?

Approach

We can use Equation (8.5) to find the engine's instantaneous power output. As an intermediate step, the rotational speed first must be converted into the consistent units of rad/s by applying the conversion factor 1 rpm = 0.1047 rad/s from Table 8.1.

Solution

In dimensionally consistent units, the engine's speed is

$$\omega = (3600 \ \text{rpm})\left(0.1047 \frac{\text{rad/s}}{\text{rpm}}\right)$$

$$= 376.9\,(\overline{\text{rpm}})\left(\frac{\text{rad/s}}{\overline{\text{rpm}}}\right)$$

$$= 376.9 \ \frac{\text{rad}}{\text{s}}$$

The power that the engine develops is

$$P = (149 \ \text{ft} \cdot \text{lb})\left(376.9 \frac{\text{rad}}{\text{s}}\right) \qquad \leftarrow [P = T\omega]$$

$$= 5.616 \times 10^4 (\text{ft} \cdot \text{lb})\left(\frac{\text{rad}}{\text{s}}\right)$$

$$= 5.616 \times 10^4 \ \frac{\text{ft} \cdot \text{lb}}{\text{s}}$$

Here we eliminated the radian unit in the angular velocity term because it is dimensionless. Although (ft · lb)/s is an acceptable dimension for power in the USCS, the derived unit of horsepower is more conventional when expressing the power in engines and automotive applications. In the final step, we apply the conversion factor from Table 7.2:

$$P = \left(5.616 \times 10^4 \frac{\text{ft} \cdot \text{lb}}{\text{s}}\right)\left(1.818 \times 10^{-3} \frac{\text{hp}}{(\text{ft} \cdot \text{lb})/\text{s}}\right)$$

$$= 102.1\left(\frac{\text{ft} \cdot \text{lb}}{\text{s}}\right)\left(\frac{\text{hp}}{(\text{ft} \cdot \text{lb})/\text{s}}\right)$$

$$= 102.1 \ \text{hp}$$

Example 8.3 | *continued*

Discussion
This value is typical for the maximum power that an automotive engine can develop. In normal driving situations, however, the engine's power output is considerably less and will be determined by the throttle setting, the engine's speed, and the torque transferred to the transmission.

$$P = 102.1 \text{ hp}$$

▶ 8.3 Design Application: Gears

With the preceding groundwork in place to describe rotational velocity, work, and power, we next discuss some aspects associated with the design of machinery. Gears are used to transmit rotation, torque, and power between shafts by engaging specially shaped teeth on rotating disks. Geartrains can be used to increase a shaft's rotation speed but decrease torque, to keep speed and torque constant, or to reduce rotation speed but increase torque. Mechanisms incorporating gears are remarkably common in the design of machinery, and they have applications as diverse as electric can openers, automatic teller machines, electric drills, and helicopter transmissions. In this section, our objective is to explore various types of gears with an emphasis on their characteristics and the terminology used to describe them.

The shape of a gear's tooth is mathematically defined and precisely machined according to codes and standards established by industrial trade groups. The American Gear Manufacturers Association, for instance, has developed guidelines for standardizing the design and production of gears. Mechanical engineers can purchase loose gears directly from gear manufacturers and suppliers, or they can obtain prefabricated gearboxes and transmissions that are suitable for the task at hand. In some cases, when standard gears do not offer sufficient performance (such as low noise and vibration levels), specialty machine shops can be contacted to custom produce gears. In the majority of machine design, however, gears and gearboxes are usually selected as off-the-shelf components.

Just as was the case for the rolling-element bearings discussed in Section 4.6, no single "best" type of gear exists; each variant must be selected based on its suitability for the application. In the following subsections, we will discuss the gear types known as spur, rack and pinion, bevel, helical, and worm. The choice that an engineer ultimately makes for a product reflects a balance between expense and the task that the gear is expected to perform.

Spur Gears
Spur gears are the simplest type of engineering-grade gear. As shown in Figure 8.8, spur gears are cut from cylindrical blanks, and their teeth have

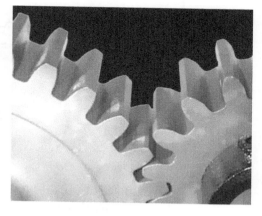

Figure 8.8

Close-up view of two spur gears in mesh.

Image courtesy of the authors.

faces that are oriented parallel to the shaft on which the gear is mounted. For the *external gears* of Figure 8.9(a), the teeth are formed on the outside of the cylinder; conversely, for an *internal or ring gear*, the teeth are located on the inside Figure 8.9(b). When two gears having complementary teeth engage and motion is transmitted from one shaft to another, the two gears are said to form a *gearset*. Figure 8.10 (see page 350) depicts a spur gearset and some of the terminology used to describe the geometry of its teeth. By convention, the smaller (driving) gear is called the *pinion*, and the other (driven) one is simply called the gear. The pinion and gear mesh at the point where the teeth approach, contact one another, and then separate.

Although a collection of many individual teeth are continuously contacting, engaging, and disengaging in a gearset, the pinion and gear can be regarded conceptually as two cylinders that are pressed against one another and that smoothly roll together. As illustrated in Figure 8.11 (see page 350), those cylinders roll on the outside of one another for two external gears, or one can roll within the other if the gearset comprises external and internal gears.

External gear
Internal or ring gear

Gearset

Pinion

Figure 8.9

(a) Two external spur gears in mesh. (b) Internal or ring gears of several sizes.

Reprinted with permission by Boston Gear Company.

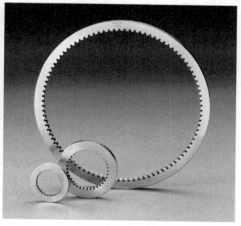

(a)

(b)

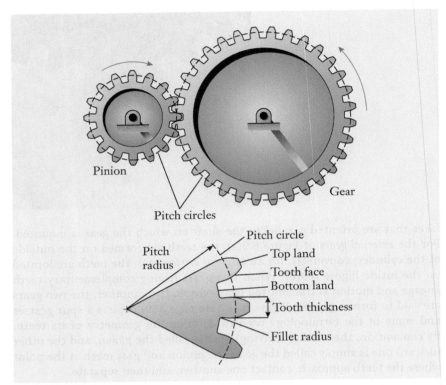

Figure 8.10

Terminology for a spur gearset and the geometry of its teeth.

Figure 8.11

Configurations of gearsets having (a) two external gears and (b) one external and one internal gear. In each case, the rotations are analogous to two cylinders that roll on one another.

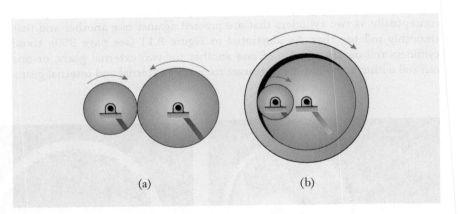

Referring to the nomenclature of Figure 8.10, the effective radius *r* of a spur gear (which is also the radius of its conceptual rolling cylinder) is called the

Pitch radius and circle

pitch radius. Continuous contact between the pinion and gear is imagined to take place at the intersection of the two *pitch circles*. The pitch radius is not the distance from the gear's center to either the top or bottom lands of a tooth. Instead, *r* is simply the radius that an equivalent cylinder would have if it rotated at the same speed as the pinion or gear.

The thickness of a tooth and the spacing between adjacent teeth are measured along the gear's pitch circle. The tooth-to-tooth spacing must be slightly larger than the tooth's thickness itself to prevent the teeth from binding against one another as the pinion and gear rotate. On the other hand, if the space between the teeth is too large, the clearance and free play could cause undesirable rattle, vibration, speed fluctuations, fatigue wear, and eventual gear failure. In the USCS, the proximity of teeth to one another is measured by a quantity called the *diametral pitch*

Diametral pitch

$$p = \frac{N}{2r} \tag{8.6}$$

where N is the number of teeth on the gear, and r is the pitch radius. The diametral pitch has the units of teeth per inch of the gear's diameter. For a pinion and gear to mesh, they must have the same diametral pitch, or the gearset will bind as the shafts begin to rotate. The product catalogs supplied by gear manufacturers will classify compatible gears according to their diametral pitch. By convention, values of $p < 20$ teeth/in. are regarded as being so-called coarse gears, and gears with $p \geq 20$ teeth/in. are said to be fine pitched. Because r and N are proportional in Equation (8.6), if the number of teeth on a gear is doubled, its radius will likewise double. In the SI, the spacing between teeth is measured not by the diametral pitch, but by a quantity called the *module*

Module

$$m = \frac{2r}{N} \tag{8.7}$$

with the units of millimeters.

In principle, pinions and gears with teeth having any complementary shape are capable of transmitting rotation between their respective shafts. However, arbitrarily shaped gear teeth may not be able to transmit consistent motion because the pinion teeth may not efficiently engage and disengage with the gear teeth. Therefore, the tooth shape on modern spur gears has been mathematically optimized so that motion is smoothly transferred between the pinion and gear. Figure 8.12 (see page 352) depicts a small portion of a gearset during three stages of its rotation. As the teeth on the pinion and gear begin to mesh, the teeth roll on the surface of one another, and their point of contact moves from one side of the pitch circles to the other. The cross-sectional shape of the tooth is called the *involute profile*, and it compensates for the fact that the tooth-to-tooth contact point moves during meshing. The term "involute" is synonymous with "intricate," and this special shape for spur gear teeth ensures that, if the pinion turns at a constant speed, the gear will also. This mathematical characteristic of the involute profile establishes what is known as the *fundamental property of gearsets*, enabling engineers to view gearsets as two cylinders rolling on one another:

Involute profile

Fundamental property of gearsets

> For spur gears with involute-shaped teeth, if one gear turns at a constant speed, so will the other.

We will use this property throughout the remainder of this chapter to analyze gearsets and geartrains.

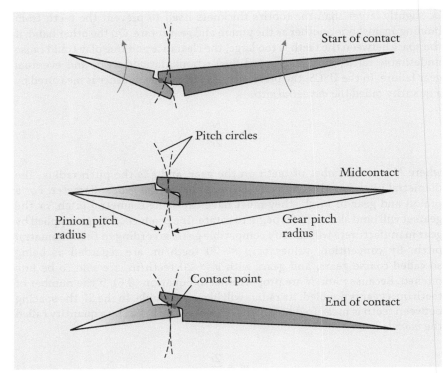

Figure 8.12

A pair of teeth in contact.

Rack and Pinion

Gears are sometimes used to convert the rotational motion of a shaft into the straight-line, or translational, motion of a slider (and vice versa). The rack-and-pinion mechanism is the limiting case of a gearset in which the gear has an infinite radius and tends toward a straight line. This configuration of a rack and pinion is shown in Figure 8.13. With the center point of the pinion fixed, the rack will move horizontally as the pinion rotates—leftward in Figure 8.13(b) as the pinion turns clockwise. The rack itself can be supported by rollers, or it can slide on a lubricated smooth surface. Racks and pinions are often used in the mechanism for steering an automobile's front wheels, and for positioning

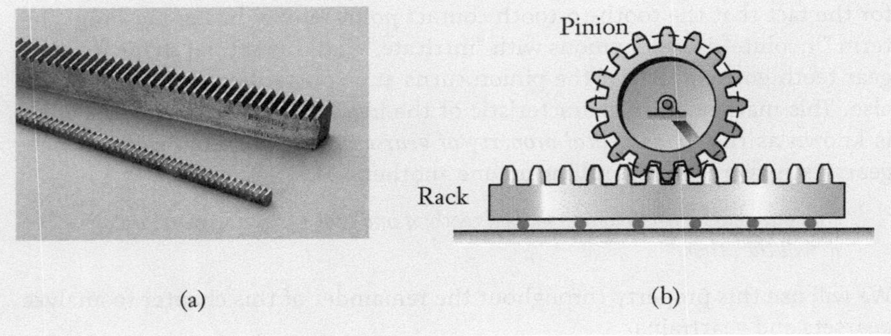

Figure 8.13

(a) Gear racks. (b) Rack-and-pinion mechanism.

Reprinted with permission by Boston Gear Company.

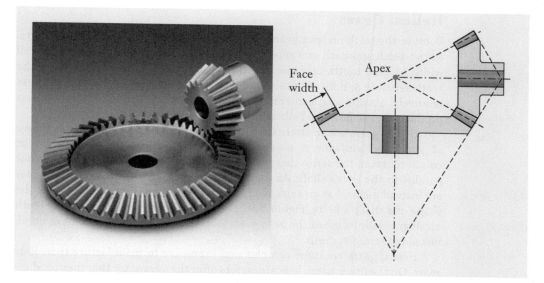

Face width

Apex

machine tables in a milling or grinding machine where the cutting head is stationary.

Bevel Gears

Whereas the teeth of spur gears are arranged on a cylinder, a bevel gear is produced by alternatively forming teeth on a blank that is shaped like a truncated cone. Figure 8.14 depicts a photograph and cross-sectional drawing of a bevel gearset. You can see how its design enables a shaft's rotation to be redirected by 90°. Bevel gears (Figure 8.15) are appropriate for applications in which two shafts must be connected at a right angle and where extensions to the shaft centerlines would intersect one another.

Figure 8.14

Two bevel gears in mesh.

Reprinted with permission by Boston Gear Company.

Figure 8.15

A collection of bevel gears, some having straight and others having spiral-shaped teeth.

Reprinted with permission by Boston Gear Company.

Helical Gears

Because the teeth on spur gears are straight with faces parallel to their shafts, as two teeth approach one another, they make contact suddenly along the full width of each tooth. Likewise, in the meshing sequence of Figure 8.12, the teeth separate and lose contact along the tooth's entire width at once. Those relatively abrupt engagements and disengagements result in spur gears producing more noise and vibration than other types of gears.

Helical gears are an alternative to spur gears, and they offer the advantage of smoother and quieter meshing. Helical gears are similar to their spur counterparts in the sense that the teeth are still formed on a cylinder but not parallel to the gear's shaft. As their name implies, the teeth on a helical gear are instead inclined at an *angle* so that each tooth wraps around the gear in the shape of a shallow helix (Figure 8.16). With the same objective of having teeth mesh gradually, some of the *bevel gears* (shown in Figure 8.15) also have *spiral*, instead of straight, teeth.

Helical gears are more complex and expensive to manufacture than spur gears. On the other hand, helical gearsets offer the advantage that they produce less noise and vibration when used in high-speed machinery. Automobile automatic transmissions, for instance, are typically constructed using both external and internal helical gears for precisely that reason. In a helical gearset, tooth-to-tooth contact starts at one edge of a tooth and proceeds gradually across its width, thus smoothing out the engagement and disengagement of teeth. Another attribute of helical gears is that they are capable of carrying greater torque and power when compared to similarly sized spur gears, because the tooth-to-tooth forces are spread over more surface area and the contact stresses are reduced.

We have seen how helix-shaped teeth can be formed on gears mounted on parallel shafts, but gears that are attached to perpendicular shafts can also incorporate helical teeth. The *crossed helical gears* of Figure 8.16(b) connect two perpendicular shafts, but, unlike the bevel gearset application, the shafts here are offset from one another, and extensions to their centerlines do not intersect. A collection of helical and crossed helical gears is shown in Figure 8.17.

Helix angle

Spiral bevel gear

Crossed helical gears

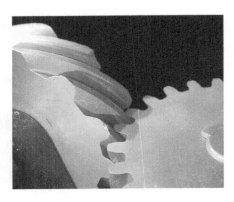

Figure 8.16

(a) A helical gear.
(b) A Pair of crossed helical gears in mesh.

Image courtesy of the authors.

(a)   (b)

Figure 8.17

A collection of helical and crossed helical gears.

Reprinted with permission by Boston Gear Company.

Worm Gearsets

If the helix angle on a pair of crossed helical gears is made large enough, the resulting pair is called a worm and worm gear. Figure 8.18 illustrates this type of gearset, in which the worm itself has only one tooth that wraps several times around a cylindrical body, similar to a thread of a screw. For each revolution of the worm, the worm gear advances by just one tooth in its rotation.

Worm gearsets are capable of large speed-reduction ratios. For instance, if the worm has only one tooth and the worm gear has 50 teeth, the speed reduction for a gearset Figure 8.18(c) would be 50 fold. The ability to package a geartrain with large speed reduction into a small physical space is an attractive feature of worm gearsets. However, the tooth profiles in worm gearsets are not involutes, and significant sliding occurs between the teeth during meshing.

Figure 8.18

(a) Worm having one continuous tooth.
(b) Worm gears.
(c) Worm gearset.

Reprinted with permission by Boston Gear Company.

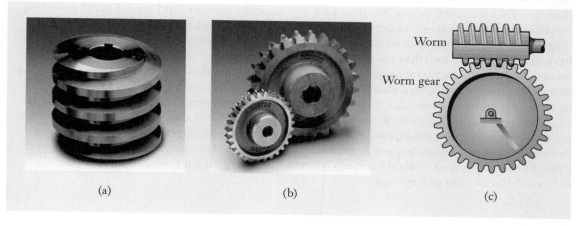

Worm

Worm gear

(a)   (b)   (c)

That friction is a major source of power loss, heating, and inefficiency when compared to other types of gears.

Another feature of worm gearsets is that they can be designed so that they are capable of being driven in only one direction, namely, from the worm to the worm gear. For such *self-locking* gearsets, the power flow cannot be reversed by having the worm gear drive the worm. This capability for motion transmission in only one direction can be exploited in such applications as hoists or jacks, where it is desirable for safety reasons to prevent the system mechanically from being back driven. Not all worm gearsets are self-locking, however, and this characteristic depends on such factors as the helix angle, the amount of friction between the worm and worm gear, and the presence of vibration.

Self-locking

Focus On   NANOMACHINES

Although the machines discussed in this chapter are typically quite large (e.g., industrial robotics, transmissions), a new breed of machines is being developed at the nanoscale and manufactured out of individual atoms. These nanomachines could perform tasks never before possible with larger machines. For example, nanomachines could someday be used to disassemble cancer cells, repair damaged bones and tissues, eliminate landfills, detect and remove impurities in drinking water, deliver drugs internally with unmatched precision, build new ozone molecules, and clean up toxins or oil spills.

In Figure 8.19(a), a conceptual design for a nanomachine that would provide fine-motion control for molecular assembly is shown. This proposed device includes cam mechanisms that use levers to drive rotating rings. Simpler machines such as nanogears have been developed that may pave the way for more complex and functional nanomachines. Researchers developed these stable nanogears by attaching benzene molecules to the outside of carbon nanotubes [Figure 8.19(b)]. Rack and pinion versions of these nanogears, developed by research teams in Germany and France, could be used to move a scalpel back and forth or to push the plunger of a syringe.

Creating these gears, racks, and pinions at the nano or even micro level has typically been an expensive and time-consuming process. However, a research team at Columbia University has developed self-assembling microgears that are created as two materials shrink at different rates when cooled. A thin metal film is deposited on top of a heat-expanded polymer. When the polymer is cooled, the metal buckles, creating regularly spaced "teeth" in the polymer. Variations in metal stiffness and cooling rates allow for differently sized gears and teeth to form.

Powering the nanomachines of the future will be an additional challenge for engineers to overcome. Currently, the power for these nanomachines must come from chemical sources such as light from a laser, imitating the photosynthetic process that plants use to create energy from sunlight. However, scientists and engineers are currently developing other sources of power, including nanogenerators powered by ultrasonic waves, magnetic fields, or electric currents. Nanomachines designed, developed, and tested by engineers could soon be having widespread impact on medical, environmental, and energy issues around the globe.

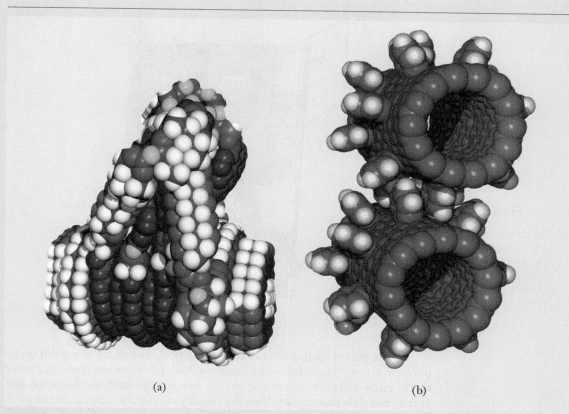

(a)   (b)

Figure 8.19

(a) A mechanism made up of approximately 2500 atoms to move and assemble individual molecules.
(b) Nanogears made up of hundreds of individual atoms.

(a) © Institute for Molecular Manufacturing. www.imm.org; (b) Courtesy of NASA.

▶ 8.4 Speed, Torque, and Power in Gearsets

A gearset is a pair of gears that mesh with one another, and it forms the basic building block of larger-scale systems, such as transmissions, that transmit rotation, torque, and power between shafts (Figure 8.20, see page 358). In this section, we examine the speed, torque, and power characteristics of two meshing gears. In later sections, we will extend those results to simple, compound, and planetary geartrains.

Speed

In Figure 8.21(a) (see page 359), the smaller of the two gears is called the pinion (denoted by p), and the larger is called the gear (denoted by g). The pitch radii of the pinion and gear are denoted by r_p and r_g, respectively.

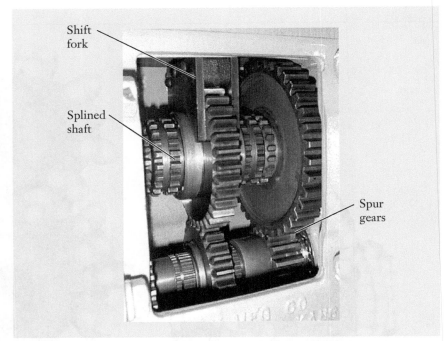

Figure 8.20

The shift fork in the transmission is used to slide a gear along the upper shaft so that the speed of the output shaft changes.

As the pinion rotates with angular velocity ω_p, the speed of a point on its pitch circle is $v_p = r_p\omega_p$, following Equation (8.2). Likewise, the speed of a point on the pitch circle of the gear is $v_g = r_g\omega_g$. Because the teeth on the pinion and gear do not slide past one another, the velocity of points in contact on the pitch circles are the same, and $r_g\omega_g = r_p\omega_p$. With the angular velocity of the pinion's shaft being known as the input to the gearset, the speed of the output shaft in Figure 8.21(a) is

$$\omega_g = \left(\frac{r_p}{r_g}\right)\omega_p = \left(\frac{N_p}{N_g}\right)\omega_p \tag{8.8}$$

Rather than perform calculations in terms of the pitch radii r_p and r_g, it is simpler to work with the numbers of teeth N_p and N_g. Although it is a simple matter to count the number of teeth on a gear, measuring the pitch radius is not so straightforward. So that the pinion and gear mesh smoothly, they must be manufactured with the same distance between teeth, as measured by either the diametral pitch or module. In both Equations (8.6) and (8.7), the number of teeth on a gear is proportional to its pitch radius. Since the diametral pitch (or module) of the pinion and gear are the same, the ratio of radii in Equation (8.8) also equals the ratio of their teeth numbers.

Velocity ratio

The *velocity ratio* of the gearset is defined as

$$VR = \frac{\text{output speed}}{\text{input speed}} = \frac{\omega_g}{\omega_p} = \frac{N_p}{N_g} \tag{8.9}$$

which is a constant that relates the output and input speeds, just as the mechanical advantage is a constant that relates the input and output forces acting on a mechanism.

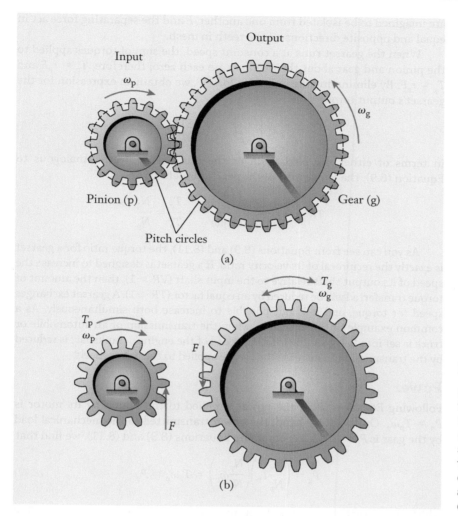

Figure 8.21

(a) A pinion and gear form a gearset.
(b) The pinion and gear are conceptually separated to expose the driving component of the tooth force and the input and output torques.

Torque

We next consider how torque transfers from the shaft of the pinion to the shaft of the gear. For example, imagine that the pinion in Figure 8.21(a) is driven by a motor and that the shaft of the gear is connected to a mechanical load such as a crane or pump. In the diagrams of Figure 8.21(b), the motor applies torque T_p to the pinion, and torque T_g is applied to the gear's shaft by the load being driven. The tooth force F that is exposed in this diagram is the physical means by which torque is transferred between the pinion and gear. The full meshing force is the resultant of F and another component that acts along the line of action passing through the centers of the two shafts. That second force component, which would be directed horizontally in Figure 8.21(b), tends to separate the pinion and gear from one another, but it does not contribute to the transfer of torque, and so it is omitted from the diagram for clarity. As the pinion and gear are imagined to be isolated from one another, F and the separating force act in equal and opposite directions on the teeth in mesh.

When the gearset runs at a constant speed, the sum of torques applied to the pinion and gear about their centers are each zero; therefore, $T_p = r_p F$ and $T_g = r_g F$. By eliminating the unknown force F, we obtain an expression for the gearset's output torque

$$T_g = \left(\frac{r_g}{r_p}\right)T_p = \left(\frac{N_g}{N_p}\right)T_p \tag{8.10}$$

Torque ratio

in terms of either the pitch radii or the numbers of teeth. Analogous to Equation (8.9), the *torque ratio* of the gearset is defined as

$$TR = \frac{\text{outcome torque}}{\text{input torque}} = \frac{T_g}{T_p} = \frac{N_g}{N_p} \tag{8.11}$$

As you can see from Equations (8.9) and (8.11), the torque ratio for a gearset is exactly the reciprocal of its velocity ratio. If a gearset is designed to increase the speed of its output shaft relative to the input shaft ($VR > 1$), then the amount of torque transferred will be reduced by an equal factor ($TR < 1$). A gearset exchanges speed for torque, and it is not possible to increase both simultaneously. As a common example of this principle, when the transmission of an automobile or truck is set into low gear, the rotation speed of the engine's crankshaft is reduced by the transmission to increase the torque applied to the drive wheels.

Power

Following Equation (8.5), the power supplied to the pinion by its motor is $P_p = T_p \omega_p$. On the other hand, the power transmitted to the mechanical load by the gear is $P_g = T_g \omega_g$. By combining Equations (8.9) and (8.11), we find that

$$P_g = \left(\frac{N_g}{N_p}T_p\right)\left(\frac{N_p}{N_g}\omega_p\right) = T_p \omega_p = P_p \tag{8.12}$$

which shows that the input and output power levels are exactly the same. The power supplied to the gearset is identical to the power that it transfers to the load. From a practical standpoint, any real gearset incurs frictional losses, but Equation (8.12) is a good approximation for gearsets made of quality gears and bearings where friction is small relative to the overall power level. In short, any reduction in power between the input and output of a gearset will be associated with frictional losses, not with the intrinsic changes of speed and torque.

▶ 8.5 Simple and Compound Geartrains

Simple Geartrain

For most combinations of a single pinion and gear, a reasonable limit on the velocity ratio lies in the range of 5 to 10. Larger velocity ratios often become impractical either because of size constraints or because the pinion would

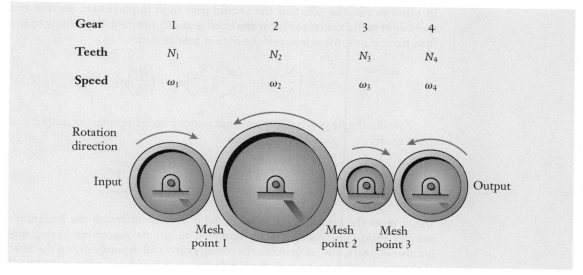

Gear	1	2	3	4
Teeth	N_1	N_2	N_3	N_4
Speed	ω_1	ω_2	ω_3	ω_4

Rotation direction

Input

Output

Mesh point 1   Mesh point 2   Mesh point 3

Figure 8.22

A simple geartrain formed as a serial connection of four gears on separate shafts. The gears' teeth are omitted for clarity.

need to have too few teeth for it to smoothly engage with the gear. One might therefore consider building a geartrain that is formed as a serial chain of more than two gears. Such a mechanism is called a *simple geartrain*, and it has the characteristic that each shaft carries a single gear.

Figure 8.22 shows an example of a simple geartrain having four gears. To distinguish the various gears and shafts, we set the convention that the input gear is labeled as gear 1, and the other gears are numbered sequentially. The numbers of teeth and rotational speeds of each gear are represented by the symbols N_i and ω_i. We are interested in determining the speed ω_4 of the output shaft for a given speed ω_1 of the input shaft. The direction that each gear rotates can be determined by recognizing that, for external gearsets, the direction reverses at each mesh point.

We can view the simple geartrain as a sequence of connected gearsets. In that sense, Equation (8.9) can be applied recursively to each pair of gears. Beginning at the first mesh point,

$$\omega_2 = \left(\frac{N_1}{N_2}\right)\omega_1 \qquad (8.13)$$

in the counterclockwise direction. At the second mesh point,

$$\omega_3 = \left(\frac{N_2}{N_3}\right)\omega_2 \qquad (8.14)$$

In designing a geartrain, we are more interested in the relationship between the input and output shaft speeds than in the speeds ω_2 and ω_3 of the intermediate shafts. By combining Equations (8.13) and (8.14), we can eliminate the intermediate variable ω_2:

$$\omega_3 = \left(\frac{N_2}{N_3}\right)\left(\frac{N_1}{N_2}\right)\omega_1 = \left(\frac{N_1}{N_3}\right)\omega_1 \qquad (8.15)$$

In other words, the effect of the second gear (and in particular, its size and number of teeth) cancels as far as the third gear is concerned. Proceeding to the final mesh point, the velocity of the output gear becomes

$$\omega_4 = \left(\frac{N_3}{N_4}\right)\omega_3 = \left(\frac{N_3}{N_4}\right)\left(\frac{N_1}{N_3}\right)\omega_1 = \left(\frac{N_1}{N_4}\right)\omega_1 \tag{8.16}$$

For this simple geartrain, the overall velocity ratio between the output and input shafts is

$$VR = \frac{\text{output speed}}{\text{input speed}} = \frac{\omega_4}{\omega_1} = \frac{N_1}{N_4} = \frac{N_{\text{input}}}{N_{\text{output}}} \tag{8.17}$$

The sizes of the intermediate gears 2 and 3 have no effect on the geartrain's velocity ratio. This result can be extended to simple geartrains having any number of gears, and, in general, the velocity ratio will depend only on the sizes of the input and output gears. In a similar manner, by applying Equation (8.12) for power conservation with pairs of ideal gears, the power supplied to the first gear must balance the power transmitted by the final gear. Since $VR \times TR = 1$ for an ideal gearset, the torque ratio is likewise unaffected by the numbers of teeth on gears 2 and 3, and

$$TR = \frac{\text{output torque}}{\text{input torque}} = \frac{T_4}{T_1} = \frac{N_4}{N_1} = \frac{N_{\text{output}}}{N_{\text{input}}} \tag{8.18}$$

Idler gear

Because the intermediate gears of a simple geartrain provide no speed or torque modifications as a whole, they are sometimes called *idler gears*. Although they may have no direct effect on *VR* and *TR*, the idler gears contribute indirectly to the extent that a designer can insert them to gradually increase or decrease the dimensions of adjacent gears. Additional idler gears also enable the input and output shafts to be separated farther from one another, but the power transmission chains and belts discussed later in this chapter could also be used for that purpose.

Compound Geartrain

As an alternative to simple geartrains, compound geartrains can be used in transmissions when larger velocity or torque ratios are needed or when the gearbox must be made physically compact. A *compound geartrain* is based on the principle of having more than one gear on each intermediate shaft. In the geartrain of Figure 8.23, the intermediate shaft carries two gears having different numbers of teeth. To determine the overall velocity ratio ω_4/ω_1 between the input and output shafts, we apply Equation (8.9) to each pair of meshing gears. Beginning with the first pair in mesh,

$$\omega_2 = \left(\frac{N_1}{N_2}\right)\omega_1 \tag{8.19}$$

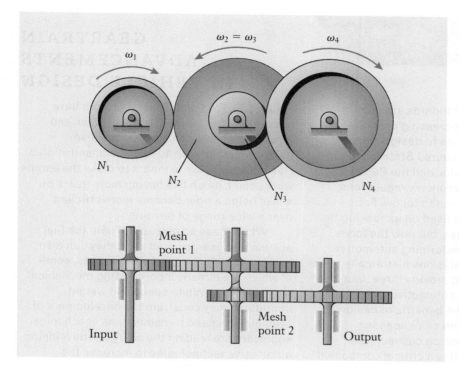

Figure 8.23

A compound geartrain in which the second and third gears are mounted on the same shaft and rotate together. The gears' teeth are omitted for clarity.

Because gears 2 and 3 are mounted on the same shaft, $\omega_3 = \omega_2$. Proceeding to the next mesh point between gears 3 and 4, we have

$$\omega_4 = \left(\frac{N_3}{N_4}\right)\omega_3 \qquad (8.20)$$

Combining Equations (8.19) and (8.20), the speed of the output shaft becomes

$$\omega_4 = \left(\frac{N_3}{N_4}\right)\left(\frac{N_1}{N_2}\right)\omega_1 \qquad (8.21)$$

and the velocity ratio is

$$VR = \frac{\text{output speed}}{\text{input speed}} = \left(\frac{\omega_4}{\omega_1}\right) = \left(\frac{N_1}{N_2}\right)\left(\frac{N_3}{N_4}\right) \qquad (8.22)$$

A compound geartrain's velocity ratio is the product of the velocity ratios between the pairs of gears at each mesh point, a result that can be generalized to compound geartrains having any number of additional stages. Unlike the gears in a simple geartrain, the sizes of the intermediate gears here influence the velocity ratio. The clockwise or counterclockwise rotation of any gear is determined by the rule of thumb: With an even number of mesh points, the output shaft rotates in the same direction as the input shaft. Conversely, if the number of mesh points is odd, the input and output shafts rotate opposite to one another.

Focus On

GEARTRAIN ADVANCEMENTS IN VEHICLE DESIGN

Rising fuel economy standards around the world are putting increasing pressure on mechanical engineers to design more efficient vehicles. The United States, Japan, China, Canada, Australia, and the Republic of Korea all have fuel economy regulations, with other countries soon to follow. For decades, engineers focused on increasing the efficiency of the engines, but now the focus has turned also to transforming automotive transmissions. Transmissions historically have housed geartrains that provide three, four, or five speeds. However, automotive companies are now recognizing the benefits of designing transmissions with even more speeds.

Since the transmission connects the engine to the wheels, it is a critical component of a vehicle. It is also one of the most complicated components, consisting of various high-precision gear trains as illustrated in

Figure 8.24. Ford and General Motors have codeveloped a 10-speed transmission, and Hyundai, Volkswagen, and Kia are also developing their own 10-speed transmissions. Having more gears is meant to make the engine work easier, much like having more gears on a bike helps a rider become more efficient over a wide range of terrains.

While these advances will help the fuel economy of gas-powered cars, they can also improve the efficiency of electric cars, some of which are already incorporating mechanical transmissions. While space and weight limitations may constrain the development of even higher speed transmissions, mechanical engineers are leading the charge in developing automotive technologies to increase the performance and fuel economy of a wide range of vehicles.

Figure 8.24

A portion of a vehicle transmission

Gajic Dragan/Shutterstock.com

Example 8.4 | *Speed, Torque, and Power in a Simple Geartrain*

The input shaft to the simple geartrain is driven by a motor that supplies
1 kW of power at the operating speed of 250 rpm. (See Figure 8.25.)
(a) Determine the speed and rotation direction of the output shaft. (b) What
magnitude of torque does the output shaft transfer to its mechanical load?

Figure 8.25

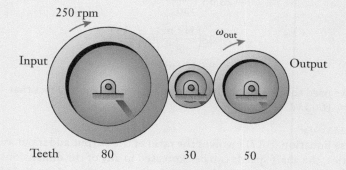

| Teeth | 80 | 30 | 50 |

Approach
To find the rotation direction of the output shaft, we recognize that the input
shaft rotates clockwise in the figure, and, at each mesh point, the direction of
rotation reverses. Thus, the 30-tooth gear rotates counterclockwise, and the
50-tooth output gear rotates clockwise. To determine the speed of the output
shaft, we will apply Equation (8.17). In part (b), for an ideal geartrain where
friction can be neglected, the input and output power levels are identical. We
can apply Equation (8.5) to relate speed, torque, and power.

Solution
(a) The speed of the output shaft is

$$\omega_{\text{out}} = \left(\frac{80}{50}\right)(250 \text{ rpm}) \qquad \leftarrow \left[VR = \frac{N_{\text{input}}}{N_{\text{output}}} \right]$$

$$= 400 \text{ rpm}$$

(b) The instantaneous power is the product of torque and rotation
speed. For the calculation to be dimensionally consistent, the speed
of the output shaft first must be converted to the units of radians
per second by using the conversion factor listed in Table 8.1:

$$\omega_{\text{out}} = 400 \text{ rpm}\left(0.1047 \frac{\text{rad/s}}{\text{rpm}}\right)$$

$$= 41.88\left(\frac{\text{rpm} \cdot \text{rad/s}}{\text{rpm}}\right)$$

$$= 41.88 \frac{\text{rad}}{\text{s}}$$

Example 8.4 | *continued*

The output torque becomes

$$T_{out} = \frac{1000 \text{ W}}{41.88 \text{ rad/s}} \quad \leftarrow \left[T = \frac{P}{\omega} \right]$$

$$= 23.88 \frac{\text{W} \cdot \text{s}}{\text{rad}}$$

$$= 23.88 \left(\frac{\text{N} \cdot \text{m}}{\text{s}} \right) (\text{s}) \left(\frac{1}{\text{rad}} \right)$$

$$= 23.88 \text{ N} \cdot \text{m}$$

Here we used the definition that $1 \text{ W} = 1 \text{ (N} \cdot \text{m)/s}$ and the fact that $1 \text{ kW} = 1000 \text{ W}$.

Discussion
Because Equation (8.17) involves the ratio of the output and input angular velocities, the shaft speeds can be expressed in any of the dimensions that are appropriate for angular velocity, including rpm. Also, the speed of the 30-tooth gear will be larger than the input or output gears because of its smaller size. When calculating the torque from the expression $P = T\omega$, the radian unit in the angular velocity can be directly canceled because it is a dimensionless measure of an angle.

$$\omega_{out} = 400 \text{ rpm}$$
$$T_{out} = 23.88 \text{ N} \cdot \text{m}$$

Example 8.5 | *Money Changer Geartrain*

A gear reduction mechanism is used in a machine that accepts bill currency, inspects it, and returns coins in change. The numbers of teeth on the pinion and gear in the first speed-reduction stage are shown in Figure 8.26. The worm has a single tooth that meshes with the 16 tooth worm gear. Determine the velocity ratio between the output shaft's speed and the speed of the direct-current motor.

Approach
To find the velocity ratio between the output shaft and the input speed, we recognize that the motor's speed is reduced in two stages: first, in the gearset comprising the 10 tooth pinion and 23 tooth gear, and second, at the mesh between the worm and worm gear. We denote the angular velocity of the motor's shaft as ω_{in}. To determine the speed of the transfer shaft, we will apply Equation (8.9). In the second speed-reduction stage,

Example 8.5 | *continued*

Figure 8.26

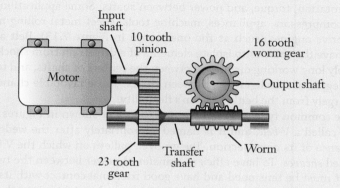

the worm gear will advance by one tooth for each rotation of the transfer shaft.

Solution

The speed of the transfer shaft is

$$\omega_{transfer} = \frac{10}{23}\omega_{in} \qquad \leftarrow \left[\omega_g = \left(\frac{N_p}{N_g}\right)\omega_p\right]$$

$$= 0.4348\,\omega_{in}$$

or $\omega_{in} = 2.3\,\omega_{transfer}$. Because the worm gear has 16 teeth and the worm has only one, the speeds of the transfer shaft and output shaft are related by

$$\omega_{out} = \left(\frac{1}{16}\right)\omega_{transfer}$$

$$= 0.0625\,\omega_{transfer}$$

The geartrain's overall velocity ratio is

$$VR = \frac{0.0625\,\omega_{transfer}}{2.3\,\omega_{transfer}} \qquad \leftarrow \left[VR = \frac{\omega_{out}}{\omega_{in}}\right]$$

$$= 2.717 \times 10^{-2}$$

Discussion

The speed of the output shaft is reduced to 2.7% of the input shaft's speed, for a speed reduction factor of 36.8 fold. This is consistent with the use of worm gears in design applications where large speed reduction ratios are required.

$$VR = 2.717 \times 10^{-2}$$

▶ 8.6 Design Application: Belt and Chain Drives

Similar to geartrains, belt and chain drives can also be used to transfer rotation, torque, and power between shafts. Some applications include compressors, appliances, machine tools, sheet-metal rolling mills, and automotive engines (such as the one shown in Figure 7.13). Belt and chain drives have the abilities to isolate elements of a drivetrain from shock, to have relatively long working distances between the centers of shafts, and to tolerate some degree of misalignment between shafts. Those favorable characteristics stem largely from the belt's or chain's flexibility.

The common type of power transmission belt shown in Figures 8.27 and 8.28 is called a *V-belt*, and it is named appropriately after the wedge-shaped appearance of its cross section. The grooved pulleys on which the V-belt rides are called *sheaves*. To have efficient transfer of power between the two shafts, the belt must be tensioned and have good frictional contact with its sheaves. In fact, the belt's cross section is designed to wedge tightly into the sheave's groove. The capability of V-belts to transfer load between shafts is determined by the belt's wedge angle and by the amount of friction between the belt and the surface of the sheaves. The exterior of a V-belt is made from a synthetic rubber to increase the amount of friction. Because elastomer materials have low elastic modulus and stretch easily, V-belts are usually reinforced internally with fiber or wire cords that carry most of the belt's tension.

Although V-belts are well suited for transmitting power, some amount of slippage invariably occurs between the belt and sheaves because the only contact between the two is friction. Slippage between gears, on the other hand, cannot occur because of the direct mechanical engagement between

V-belt

Sheave

Figure 8.27

A V-belt and its sheaves.

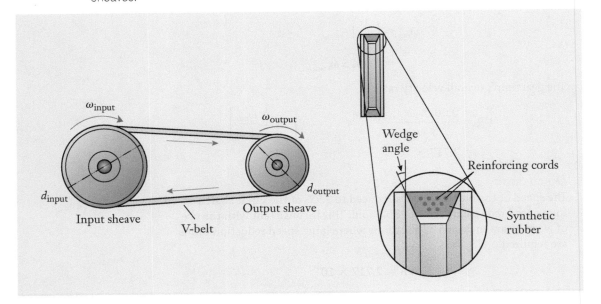

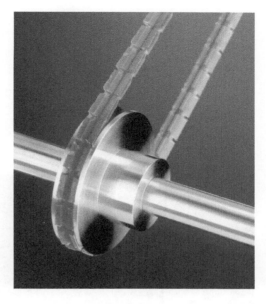

Figure 8.28

A segmented V-belt in contact with its sheave.

Reprinted with permission by W. M. Berg, Inc.

their teeth. Geartrains are a *synchronous* method of rotation; that is, the input and output shafts are synchronized, and they rotate together exactly. Belt slippage is not a concern if the engineer is interested only in transmitting power, as in a gasoline engine that drives a compressor or generator. On the other hand, for such precision applications as robotic manipulators and valve timing in automotive engines, the rotation of the shafts must remain perfectly synchronized. Timing belts address that need by having molded teeth that mesh in matching grooves on their sheaves. *Timing belts*, such as the one shown in Figure 8.29, combine some of the best features of belts—mechanical isolation and long working distances between shaft centers—with a gearset's ability to provide synchronous motion.

Synchronous rotation

Timing belt

Chain drives (Figures 8.30 and 8.31, see page 370) are another design option when synchronous motion is required, particularly when high torque or power must be transmitted. Because of their metallic link construction, chain and sprocket mechanisms are capable of carrying greater forces than belts, and they are also able to withstand high-temperature environments.

Chain drives

For timing belts and chains (and for V-belts when slippage on the sheaves can be neglected), the angular velocities of the two shafts are proportional to

Figure 8.29

A magnified view of the teeth on a timing belt. The reinforcing cords are exposed at the belt's cross section.

Reprinted with permission by W. M. Berg, Inc.

Figure 8.30

A chain and sprocket used in a power-transmission drive.

Image courtesy of the authors.

Figure 8.31

Engineers and scientists have manufactured micro mechanical chain drives by adapting techniques that are used in the production of integrated electronic circuits. The distance between each link in this chain is 50 microns, smaller than the diameter of a human hair.

Reprinted with permission of Sandia National Laboratories.

one another in a manner similar to a gearset. In Figure 8.27, for instance, the speed of the belt when it wraps around the first sheave $[(d_{\text{input}}/2)\, \omega_{\text{input}}]$ must be the same as when it wraps around the second sheave $[(d_{\text{output}}/2)\, \omega_{\text{output}}]$. In much the same manner as Equation (8.9), the velocity ratio between the output and input shafts is given by

$$VR = \frac{\text{output speed}}{\text{input speed}} = \frac{\omega_{\text{output}}}{\omega_{\text{input}}} = \frac{d_{\text{input}}}{d_{\text{output}}} \tag{8.23}$$

where the pitch diameters of the sheaves are denoted by d_{input} and d_{output}.

Example 8.6 | *Computer Scanner*

The mechanism in a desktop computer scanner converts the rotation of a motor's shaft into the side-to-side motion of the scanning head (Figure 8.32). During a portion of a scan operation, the drive motor turns at 180 rpm. The gear attached to the motor's shaft has 20 teeth, and the timing belt has 20 teeth/in. The two other gears are sized as indicated. In the units of in./s, calculate the speed at which the scanning head moves.

Example 8.6 | *continued*

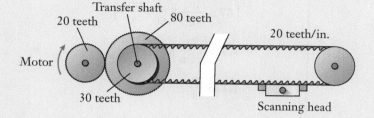

Figure 8.32

Approach
To find the speed of the scanning head, we must first determine the speed of the transfer shaft by applying Equation (8.9). Equation (8.2) can then be used to relate the rotational speed of the transfer shaft to the belt's speed.

Solution
The angular velocity of the transfer shaft is

$$\omega = \left(\frac{20}{80}\right)(180 \text{ rpm}) \quad \leftarrow \left[\omega_g = \left(\frac{N_p}{N_g}\right)\omega_p\right]$$

$$= 45 \text{ rpm}$$

Because the 30-tooth and 80-tooth gears are mounted on the same shaft, the 30-tooth gear also rotates at 45 rpm. With each rotation of the transfer shaft, 30 teeth of the timing belt mesh with the gear, and the timing belt advances by the distance

$$x = \frac{30 \text{ teeth/rev}}{20 \text{ teeth/in.}}$$

$$= 1.5\left(\frac{\text{teeth}}{\text{rev}}\right)\left(\frac{\text{in.}}{\text{teeth}}\right)$$

$$= 1.5 \frac{\text{in.}}{\text{rev}}$$

The speed v of the timing belt is the product of the shaft's speed ω and the amount x by which the belt advances with each revolution. The scanning head's speed becomes

$$v = \left(1.5 \frac{\text{in.}}{\text{rev}}\right)(45 \text{ rpm})$$

$$= 67.5\left(\frac{\text{rev}}{\text{min}}\right)\left(\frac{\text{in.}}{\text{rev}}\right)$$

$$= 67.5 \frac{\text{in.}}{\text{min}}$$

Example 8.6 | *continued*

In the units of inches per second, the velocity is

$$v = \left(67.5\ \frac{\text{in.}}{\text{min}}\right)\left(\frac{1}{60}\ \frac{\text{min}}{\text{s}}\right)$$

$$= 1.125\left(\frac{\text{in.}}{\text{min}}\right)\left(\frac{\text{min}}{\text{s}}\right)$$

$$= 1.125\ \frac{\text{in.}}{\text{s}}$$

Discussion
This mechanism achieves a speed reduction and a conversion between rotational and straight-line motion in two stages. First, the (input) angular velocity of the motor's shaft is reduced to the speed of the transfer shaft. Second, the timing belt converts the transfer shaft's rotational motion to the (output) straight-line motion of the scanning head.

$$v = 1.125\ \frac{\text{in.}}{\text{s}}$$

Example 8.7 | *Treadmill's Belt Drive*

A timing belt is used to transfer power in the exercise treadmill from the motor to the roller that supports the walking/running belt (Figure 8.33). The pitch diameters of the sheaves on the two shafts are 1.0 in. and 2.9 in., and the roller has a diameter of 1.75 in. For a running pace of 7 min/mi, at what speed should the motor be set? Express your answer in the units of rpm.

Figure 8.33

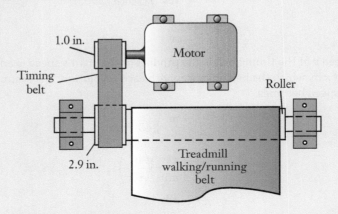

Example 8.7 | *continued*

Approach
To find the appropriate speed of the motor, we first find the roller's angular velocity and then relate this to the motor speed. The speed of the walking/running belt is related to the roller's angular velocity by Equation (8.2). Since the timing belt's speed is the same whether it is in contact with the sheave on the roller or with the sheave on the motor, the angular velocities of those two shafts are related by Equation (8.23).

Solution
In dimensionally consistent units, the speed of the walking/running belt is

$$v = \left(\frac{1 \text{ mi}}{7 \text{ min}}\right)\left(5280\frac{\text{ft}}{\text{mi}}\right)\left(12\frac{\text{in.}}{\text{ft}}\right)\left(\frac{1}{60}\frac{\text{min}}{\text{s}}\right)$$

$$= 150.9\left(\frac{\text{mi}}{\text{min}}\right)\left(\frac{\text{ft}}{\text{mi}}\right)\left(\frac{\text{in.}}{\text{ft}}\right)\left(\frac{\text{min}}{\text{s}}\right)$$

$$= 150.9\frac{\text{in.}}{\text{s}}$$

The angular velocity of the roller is

$$\omega_{\text{roller}} = \frac{150.9 \text{ in./s}}{(1.75 \text{ in.})/2} \quad \leftarrow [v = rw]$$

$$= 172.4\left(\frac{\text{in.}}{\text{s}}\right)\left(\frac{1}{\text{in.}}\right)$$

$$= 172.4\frac{\text{rad}}{\text{s}}$$

where the dimensionless radian unit has been used for the roller's angle. In the engineering units of rpm, the roller's speed is

$$\omega_{\text{roller}} = \left(172.4\frac{\text{rad}}{\text{s}}\right)\left(9.549\frac{\text{rpm}}{\text{rad/s}}\right)$$

$$= 1646\left(\frac{\text{rad}}{\text{s}}\right)\left(\frac{\text{rpm}}{\text{rad/s}}\right)$$

$$= 1646 \text{ rpm}$$

Since the 2.9 in. diameter sheave turns at the same speed as the roller, the speed of the motor's shaft is

$$\omega_{\text{roller}} = \left(\frac{2.9 \text{ in.}}{1.0 \text{ in.}}\right)(1646 \text{ rpm}) \quad \leftarrow \left[VR = \frac{d_{\text{input}}}{d_{\text{output}}}\right]$$

$$= 4774\left(\frac{\text{in.}}{\text{in.}}\right)(\text{rpm})$$

$$= 4774 \text{ rpm}$$

Example 8.7 | *continued*

Discussion

This belt drive is similar to a compound geartrain in the sense that two belts (the timing belt and the walking/running belt) are in contact with sheaves on the same shaft. Because the roller and the 2.9 in. sheave have different diameters, the two belts move at different speeds v and are assumed not to slip. Many treadmill motors are rated for a maximum of 4000–5000 rpm. So our answer, though on the high end of the range, is certainly reasonable for the speed of the runner.

$$\omega_{\text{motor}} = 4774 \text{ rpm}$$

▶ 8.7 Planetary Geartrains

Sun gear

Planet gear

Carrier

Figure 8.34

(a) A simple gearset in which the pinion and gear are connected by a fixed-ground link. (b) A planetary gearset where the planet and carrier link can rotate about the center of the sun gear.

Up to this point, our geartrain shafts have been connected to the housing of a gearbox by bearings, and the centers of the shafts themselves did not move. The gearsets, simple geartrains, compound geartrains, and belt drives of the previous sections were all of this type. In some geartrains, however, the centers of certain gears may be allowed to move. Such mechanisms are called planetary geartrains because the motion of their gears is (in many ways) analogous to the orbit of a planet around a star.

Simple and planetary gearsets are contrasted in Figure 8.34. Conceptually, we can view the center points of gears in the simple gearset as being connected by a link that is fixed to the ground. In the planetary system of Figure 8.34(b), on the other hand, although the center of the *sun gear* is stationary, the center of the *planet gear* can orbit around the sun gear. The planet gear rotates about its own center, meshes with the sun gear, and then orbits as a whole about the center of the sun gear. The link that connects the centers of the two gears is called the *carrier*. Planetary geartrains are often used as speed reducers, and one application is the geartrain of Figure 8.35 that is used in the transmission of a light-duty helicopter.

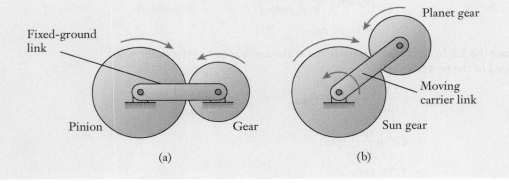

Fixed-ground link

Pinion   Gear

Planet gear

Moving carrier link

Sun gear

(a)   (b)

Figure 8.35

A planetary geartrain used in a helicopter's transmission. The diameter of the ring gear is approximately one foot.

NASA-Lewis Research Center

In Figure 8.34(b), it is straightforward enough to connect the shaft of the sun gear to a power source or to a mechanical load. However, because of the planet gear's orbital motion, connecting the planet gear to the shaft of another machine directly is not feasible. To construct a more functional geartrain, the *ring gear* shown in Figure 8.36 is used to convert the motion of the planet gear into rotation of the ring gear and its shaft. The ring gear is an internal gear, whereas the sun and planet gears are external gears.

In this configuration, a planetary geartrain has three input–output connection points, as indicated in Figure 8.37 (see page 376): the shafts of the sun gear, carrier, and ring gear. Those connections can be configured to form a geartrain having two input shafts (for instance, the carrier and the sun gear) and one output shaft (the ring gear in this case), or a geartrain having one input and two output shafts. A planetary geartrain can therefore combine power from two sources into one output, or it can split the power from one source into two outputs. In a rear-wheel-drive automobile, for instance, power supplied by the engine is split between the two drive wheels by a special type of planetary geartrain called a differential (described later in this section). It is also possible to fix one of the planetary geartrain's connections to ground (say, the ring gear) so that there are only one input shaft and one output shaft (in this case, the carrier and sun gear). Alternatively, two of the shafts can be coupled, again reducing the number of connection points from three to two. Versatile configurations such as these are exploited in the operation of

Ring gear

Figure 8.36

Front and cross-sectional views of a planetary geartrain.

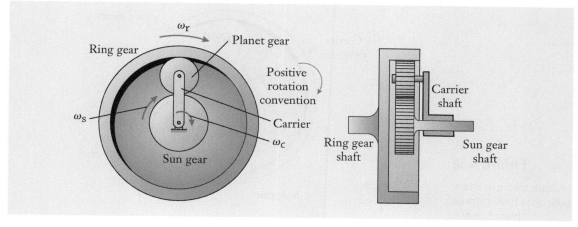

Figure 8.37

Three input–output connection points to a planetary geartrain.

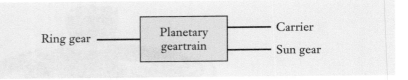

automotive automatic transmissions. By the proper sizing of the sun, planet, and ring gears and by selecting the input and output connections, engineers can obtain either very small or very large velocity ratios in a geartrain of compact size.

Planetary geartrains are usually constructed with more than one planet gear to reduce noise, vibration, and the forces applied to the gear teeth. A *balanced planetary geartrain* is depicted in Figure 8.38. When multiple planet gears are present, the carrier is sometimes called the *spider*, because it has several (although perhaps not as many as eight) legs that evenly separate the planet gears around the circumference.

Balanced geartrain
Spider

The rotations of shafts and the flow of power through simple and compound geartrains are usually straightforward enough to visualize. Planetary systems are more complicated to the extent that power can flow through the geartrain in multiple paths, and our intuition may not always be sufficient to determine the rotation directions. For instance, if the carrier and sun gears in Figure 8.36 are both driven clockwise, and the carrier's speed is greater than the sun's, then the ring gear will rotate clockwise. However, as the speed of the sun gear is gradually increased, the ring gear will slow down, stop, and then actually reverse its direction and rotate counterclockwise. With that in mind, instead of relying on our intuition, we can apply a design equation that relates the rotational velocities of the sun gear (ω_s), carrier (ω_c), and ring gear (ω_r):

$$\omega_s + n\omega_r - (1 + n)\,\omega_c = 0 \qquad (8.24)$$

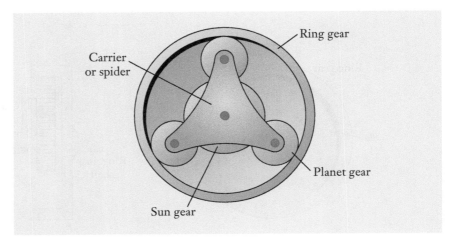

Figure 8.38

A balanced planetary geartrain having three planet gears.

With the numbers of teeth on the sun and ring gears denoted by N_s and N_r, the geartrain's *form factor n* in Equation (8.24) is

$$n = \frac{N_r}{N_s} \qquad (8.25)$$

This parameter is a size ratio that enters into the speed calculations for planetary geartrains and makes them more convenient; specifically, n is not the number of teeth on any gear. When n is a large number, then the planetary geartrain has a relatively small sun gear, and vice versa. The numbers of teeth on the sun, planet, and ring gears are related by

$$N_r = N_s + 2N_p \qquad (8.26)$$

The direction in which each shaft rotates can be determined from the positive or negative signs of the angular velocity terms in Equation (8.24). As shown in Figure 8.36, we apply the convention that rotations of the shafts are positive when they are directed clockwise and negative when counterclockwise. By consistently applying the *sign convention*, we rely on the outcome of the calculation to indicate the directions that the shafts rotate. Of course, the clockwise convention is arbitrary, and we could just as well have chosen the counterclockwise direction as positive. In any event, once we define the convention for positive rotations, we use it consistently throughout a calculation.

A *differential* is a special type of planetary geartrain used in automobiles. The layout of the drivetrain for a rear-wheel-drive vehicle is shown in Figure 8.39. The engine is located at the front of the automobile, and the crankshaft feeds into the transmission. The speed of the engine's crankshaft is reduced by the transmission, and the driveshaft extends down the length of the vehicle to the

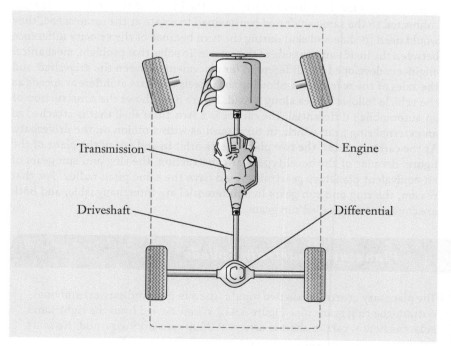

Transmission

Engine

Driveshaft

Differential

Figure 8.39

Drivetrain in a rear-wheel-drive automobile.

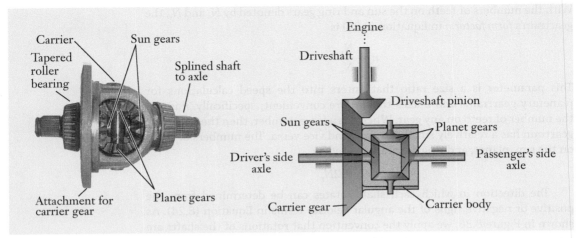

Figure 8.40

A differential for a small sedan automobile, which can be viewed as a planetary geartrain constructed from bevel-type gears.

Image courtesy of the authors.

rear wheels. The transmission adjusts the velocity ratio between the rotation speeds of the engine's crankshaft and the driveshaft. In turn, the differential transfers torque from the driveshaft and splits it between the wheels on the driver's and passenger's sides. The differential therefore has one input (the driveshaft) and two outputs (the wheel axles). Look underneath a rear-wheel-drive automobile, and see if you can identify the transmission, driveshaft, and differential.

The differential enables the rear wheels to rotate at different speeds when the vehicle turns a corner, all the while being powered by the engine. As an automobile turns, the drive wheel on the outside of the curve rolls through a greater distance than does the wheel on the inside. If the drive wheels were connected to the same shaft and constrained to rotate at the same speed, they would need to slide and skid during the turn because of the velocity difference between the inside and outside of the corner. To solve that problem, mechanical engineers developed the differential for placement between the driveshaft and the axles of the rear wheels, allowing the wheels to rotate at different speeds as the vehicle follows curves along a road. Figure 8.40 shows the construction of an automobile's differential. The carrier is a structural shell that is attached to an external ring gear, which, in turn, meshes with a pinion on the driveshaft. As the carrier rotates, the two planet gears orbit in and out of the plane of the figure. Because of the bevel-type gear construction, the ring and sun gears of an equivalent planetary geartrain would have the same pitch radius. For that reason, the ring and sun gears in a differential are interchangeable, and both are conventionally called sun gears.

Example 8.8 | *Planetary Geartrain Speeds*

The planetary geartrain has two inputs (the sun gear and carrier) and one output (the ring gear). (See Figure 8.41.) When viewed from the right-hand side, the hollow carrier shaft is driven at 3600 rpm clockwise, and the shaft for the sun gear turns at 2400 rpm counterclockwise. (a) Determine the

Example 8.8 | continued

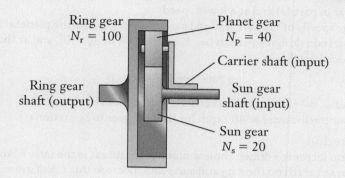

Figure 8.41

Ring gear
$N_r = 100$

Planet gear
$N_p = 40$

Carrier shaft (input)

Ring gear
shaft (output)

Sun gear
shaft (input)

Sun gear
$N_s = 20$

speed and direction of the ring gear. (b) The carrier's shaft is driven by an electric motor capable of reversing its direction. Repeat the calculation for the case in which it is instead driven at 3600 rpm counterclockwise. (c) For what speed and direction of the carrier's shaft will the ring gear not rotate?

Approach

To find the speed and direction of the appropriate gear in each operating condition, we apply Equations (8.24) and (8.25), together with the sign convention shown in Figure 8.36. With clockwise rotations being positive, the known speeds in part (a) are $\omega_c = 3600$ rpm and $\omega_s = -2400$ rpm. In part (b), when the carrier's shaft reverses direction, $\omega_c = -3600$ rpm.

Solution

(a) Based on the numbers of teeth shown in the diagram, the form factor is

$$n = \frac{100 \text{ teeth}}{20 \text{ teeth}} \qquad \leftarrow \left[n = \frac{N_r}{N_s} \right]$$

$$= 5 \frac{\text{teeth}}{\text{teeth}}$$

$$= 5$$

Substituting the given quantities into the planetary geartrain design equation, $[\omega_s + n\omega_r - (1 + n)\, \omega_c = 0]$

$$(-2400 \text{ rpm}) + 5\omega_r - 6(3600 \text{ rpm}) = 0$$

and $\omega_r = 4800$ rpm. Because ω_r is positive, the ring gear rotates clockwise.

(b) When the carrier's shaft reverses, $\omega_c = -3600$ rpm. The speed of the ring gear using the planetary geartrain design equation $[\omega_s + n\omega_r - (1 + n)\, \omega_c = 0]$ becomes

$$(-2400 \text{ rpm}) + 5\omega_r - 6(-3600 \text{ rpm}) = 0$$

Example 8.8 | *continued*

and $\omega_r = -3840$ rpm. The ring gear rotates in the direction opposite that in part (a) and at a lower speed.

(c) The speeds of the sun gear and carrier are related by the planetary geartrain design equation $[\omega_s + n\omega_r - (1 + n)\omega_c = 0]$, and at that special condition

$$(-2400 \text{ rpm}) - 6\omega_c = 0$$

and $\omega_c = -400$ rpm. The carrier's shaft should be driven counterclockwise at 400 rpm for the ring gear to be stationary.

Discussion

The form factor is a dimensionless number because it is the ratio of the numbers of teeth on the ring and sun gears. Because this calculation involves only rotational speeds, not the velocity of a point as in Equation (8.2), it is acceptable to use the dimensions of rpm rather than convert the units for angular velocity into radians per unit time. These results demonstrate the complicated yet flexible nature of planetary geartrains to produce a wide range of output motions.

Clockwise carrier: $\omega_r = 4800$ rpm (clockwise)
Counterclockwise carrier: $\omega_r = 3840$ rpm (counterclockwise)
Stationary ring: $\omega_c =\ \ 400$ rpm (counterclockwise)

Example 8.9 | *Torque in a Planetary Geartrain*

In part (a) of the previous example, the carrier and sun gear are driven by engines producing 2 hp and 5 hp, respectively (Figure 8.42). Determine the torque applied to the shaft of the output ring gear.

Approach

To calculate the torque applied to the output shaft, we recognize that, to balance the power supplied to the geartrain, the output shaft must transfer a total of 7 hp to a mechanical load. To apply Equation (8.5) in dimensionally consistent units, first convert the units for power from horsepower to (ft · lb)/s with the conversion factor in Table 7.2.

Figure 8.42

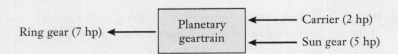

Ring gear (7 hp) ← | Planetary geartrain | ← Carrier (2 hp)
← Sun gear (5 hp)

Example 8.9 | *continued*

Solution

In dimensionally consistent units, the speed of the ring gear's output shaft is

$$\omega_r = (4800 \text{ rpm})\left(0.1047\frac{\text{rad/s}}{\text{rpm}}\right)$$

$$= 502.6 (\text{rpm})\left(\frac{\text{rad/s}}{\text{rpm}}\right)$$

$$= 502.6\frac{\text{rad}}{\text{s}}$$

The total power supplied to the geartrain is $P = 7$ hp, and, in dimensionally consistent units, this becomes

$$P = (7 \text{ hp})\left(550\frac{(\text{ft} \cdot \text{lb})/\text{s}}{\text{hp}}\right)$$

$$= 3850 \text{ (hp)}\left(\frac{(\text{ft} \cdot \text{lb})/\text{s}}{\text{hp}}\right)$$

$$= 3850\frac{\text{ft} \cdot \text{lb}}{\text{s}}$$

The output torque is therefore

$$T = \frac{3850 \text{ (ft} \cdot \text{lb)}/\text{s}}{502.6 \text{ rad/s}} \qquad \leftarrow [P = T\omega]$$

$$= 7.659\left(\frac{\text{ft} \cdot \text{lb}}{\text{s}}\right)\left(\frac{\text{s}}{\text{rad}}\right)$$

$$= 7.659 \text{ ft} \cdot \text{lb}$$

Discussion

As before, we can directly cancel the radian unit when calculating the torque because the radian is a dimensionless measure of an angle. For this geartrain as a whole, the input power is equivalent to the power transferred to the mechanical load to the extent that friction within the geartrain can be neglected.

$$T = 7.659 \text{ ft} \cdot \text{lb}$$

Summary

I n this chapter, we discussed the topics of motion and power transmission in machinery in the context of geartrains and belt drives. The important quantities introduced in this chapter, common symbols representing them and their units are summarized in Table 8.2, and Table 8.3 reviews the key equations used. The motion of geartrains and belt or chain drives encompasses mechanical components, forces and torques, and energy and power.

Gearsets, simple geartrains, compound geartrains, planetary geartrains, and belt and chain drives are used to transmit power, to change the rotation speed of a shaft, and to modify the torque applied to a shaft. More broadly, geartrains and belt and chain drives are examples of mechanisms commonly encountered in mechanical engineering. Mechanisms are combinations of gears, sheaves, belts, chains, links, shafts, bearings, springs, cams, lead screws, and other building-block components that can be assembled to convert one type of motion into another. Thousands of recipes for various mechanisms are available in printed and online mechanical engineering resources with applications including robotics, engines, automatic-feed mechanisms, medical devices, conveyor systems, safety latches, ratchets, and self-deploying aerospace structures.

Table 8.2

Quantities, Symbols, and Units That Arise when Analyzing Motion and Power-Transmission Machinery

Quantity	Conventional Symbols	Conventional Units	
		USCS	SI
Velocity	v	in./s, ft/s	mm/s, m/s
Angle	θ	deg, rad	
Angular velocity	ω	rpm, rps, deg/s, rad/s	
Torque	T	in · lb, ft · lb	N · m
Work	W	ft · lb	J
Instantaneous power	P	hp, (ft · lb)/s	W
Diametral pitch	p	teeth/in.	—
Module	m	—	mm
Number of teeth	N	—	—
Velocity ratio	VR	—	—
Torque ratio	TR	—	—
Form factor	n	—	—

Velocity	$v = r\omega$
Work of a torque	$W = T\Delta\theta$
Instantaneous power	
Force	$P = Fv$
Torque	$P = T\omega$
Gear design	
Module	$m = \dfrac{2r}{N}$
Diametral pitch	$p = \dfrac{N}{2r}$
Velocity ratio	
Gearset	$VR = \dfrac{N_p}{N_g}$
Simple geartrain	$VR = \dfrac{N_{input}}{N_{output}}$
Compound geartrain	$VR = \left(\dfrac{N_1}{N_2}\right)\left(\dfrac{N_3}{N_4}\right)\cdots$
Belt/chain drive	$VR = \dfrac{d_{input}}{d_{output}}$
Torque ratio	$TR = \dfrac{1}{VR}$
Planetary geartrain	
Form factor	$n = \dfrac{N_r}{N_s}$
Velocity	$\omega_s + n\omega_r - (1+n)\,\omega_c = 0$
Teeth	$N_r = N_s + 2N_p$

Table 8.3

Key Equations
That Arise When
Analyzing Motion and
Power-Transmission
Machinery

Self-Study and Review

8.1. List several of the units that engineers use for angular velocity.

8.2. In what types of calculations must the unit of rad/s be used for angular velocity?

8.3. What is the difference between average and instantaneous power?

8.4. Sketch the shape of a spur gear's teeth.

8.5. What is the fundamental property of gearsets?

8.6. Define the terms "diametral pitch" and "module."

8.7. What is a rack and pinion?

8.8. How do helical gears differ from spur gears?

8.9. Make a sketch to show the difference in shaft orientations when bevel gears and crossed helical gears are used.

8.10. How do simple and compound geartrains differ?

8.11. How are the velocity and torque ratios of a geartrain defined?

8.12. What relationship exists between the velocity and torque ratios for an ideal geartrain?

8.13. What are some of the differences between a V-belt and a timing belt?

8.14. Sketch a planetary geartrain, label its main components, and explain how it operates.

8.15. Describe the main components of an automobile's drivetrain.

8.16. What function does the differential in an automobile serve?

Problems

P8.1

An automobile travels at 30 mph, which is also the speed of the center C of the tire (Figure P8.1). If the outer radius of the tire is 15 in., determine the rotational velocity of the tire in the units of rad/s, deg/s, rps, and rpm.

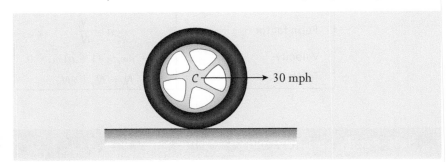

Figure P8.1

P8.2

Assume that the tire in P8.1 is now sitting on ice, spinning with the same rotational velocity but not translating forward. Determine the velocities of points at the top of the tire and at the bottom of the tire where it contacts the ice. Also, identify the direction of these velocities.

P8.3

The disk in a computer hard drive spins at 7200 rpm (Figure P8.3). At the radius of 30 mm, a stream of data is magnetically written on the disk, and the spacing between data bits is 25 μm. Determine the number of bits per second that pass by the read/write head.

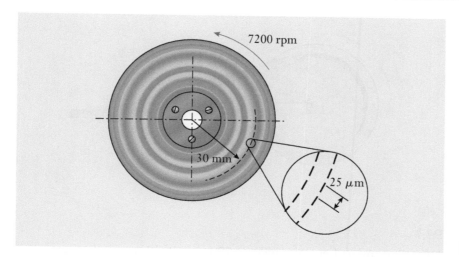

Figure P8.3

P8.4

The lengths of the two links in the industrial robot are $AB = 22$ in. and $BC = 18$ in. (Figure P8.4). The angle between the two links is held constant at $40°$ as the robot's arm rotates about base A at $300°/s$. Calculate the velocity of the center C of the robot's gripper.

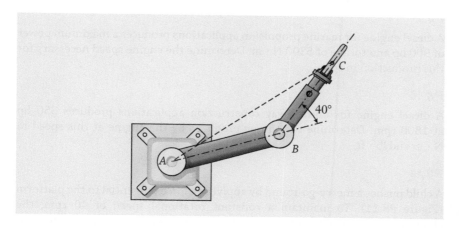

Figure P8.4

P8.5

For the industrial robot in P8.4, assume that the robot arm stops rotating about base A. However, the link BC begins to rotate about joint B. If link BC rotates $90°$ in 0.2 s, calculate the velocity of point C.

P8.6

The driving disk spins at a constant 280 rpm clockwise (Figure P8.6). Determine the velocity of the connecting pin at A. Also, does the collar at B move with constant velocity? Explain your answer.

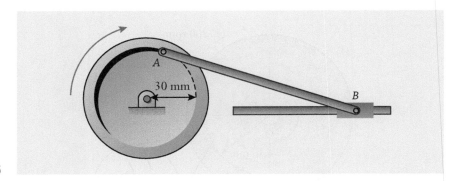

Figure P8.6

P8.7

A gasoline-powered engine produces 15 hp as it drives a water pump at a construction site. If the engine's speed is 450 rpm, determine the torque T that is transmitted from the output shaft of the engine to the pump. Express your answer in the units of ft · lb and N · m.

P8.8

A small automobile engine produces 260 N · m of torque at 2100 rpm. Determine the engine's power output in the units of kW and hp.

P8.9

A diesel engine for marine propulsion applications produces a maximum power of 900 hp and torque of 5300 N · m. Determine the engine speed necessary for this production in rpm.

P8.10

A diesel engine for on-highway construction applications produces 350 hp at 1800 rpm. Determine the torque produced by the engine at this speed in N · m and lb · ft.

P8.11

A child pushes a merry-go-round by applying a force tangential to the platform (Figure P8.11). To maintain a constant rotational speed of 40 rpm, the child must exert a constant force of 90 N to overcome the slowing effects of friction in the bearings and platform. Calculate the power exerted by the child in horsepower to operate the merry-go-round. The diameter of the platform is 8 ft.

P8.12

The torque produced by a 2.5 L automobile engine operating at full throttle was measured over a range of speeds (Figure P8.12). By using this graph of torque as a function of the engine's speed, prepare a second graph to show how the power output of the engine (in the units of hp) changes with speed (in rpm).

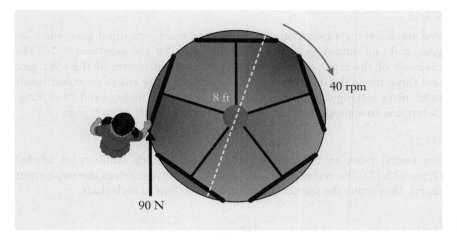

Figure P8.11

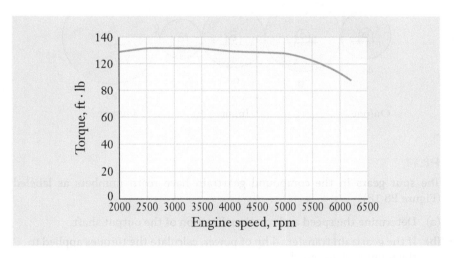

Figure P8.12

P8.13

A spur gearset has been designed with the following specifications:

> Pinion gear: number of teeth = 32, diameter = 3.2 in.

> Output gear: number of teeth = 96, diameter = 8.0 in.

Determine whether this gearset will operate smoothly.

P8.14

The radius of an input pinion is 3.8 cm, and the radius of an output gear is 11.4 cm. Calculate the velocity and torque ratios of the gearset.

P8.15

The torque ratio of a gearset is 0.75. The pinion gear has 36 teeth and a diametral pitch of 8. Determine the number of teeth on the output gear and the radii of both gears.

P8.16

You are designing a geartrain with three spur gears: one input gear, one idler gear, and one output gear. The diametral pitch for the geartrain is 16. The diameter of the input gear needs to be twice the diameter of the idler gear and three times the diameter of the output gear. The entire geartrain needs to fit into a rectangular footprint of no larger than 16 in. high and 24 in. long. Determine an appropriate number of teeth and diameter of each gear.

P8.17

The helical gears in the simple geartrain have teeth numbers as labeled (Figure P8.17). The central gear rotates at 125 rpm and drives the two output shafts. Determine the speeds and rotation directions of each shaft.

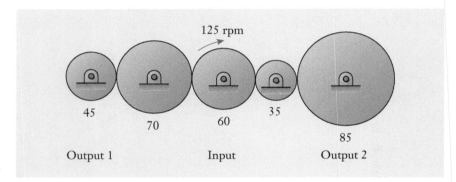

Figure P8.17

P8.18

The spur gears in the compound geartrain have teeth numbers as labeled (Figure P8.18).

(a) Determine the speed and rotation direction of the output shaft.

(b) If the geartrain transfers 4 hp of power, calculate the torques applied to the input and output shafts.

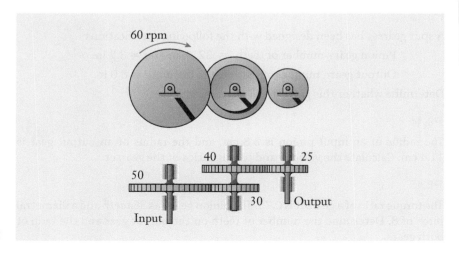

Figure P8.18

P8.19

The disk in a computer's magnetic hard drive spins about its center spindle while the recording head C reads and writes data (Figure P8.19). The head is positioned above a specific data track by the arm, which pivots about its bearing at B. As the actuator motor A turns through a limited angle of rotation, its 6 mm radius pinion rotates along the arc-shaped segment, which has a radius of 52 mm about B. If the actuator motor turns at 3000 deg/s over a small range of motion during a track-to-track seek operation, calculate the speed of the read/write head.

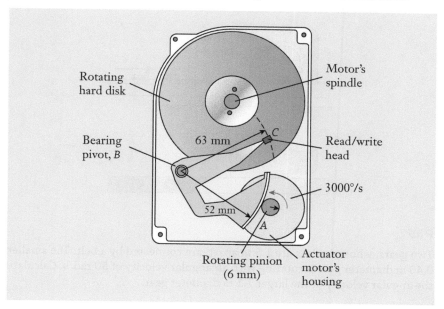

Figure P8.19

P8.20

For the compound geartrain, obtain an equation for the velocity and torque ratios in terms of the numbers of teeth labeled in Figure P8.20.

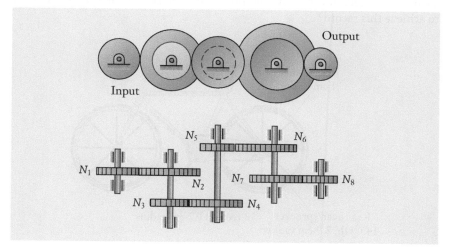

Figure P8.20

P8.21

In the motorized rack-and-pinion system in a milling machine, the pinion has 50 teeth and a pitch radius of 0.75 in. (Figure P8.21). Over a certain range of its motion, the motor turns at 800 rpm.

(a) Determine the horizontal velocity of the rack.

(b) If a peak torque of 10 ft · lb is supplied by the motor, determine the pulling force that is produced in the rack and transferred to the milling machine.

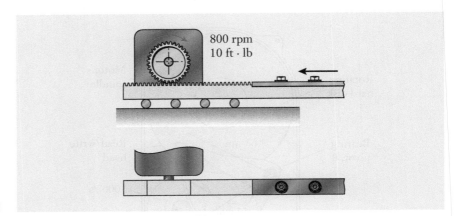

Figure P8.21

P8.22

Two gears, whose centers are 1.5 m apart, are connected by a belt. The smaller 0.40 m diameter gear is rotating with an angular velocity of 50 rad/s. Calculate the angular velocity of the larger 1.1 m diameter gear.

P8.23

The world distance record for riding a standard bicycle in 1 h is 54.526 km, set in 2015 in London (Figure P8.23). A fixed-gear bike was used, which means that the bike effectively had only one gear. Assuming that the rider pedaled at a constant speed, how fast did the rider pedal the front gear/sprocket in rpm to achieve this record?

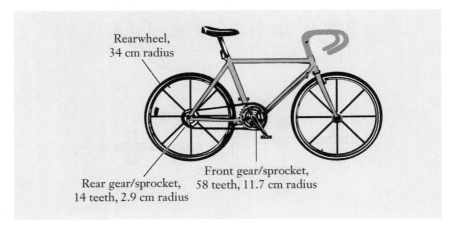

Rearwheel,
34 cm radius

Rear gear/sprocket,
14 teeth, 2.9 cm radius

Front gear/sprocket,
58 teeth, 11.7 cm radius

Figure P8.23

P8.24

(a) By directly examining sprockets and counting their numbers of teeth, determine the velocity ratio between the (input) pedal crank and the (output) rear wheel for your multiple-speed bicycle, a friend's, or a family member's. Make a table to show how the velocity ratio changes depending on which sprocket is selected in the front and back of the chain drive. Tabulate the velocity ratio for each speed setting.

(b) For a bicycle speed of 15 mph, determine the rotational speeds of the sprockets and the speed of the chain for one of the chain drive's speed settings. Show in your calculations how the units of the various terms are converted to obtain a dimensionally consistent result.

P8.25

Estimate the velocity ratio between the (input) engine and the (output) drive wheels for your automobile (or a friend's or a family member's) in several different transmission settings. You will need to know the engine's speed (as read on the tachometer), the vehicle's speed (on the speedometer), and the outer diameter of the wheels. Show in your calculations how the units of the various terms are converted to obtain a dimensionally consistent result.

P8.26

The geartrain in the transmission for the Segway® Personal Transporter uses helical gears to reduce noise and vibration (Figure P8.26). The vehicle's wheels have a diameter of 48 cm, and it has a top speed of 12.5 mph.

(a) Each gearset was designed to have a noninteger gear ratio so that the teeth mesh at different points from revolution to revolution, therefore reducing wear and extending the transmission's life. What is the velocity ratio at each mesh point?

(b) What is the velocity ratio for the entire geartrain?

(c) In the units of rpm, how fast does the motor's shaft turn at top speed?

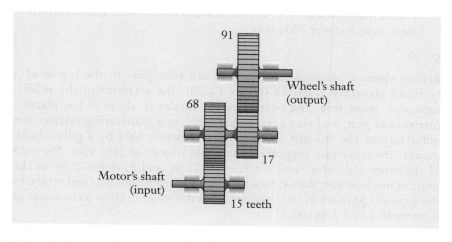

Figure P8.26

P8.27

The mechanism that operates the load/unload tray for holding the disk in a tabletop Blu-ray^{TM} player uses nylon spur gears, a rack, and a belt drive (Figure P8.27). The gear that meshes with the rack has a module of 2.5 mm. The gears have the numbers of teeth indicated, and the two sheaves have diameters of 7 mm and 17 mm. The rack is connected to the tray that holds the disk. For the tray to move at 0.1 m/s, how fast must the motor turn?

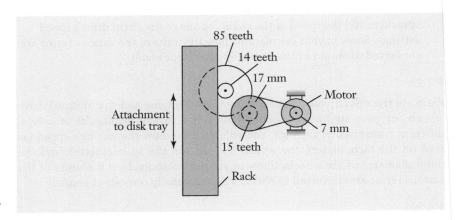

Figure P8.27

P8.28

Explain why the number of teeth on the ring gear in a planetary geartrain is related to the sizes of the sun and planet gears by the equation $N_r = N_s + 2N_p$.

P8.29

A planetary geartrain with $N_s = 48$ and $N_p = 30$ uses the carrier and ring gear as inputs and the sun gear as output. When viewed from the right-hand side in Figure 8.36, the hollow carrier shaft is driven at 1200 rpm clockwise, and the shaft for the ring gear is driven at 1000 rpm counterclockwise.

(a) Determine the speed and rotation direction of the sun gear.

(b) Repeat the calculation for the case in which the carrier is instead driven counterclockwise at 2400 rpm.

P8.30

Rolling element bearings (Section 4.6) are analogous to the layout of a balanced planetary geartrain (Figure P8.30). The rotations of the rollers, separator, inner race, and outer race are similar to those of the planets, carrier, sun gear, and ring gear, respectively, in a planetary geartrain. The outer race of the straight roller bearing shown is held by a pillow-block mount. The inner race supports a shaft that rotates at 1800 rpm. The radii of the inner and outer races are $R_i = 0.625$ in. and $R_o = 0.875$ in. In the units of ms, how long does it take for roller 1 to orbit the shaft and return to the topmost position in the bearing? Does the roller orbit in a clockwise or counterclockwise direction?

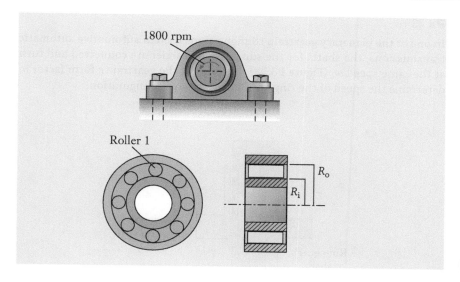

P8.31

The ring gear on a planetary geartrain similar to Figure 8.38 has 60 teeth and is rotating clockwise at 120 rpm. The planet and sun gears are identically sized. The sun gear is rotating clockwise at 150 rpm. Calculate the speed and rotation direction of the carrier.

P8.32

(a) The shaft of the sun gear in the planetary geartrain is held stationary by a brake (Figure P8.32). Determine the relationship between the rotational speeds of the shafts for the ring gear and carrier. Do those shafts rotate in the same direction or opposite directions?

(b) Repeat the exercise for the case in which the ring gear shaft is instead held stationary.

(c) Repeat the exercise for the case in which the carrier shaft is instead held stationary.

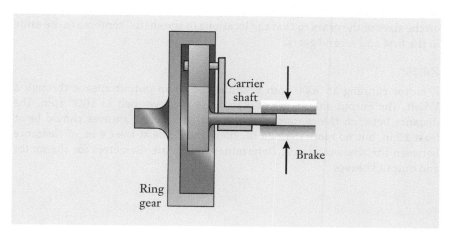

Figure P8.32

P8.33

In one of the planetary geartrain configurations used in automotive automatic transmissions, the shafts for the sun gear and carrier are connected and turn at the same speed ω_o (Figure P8.33). In terms of the geartrain's form factor n, determine the speed of the ring gear's shaft for this configuration.

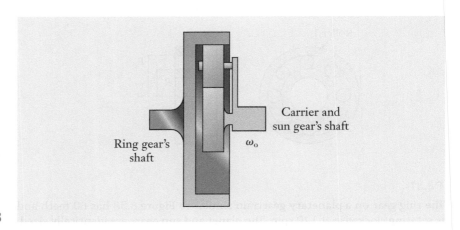

Figure P8.33

P8.34*

A gearbox is to be designed to provide an overall velocity ratio of exactly 24:1, while minimizing the overall size of the gearbox. Also, the rotational direction of the input and output shafts should be the same. Determine appropriate values for the number of teeth for each gear.

P8.35*

In a two-speed gearbox (Figure P8.35), the lower shaft has splines, and it can slide horizontally as an operator moves the shift fork. Design the transmission and choose the number of teeth on each gear to produce velocity ratios of approximately 0.8 (in first gear) and 1.2 (in second gear). Select gears having only even numbers of teeth between 40 and 80. Note that a constraint exists on the sizes of the gears so that the locations of the shafts' centers are the same in the first and second gears.

P8.36*

A motor running at 3000 rpm is connected to an output sheave through a V-belt. The output sheave drives a separate conveyer belt at 1000 rpm. The distance between the centers of the motor and output sheaves should be at least 12 in. but no more than 24 in. There should be at least 4 in. of clearance between the sheaves as well. Determine appropriate diameters for the motor and output sheaves.

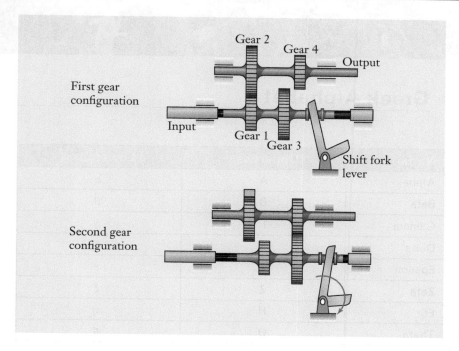

First gear configuration

Input

Gear 2
Gear 4
Output

Gear 1
Gear 3

Shift fork lever

Second gear configuration

Figure P8.35

References

Drexler, K. E., *Nanosystems: Molecular Machinery, Manufacturing, and Computation*, Hoboken, NJ: Wiley Interscience, 1992.

Lang, G. F., "S&V Geometry 101," *Sound and Vibration*, 1999, **33**(5), pp. 16–26.

Norton, R. L., *Design of Machinery: An Introduction to the Synthesis and Analysis of Mechanisms and Machines*, 3rd ed. New York: McGraw-Hill, 2004.

"Undercover Gears," *Gear Technology*, March/April 2002, p. 56.

Wilson, C. E., and Sadler, J. P., *Kinematics and Dynamics of Machinery*, 3rd ed. Upper Saddle River, NJ: Prentice Hall, 2003.

▶ Greek Alphabet

Name	Uppercase	Lowercase
Alpha	A	α
Beta	B	β
Gamma	Γ	γ
Delta	Δ	δ
Epsilon	E	ε
Zeta	Z	ζ
Eta	H	η
Theta	Θ	θ
Iota	I	ι
Kappa	K	κ
Lambda	Λ	λ
Mu	M	μ
Nu	N	ν
Xi	Ξ	ξ
Omicron	O	o
Pi	Π	π
Rho	P	ρ
Sigma	Σ	σ
Tau	T	τ
Upsilon	Υ	υ
Phi	Φ	ϕ
Chi	X	χ
Psi	Ψ	ψ
Omega	Ω	ω

Table A.1

▶ Trigonometry Review

B.1 Degrees and Radians

The magnitude of an angle can be measured by using either degrees or radians. One full revolution around a circle corresponds to 360° or 2π radians. The radian unit is abbreviated as rad. Likewise, half of a circle corresponds to an angle of 180° or π rad. You can convert between radians and degrees by using the following factors:

$$1 \text{ deg} = 1.7453 \times 10^{-2} \text{ rad} \qquad \text{(B.1)}$$
$$1 \text{ rad} = 57.296 \text{ deg} \qquad \text{(B.2)}$$

B.2 Right Triangles

A right angle has a measure of 90°, or equivalently, $\pi/2$ rad. As shown in Figure B.1, a right triangle is composed of one right angle and two acute angles. An acute angle is one that is smaller than 90°. In this case, the acute angles of the triangle are denoted by the lowercase Greek characters phi (ϕ) and theta (θ), as listed in Appendix A. Since the magnitudes of the triangle's three angles sum to 180°, the two acute angles in the triangle are related by

$$\phi + \theta = 90° \qquad \text{(B.3)}$$

The lengths of the two sides that meet and form the right angle are denoted by x and y. The remaining longer side is called the hypotenuse, and it has length z. An acute angle is formed between the hypotenuse and either side adjacent to it. The three side lengths are related to one another by the Pythagorean theorem

$$x^2 + y^2 = z^2 \qquad \text{(B.4)}$$

If the lengths of two sides in a right triangle are known, the third length can be determined from this expression.

The lengths and angles in a right triangle are also related to one another by properties of the trigonometric functions called sine, cosine, and tangent.

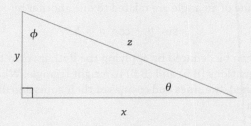

Figure B.1

Right triangle with side lengths x, y, z and acute angles ϕ and θ.

	0°	30°	45°	60°	90°
sin	0	0.5	$\dfrac{\sqrt{2}}{2}$	$\dfrac{\sqrt{3}}{2}$	1
cos	1	$\dfrac{\sqrt{3}}{2}$	$\dfrac{\sqrt{2}}{2}$	0.5	0
tan	0	$\dfrac{\sqrt{3}}{3}$	1	$\sqrt{3}$	∞

Table B.1

Some Values of the Sine, Cosine, and Tangent Functions

Each of these functions is defined as the ratio of one side's length to another. Referring to the angle θ in Figure B.1, the sine, cosine and tangent of θ involve the following ratios of the adjacent side's length (x), the opposite side's length (y), and the hypotenuse's length (z):

$$\sin \theta = \frac{y}{z} \left(\frac{\text{opposite side's length}}{\text{hypotenuse's length}} \right) \tag{B.5}$$

$$\cos \theta = \frac{x}{z} \left(\frac{\text{adjacent side's length}}{\text{hypotenuse's length}} \right) \tag{B.6}$$

$$\tan \theta = \frac{y}{x} \left(\frac{\text{opposite side's length}}{\text{adjacent side's length}} \right) \tag{B.7}$$

Similarly, for the other acute angle (ϕ) in the triangle,

$$\sin \phi = \frac{x}{z} \tag{B.8}$$

$$\cos \phi = \frac{y}{z} \tag{B.9}$$

$$\tan \phi = \frac{x}{y} \tag{B.10}$$

From these definitions, you can see such characteristics of these functions as $\sin(45°) = \cos(45°) = \dfrac{\sqrt{2}}{2}$ and $\tan(45°) = 1$. Other numerical values of the sine, cosine, and tangent functions are listed in Table B.1.

B.3 Identities

The sine and cosine of an angle are related to one another by

$$\sin^2 \theta + \cos^2 \theta = 1 \tag{B.11}$$

This expression can be deduced by applying the Pythagorean theorem and the definitions in Equations (B.5) and (B.6) to a right triangle. When two angles θ_1 and θ_2 are combined, the sines and cosines of their sums and differences can be determined from

$$\sin (\theta_1 \pm \theta_2) = \sin \theta_1 \cos \theta_2 \pm \cos \theta_1 \sin \theta_2 \tag{B.12}$$

$$\cos (\theta_1 \pm \theta_2) = \cos \theta_1 \cos \theta_2 \pm \sin \theta_1 \sin \theta_2 \tag{B.13}$$

In particular, when $\theta_1 = \theta_2$, these expressions are used to deduce the double angle formulas

$$\sin 2\theta = 2 \sin \theta \cos \theta \qquad \text{(B.14)}$$
$$\cos 2\theta = \cos^2 \theta - \sin^2 \theta \qquad \text{(B.15)}$$

B.4 Oblique Triangles

Simply put, an oblique triangle is any triangle that is not a right triangle. Therefore, an oblique triangle does not contain a right angle. There are two types of oblique triangles. In an acute triangle, all three angles have magnitudes of less than 90°. In an obtuse triangle, one of the angles is greater than 90°, and the other two are each less than 90°. In all cases, the sum of the angles in a triangle is 180°.

Two theorems of trigonometry can be applied to determine an unknown side length or angle in an oblique triangle. These theorems are called the laws of sines and cosines. The law of sines is based on the ratio of a triangle's side length to the sine of the opposing angle. That ratio is the same for the three pairs of lengths and opposite angles in the triangle. Referring to the side lengths and angles labeled in Figure B.2, the law of sines is

$$\frac{a}{\sin A} = \frac{b}{\sin B} = \frac{c}{\sin C} \qquad \text{(B.16)}$$

When we happen to know the lengths of two sides in a triangle and their included angle, the law of cosines can be used to find the length of the triangle's third side:

$$c^2 = a^2 + b^2 - 2ab \cos C \qquad \text{(B.17)}$$

When $C = 90°$, this expression reduces to $c^2 = a^2 + b^2$, which is a restatement of the Pythagorean theorem (B.4).

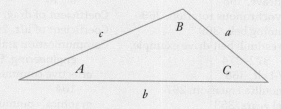

Figure B.2

An oblique triangle with side lengths a, b, c and opposing angles A, B, and C.

Index